Student's Solutions Manual

Doreen Kelly

Mesa Community College

Prealgebra

Third Edition

Tom Carson

Columbia State Community College

PEARSON

Addison
Wesley

Boston San Francisco New York
London Toronto Sydney Tokyo Singapore Madrid
Mexico City Munich Paris Cape Town Hong Kong Montreal

Reproduced by Pearson Addison-Wesley from electronic files supplied by the author.

Copyright © 2009 Pearson Education, Inc.
Publishing as Pearson Addison-Wesley, 75 Arlington Street, Boston, MA 02116.

ISBN-13: 978-0-321-52380-8
ISBN-10: 0-321-52380-6

7   BRR 12 11

CONTENTS

Chapter 1 Whole Numbers

Exercises Set 1.1

1. 1, 2, 3, 4

3. No, zero is not a natural number.

5. 1. Write the name of the digits in the left-most period.
   2. Write the period name followed by a comma.
   3. Repeat steps 1 and 2 until you get to the ones period. Do not follow the ones period with its name.

7. Consider the digit to the right of the desired place value. If this digit is 5 or greater, round up. If it is 4 or less, round down.

9. 0 is in the hundreds place.

11. 5 is in the ten millions place.

13. 7 is in the thousands place.

15. 7 is in the hundred thousands place.

17. $24{,}319 = 2 \times 10{,}000 + 4 \times 1000 + 3 \times 100 + 1 \times 10 + 9 \times 1$

19. $5{,}213{,}304 = 5 \times 1{,}000{,}000 + 2 \times 100{,}000 + 1 \times 10{,}000 + 3 \times 1000 + 3 \times 100 + 4 \times 1$

21. $93{,}014{,}008 = 9 \times 10{,}000{,}000 + 3 \times 1{,}000{,}000 + 1 \times 10{,}000 + 4 \times 1000 + 8 \times 1$

23. 8792

25. 6,039,020: We replaced the missing hundred thousand, hundred, and one place values with zeros.

27. 40,980,109: We replaced the missing million, thousand, and ten place values with zeros.

29. seven thousand, seven hundred sixty-eight

31. two hundred ninety-nine million, four hundred ninety-nine thousand, seven hundred thirty-eight

33. one hundred eighty-six thousand, one hundred seventy-one

35. $599 < 899$, because 599 is less than 899.

37. $4{,}299{,}308 > 4{,}298{,}308$, because 4,299,308 is greater than 4,298,308.

39. $609{,}001 = 609{,}001$, because 609,001 is equal to 609,001.

41.

43.

45.

47. a. $56,283   b. $38,246

   c. Massachusetts, New York, Illinois, and California

   d. cost of living

49. 5,652,992,000: 2 is in the thousands place. Look to the 4 on the right. Since 4 is 4 or less, we round the 2 down.

51. 5,653,000,000: 2 is in the millions place. Look to the 9 on the right. Since 9 is greater than 5, we round the 2 up.

53. 6,000,000,000: 5 is in the billions place. Look to the 6 on the right. Since 6 is greater than 5, we round the 5 up.

55. 5,652,992,500: 4 is in the hundreds place. Look to the 8 on the right. Since 8 is greater than 5, we round the 4 up.

57. 30,000: 3 is furthest to the left. Look to the 2 on the right. Since 2 is 4 or less, we round the 3 down.

59. 900,000: 8 is furthest to the left. Look to the 5 on the right. Since 5 is 5 or greater, we round the 8 up.

61. 9,000: 8 is furthest to the left. Look to the 7 on the right. Since 7 is greater than 5, we round the 8 up.

63. $53,000: The average salary is $52,600. The 2 is in the thousands place. Look to the 6 on the right. Since 6 is greater than 5, we round the 2 up.

65. $514,000,000,000: The total social security income in 2001 is 513,834,000,000. The 3 is in the billions place. Look to the 8 on the right. Since 8 is greater than 5, we round the 3 up.

67. 93,000,000 miles: We chose the millions place to round because 93,00,000 is reasonably accurate yet simpler. (Answers may vary.)

69. We arrange the numbers in order from least to greatest, then locate the middle number.

   $32,416, $32,788, $34,030, $38,703, $39,800, $39,900, $39,975, $40,400, $58,500

   The median is $39,800.

71. We arrange the scores in order from least to greatest, then locate the middle number.

    54, 56, 64, 70, 72, 74, 79, 80, 82, 86, 86, 88. 90, 98, 100

    The median is 80.

73. We arrange the heights in order from least to greatest, then locate the middle number.

    68, 68, 71, 72, 75, 75, 77, 77, 77, 77, 79

    The median height is 75 in.

Exercises Set 1.2

1. Changing the order of the addends does not affect the sum.

3. Perimeter is the total distance around a shape.

5. Write a related subtraction equation, subtracting the known addend from the sum.

7. Estimate: 6000   Actual: 6051
                 +3000   +2798
                 9000   8849

9. Estimate: 92,000   Actual: 91,512
              +9,000   +8,756
              101,000   100,268

11. Estimate: 10,500   Actual: 10,516
                 800   782
              +4,500   +4,516
              15,800   15,814

13. Estimate: 9,300   Actual: 9,319
                 500   519
              +5,400   5,408
              15,200   15,246

15. Estimate: 43,200   Actual: 43,210
              135,600   135,569
                2,100   2,088
                +500   +516
              181,400   181,383

17. Estimate: 5900   Actual: 5873
               −500   −521
               5400   5352

19. Estimate: 40,000   Actual: 40,302
              −6,000   −6,141
              34,000   34,161

21. Estimate: 50,000   Actual: 50,016
              −5,000   −4,682
              45,000   45,334

23. Estimate: 50,000   Actual: 51,980
              −30,000   −25,461
              20,000   26,519

25. Estimate: 210,000   Actual: 210,007
              −40,000   −43,519
              170,000   166,488

27. $8 + x = 12$ Check: $8 + 4 = 12$
    $x = 12 - 8$ $12 = 12$
    $x = 4$

29. $t + 9 = 30$ Check: $21 + 9 = 30$
    $t = 30 - 9$ $30 = 30$
    $t = 21$

31. $17 + n = 35$ Check: $17 + 18 = 35$
    $n = 35 - 17$ $35 = 35$
    $n = 18$

33. $125 + u = 280$ Check: $125 + 155 = 280$
    $u = 280 - 125$ $280 = 280$
    $u = 155$

35. $b + 76 = 301$ Check: $225 + 76 = 301$
    $b = 301 - 76$ $301 = 301$
    $b = 225$

37. The word "total" tells us to add the ticket sales.
    $548 + 354 + 481 + 427 = 1810$ tickets

39. To find the profit, we must subtract all costs from the revenue.
    $P = 548,780 - 75,348 - 284,568 - 54,214$
    $= \$134,650$

41. "How much warmer" tells us to find the difference between the temperatures.
    $5500 - 3400 = 2100$ K

43. We must subtract Serene's total purchases from the amount she started with.

    total purchases $= 2 + 8 + 21 = 31$

    starting amount − total purchases = ending amount

    $47 - 31 = \$16$

45. We must subtract the areas of the garage and the bonus room from the given area.
    $2400 - 440 - 240 = 1720$ sq. ft. is the amount of actual living space.

47. We must calculate the total current entering the node.

$25 + 19 = 44$ A

Now subtract the known current leaving the node to find the unknown current.

$44 - 11 = 33$ A

49. We must calculate the total expenses and total income/assets.

Expenses	Income/Assets
18,708	
678	
126	10,489
1,245	+32,500
+15,621	$42,989
$36,378	

To determine the balance, find the difference.

$$\begin{array}{r} 42,989 \\ -36,378 \\ \hline \$6,611 \end{array}$$

51. a. We must add the following numbers:
$14,044,000 + 7,871,000 + 7,306,000$

$+8,621,000 + 12,720,000 + 5,311,000$

$= 55,873,000$

b. We must add the following numbers:
$14,044,000 + 7,871,000 + 7,306,000$

$+8,621,000 = 37,842,000$

c. We must add the following numbers:
$12,720,000 + 5,311,000$

$= 18,031,000$

d. We must subtract the following numbers:
$7,871,000 - 7,306,000$

$= 565,000$

e. We must subtract the following numbers:
$8,621,000 - 5,311,000$

$= 3,310,000$

f. We must take subtract the number of people who attended California's Disney parks which we found in c. above from the number of people who attended the Florida's Disney parks which we found in b. above.
$37,842,000 - 18,031,000$

$= 19,811,000$

Review Exercises

1. $307,491,024$
$= 3 \times 100,000,000 + 7 \times 1,000,000$
$+ 4 \times 100,000 + 9 \times 10,000 + 1 \times 1000$
$+ 2 \times 10 + 4 \times 1$

2. one million, four hundred seventy-two thousand, three hundred fifty-nine

3. $12,305 < 12,350$, because $12,305$ is less than $12,350$

4. $23,410,000$: 0 is in the ten thousands place. Look to the 5 on the right. Since 5 is 5 or more, we round the 0 up.

5. We must put the scores in order from least to greatest and choose the score in the middle.
$64, 72, 76, 85, 88, 90, 92$

The median test score is 85.

Exercise Set 1.3

1. Changing the order of the factor does not affect the product.

3. $a(b+c) = ab + ac$

5. The area of a shape is the total number of square units that completely fill the shape.

7. By the Multiplicative Property of 1; the product of 1 and a number is always the number.

So, $352 \times 1 = 352$.

9. By the Multiplicative Property of 0; the product of 0 and a number is always 0.

So, $8 \times 9 \times 0 \times 7 = 0$.

11. Estimate: $20 \times 20 = 400$ Think 2×2 and add 2 zeros.

Actual:
$$\begin{array}{r} 24 \\ \times 18 \\ \hline 192 \\ 24 \\ \hline 432 \end{array}$$

13. Estimate: $50 \times 90 = 4500$ Think 5×9 and add 2 zeros.

Actual:
$$\begin{array}{r} 54 \\ \times 91 \\ \hline 54 \\ 486 \\ \hline 4914 \end{array}$$

15. Estimate: $200 \times 400 = 80,000$ Think 2×4 and add 4 zeros.

Actual:
$$\begin{array}{r} 246 \\ \times 381 \\ \hline 246 \\ 1968 \\ 738 \\ \hline 93,726 \end{array}$$

17.
$$\begin{array}{r} 642 \\ \times 70 \\ \hline 000 \\ 4494 \\ \hline 44,940 \end{array}$$

19.
$$\begin{array}{r} 2065 \\ \times 482 \\ \hline 4130 \\ 16520 \\ 8260 \\ \hline 995,330 \end{array}$$

21.
$$\begin{array}{r} 41,308 \\ \times 207 \\ \hline 289156 \\ 00000 \\ 82616 \\ \hline 8,550,756 \end{array}$$

23.
$$\begin{array}{r} 60,309 \\ \times 4002 \\ \hline 120618 \\ 00000 \\ 00000 \\ 241236 \\ \hline 241,356,618 \end{array}$$

25.
$$\begin{array}{r} 205 \\ \times 23 \\ \hline 615 \\ 410 \\ \hline 4715 \end{array} \qquad \begin{array}{r} 4715 \\ \times 70 \\ \hline 0000 \\ 33005 \\ \hline 330,050 \end{array}$$

27. $2^4 = 2 \cdot 2 \cdot 2 \cdot 2$
$= 4 \cdot 2 \cdot 2$
$= 8 \cdot 2$
$= 16$

29. $1^6 = 1 \cdot 1 \cdot 1 \cdot 1 \cdot 1 \cdot 1$
$= 1$

31. $3^5 = 3 \cdot 3 \cdot 3 \cdot 3 \cdot 3$
$= 9 \cdot 3 \cdot 3 \cdot 3$
$= 27 \cdot 3 \cdot 3$
$= 81 \cdot 3$
$= 243$

33. $10^7 = 10 \cdot 10 \cdot 10 \cdot 10 \cdot 10 \cdot 10 \cdot 10$
$= 100 \cdot 10 \cdot 10 \cdot 10 \cdot 10 \cdot 10$
$= 1000 \cdot 10 \cdot 10 \cdot 10 \cdot 10$
$= 10,000 \cdot 10 \cdot 10 \cdot 10$
$= 100,000 \cdot 10 \cdot 10$
$= 1,000,000 \cdot 10$
$= 10,000,000$

35. $6536^2 = 6536 \cdot 6536$
$= 42,719,296$

37. $42^5 = 42 \cdot 42 \cdot 42 \cdot 42 \cdot 42$
$= 130,691,232$

39. 9^4 : Since there are four 9's multiplied the exponent is 4.

41. 7^5 : Since there are five 7's multiplied the exponent is 5.

43. 14^6 : Since there are six 14's multiplied, the exponent is 6.

45. 24,902
$= 2 \times 10,000 + 4 \times 1000 + 9 \times 100 + 0 \times 10 + 2 \times 1$
$= 2 \times 10^4 + 4 \times 10^3 + 9 \times 10^2 + 0 \times 10^1 + 2 \times 10^0$
$= 2 \times 10^4 + 4 \times 10^3 + 9 \times 10^2 + 2 \times 1$

47. 9,128,020
$= 9 \times 1,000,000 + 1 \times 100,000 + 2 \times 10,000$
$+ 8 \times 1,000 + 0 \times 100 + 2 \times 10 + 0 \times 10^0$
$= 9 \times 10^6 + 1 \times 10^5 + 2 \times 10^4 + 8 \times 10^3 + 0 \times 10^2$
$+ 2 \times 10^1 + 0 \times 10^0$
$= 9 \times 10^6 + 1 \times 10^5 + 2 \times 10^4 + 8 \times 10^3 + 2 \times 10$

49. 407,210,925
$= 4 \times 100,000,000 + 0 \times 10,000,000$
$+ 7 \times 1,000,000 + 2 \times 100,000 + 1 \times 10,000$
$+ 0 \times 1000 + 9 \times 100 + 2 \times 10 + 5 \times 1$
$= 4 \times 10^8 + 0 \times 10^7 + 7 \times 10^6 + 2 \times 10^5$
$+ 1 \times 10^4 + 0 \times 10^3 + 9 \times 10^2 + 2 \times 10^1 + 5 \times 10^0$
$= 4 \times 10^8 + 7 \times 10^6 + 2 \times 10^5 + 1 \times 10^4 + 9 \times 10^2$
$+ 2 \times 10 + 5 \times 1$

51. For the number of intersections, we need to think in terms of a rectangular array. Multiply 97×82 = 7954 intersections. We now need to multiply 7954 intersections by 8 lights at each intersection. $7954 \times 8 = 63,632$ total traffic lights are required.

53. For the number of heartbeats in a year, we must begin with the number of heartbeats in one hour. $60 \times 70 = 4200$ heartbeats each hour.

Since there are 24 hours in a day, we must multiply. There are $4,200 \times 24 = 100,800$ heartbeats each day.

Since there are 365 days a year, we must multiply. There are $100,800 \times 365 = 36,792,000$ heartbeats each year.

If the average man lives to be 75, the estimated heartbeats in his lifetime would be $36,792,000 \times 75 = 2,759,400,000$ heartbeats.

55. Since the person will receive 10 mg of aminophylline for each kg the person weighs, this indicated repeated addition. We multiply $10 \times 74 = 740$ mg.

57. There are 52 weeks in one year. We must multiply the amount per week by the number of weeks. A telemarketer would earn up to $800 \times 52 = \$41,600$ in a year.

59. Eight 8-oz. glasses of water is $8 \times 8 = 64$ ounces. Now we must multiply $64 \times 300 = 19,200,000,000$ oz.

61. There are three slots. The first is turned left or right (2), the second is filled with the days of the week (7) and the third slot is for letter A to H (8). Multiply the 3 slots for the total combinations.

$2 \times 7 \times 8 = 112$ combinations are possible.

63. There are 5 boxes each with a 0 or a 1 choice. In other words, 5 slots each containing 2 choices. Multiply $2 \times 2 \times 2 \times 2 \times 2 = 32$ or $2^5 = 32$. No, it cannot recognize $26 + 10 = 36$ binary numbers.

65. There are 7 slots with 2 choices each.

There are $2^7 = 128$ locations.

67. The man is overestimating the amount of carpet he will need. He will order too much. His estimate is not reasonable. He needs exactly

$24 \times 26 = 624$ sq. ft.

69. $A = lw$
$A = 20 \times 15$
$A = 300$ ft.^{2}

Review Exercises

1. sixteen million, five hundred seven thousand, three hundred nine

2. $23,506 = 2 \times 10,000 + 3 \times 1000 + 5 \times 100 + 6 \times 1$

3. $54,391$
   $2,079$
   $+518$
   $\overline{56,988}$

4. $901,042$
   $-69,318$
   $\overline{831,724}$

5. $n + 19 = 32$ Check: $13 + 19 = 32$
   $n = 32 - 19$ $32 = 32$
   $n = 13$

Exercise Set 1.4

1. the number   3. 1

5. A square root of a given number is a number whose square is the given number.

7. 1: Because $26 \cdot 1 = 26$

9. 0: Because $49 \cdot 0 = 0$

11. Indeterminate, because $x \cdot 0 = 0$ is true for any number.

13. Undefined because there is no number that can make $x \cdot 0 = 22$ true.

15. No, 19,761 is not an even number. The ones digit is not even.

17. Yes, 143,706 is an even number. There is a 6 in the ones place.

19. Yes, 431,970 is an even number. There is a 0 in the ones place.

21. Yes. Add the digits $1 + 9 + 7 + 0 + 4 = 21$. 21 is divisible by 3 so 19,704 is divisible by 3.

23. No. Add the digits $2 + 6 + 0 + 9 + 3 = 20$. 20 is not divisible by 3 so 26,093 is not divisible by 3.

25. Yes. Add the digits $9 + 8 + 7 + 5 + 7 = 36$. 36 is divisible by 3 so 98,757 is divisible by 3.

27. Yes, there is a 5 in the ones place.

29. Yes, there is a 0 in the ones place.

31. No, there is not a 0 or 5 in the ones place.

33.
$$\begin{array}{r} 426 \\ 9\overline{)3834} \\ \underline{36} \\ 23 \\ \underline{18} \\ 54 \\ \underline{54} \\ 0 \end{array}$$

35.
$$\begin{array}{r} 207 \\ 5\overline{)1038} \\ \underline{10} \\ 038 \\ \underline{35} \\ 3 \end{array}$$
Answer: 207 r3

37.
$$\begin{array}{r} 217 \\ 16\overline{)3472} \\ \underline{32} \\ 27 \\ \underline{16} \\ 112 \\ \underline{112} \\ 0 \end{array}$$

39.
$$\begin{array}{r} 246 \\ 26\overline{)6399} \\ \underline{52} \\ 119 \\ \underline{104} \\ 159 \\ \underline{156} \\ 3 \end{array}$$
Answer: 246 r3

41.
$$\begin{array}{r} 307 \\ 41\overline{)12587} \\ \underline{123} \\ 287 \\ \underline{287} \\ 0 \end{array}$$

43.
$$\begin{array}{r} 230 \\ 120\overline{)27600} \\ \underline{240} \\ 360 \\ \underline{360} \\ 0 \end{array}$$

45.
$$\begin{array}{r} 1900 \\ 124\overline{)235613} \\ \underline{124} \\ 11161 \\ \underline{11160} \\ 13 \\ \underline{0} \\ 13 \end{array}$$
Answer: 1900 r13

47.
$$\begin{array}{r} 1601 \\ 207\overline{)331419} \\ \underline{207} \\ 1244 \\ \underline{1242} \\ 219 \\ \underline{207} \\ 12 \end{array}$$
Answer: 1601 r12

49. Undefined because you cannot divide by zero.

51. $9 \cdot x = 54$ Check: $9 \cdot 6 = 54$
          $x = 54 \div 9$ $54 = 54$
          $x = 6$

53. $m \cdot 11 = 88$ Check: $8 \cdot 11 = 88$
          $m = 88 \div 11$ $88 = 88$
          $m = 8$

55. $17 \cdot n = 119$ Check: $17 \cdot 7 = 119$
          $n = 119 \div 17$ $119 = 119$
          $n = 7$

57. $24 \cdot t = 0$ Check: $24 \cdot 0 = 0$
          $t = 0 \div 24$ $0 = 0$
          $t = 0$

59. $v \cdot 2 \cdot 13 = 1092$ Check: $42 \cdot 2 \cdot 13 = 1092$
          $v \cdot 26 = 1092$ $84 \cdot 13 = 1092$
              $v = 1092 \div 26$ $1092 = 1092$
              $v = 42$

61. $29 \cdot 6 \cdot h = 3480$ Check: $29 \cdot 6 \cdot 20 = 3480$
          $174 \cdot h = 3480$ $174 \cdot 20 = 3480$
              $h = 3480 \div 174$ $3480 = 3480$
              $h = 20$

63. Dedra is paid once per month, so she is paid 12 times per year. We must divide the total amount by the number of months.
     $34{,}248 \div 12 = \$2{,}854$

65. There are 60 minutes in an hour. We must divide the total ml by the number of minutes.

     $840 \div 60 = 14$ ml

67. We must divide the total number of flyers by the number on each paper.

     $500 \div 4 = 125$ pieces of paper

69. There are 100 cents in each dollar. So, $20 \times 100 = 2000$ cents. We must divide the total by the amount of each stamp. $2000 \div 39 = 51$ r11. Because there is a remainder, we round down since we cannot buy part of a stamp. Fifty-one stamps can be bought.

71. a. We must divide the total cereal produced by the amount in each box.

     $153{,}600 \div 16 = 9600$ boxes

     b. We must divide the total number of boxes by the amount in each bundle.

     $9600 \div 8 = 1200$ bundles

     c. We must divide the total number of bundles by the amount to a pallet.

     $1200 \div 24 = 50$ pallets

d. We must divide the total number of pallets by the pallets on a truck. $50 \div 28 = 1$r22. Since all pallets cannot fit on one truck, 2 trucks are needed. Twenty-eight in the first truck and 22 in the second.

73.

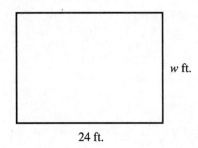

24 ft.

          $A = lw$
       $432 = 24 \cdot w$
     $432 \div 24 = w$
        18 ft. $= w$

75. We must divide the total needed by the length per section. $170 \div 8 = 21$ r2. We must round up. To have enough fencing 22 sections must be purchased.

77. 10   79. 13   81. 0   83. 1   85. 7

87. 14   89. 37   91. 214

93. We must find a number that squares to give 36. $\sqrt{36} = 6$. A 6 in. by 6 in. plate will cover the hole.

95. a. We must divide the amount of savings by the price per square foot.
     $1872 \div 13 = 144$ sq. ft.
     The Jacksons can afford 144 sq. ft. of fencing.

     b. We must find a number that squares to give 144. $\sqrt{144} = 12$. The patio can be 12 ft. × 12 ft. square.

Review Exercises

1. $184 + t = 361$ Check: $184 + 177 = 361$
          $t = 361 - 184$ $361 = 361$
          $t = 177$

2. Finding the length of the decorative border requires us to find the perimeter.

     $P = 2l + 2w$
     $P = 2 \cdot 14 + 2 \cdot 16$
     $P = 28 + 32$
     $P = 60$ ft.

Connie will need 60 ft. of decorative border.

3. Estimate: $370 \times 40 = 14{,}800$. Think 37×4 and add two zeros.

Actual:
$$
\begin{array}{r}
367 \\
\times 42 \\
\hline
734 \\
1468 \\
\hline
15{,}414
\end{array}
$$

4. $5^4 = 5 \cdot 5 \cdot 5 \cdot 5$
   $\quad = 25 \cdot 5 \cdot 5$
   $\quad = 125 \cdot 5$
   $\quad = 625$

5. $49{,}602$
   $= 4 \times 10{,}000 + 9 \times 1000 + 6 \times 100 + 2 \times 1$
   $= 4 \times 10^4 + 9 \times 10^3 + 6 \times 10^2 + 2 \times 1$

Exercise Set 1.5

1. 1. Grouping symbols
   2. Exponents and roots
   3. Multiply or divide from left to right
   4. Add or subtract from left to right

3. Subtract 4 from 10 within the parentheses. We work within grouping symbols before simplifying any other operations.

5. $7 + 5 \cdot 3 = 7 + 15$
   $\quad = 22$

7. $18 + 36 \div 4 \cdot 3 = 18 + 9 \cdot 3$
   $\quad = 18 + 27$
   $\quad = 45$

9. $12^2 - 6 \cdot 4 \div 8 = 144 - 6 \cdot 4 \div 8$
   $\quad = 144 - 24 \div 8$
   $\quad = 144 - 3$
   $\quad = 141$

11. $36 - 3^2 \cdot 4 + 14 = 36 - 9 \cdot 4 + 14$
    $\quad = 36 - 36 + 14$
    $\quad = 0 + 14$
    $\quad = 14$

13. $\dfrac{39}{13} + \sqrt{36} - 2^3 = \dfrac{39}{13} + 6 - 8$
    $\quad = 3 + 6 - 8$
    $\quad = 9 - 8$
    $\quad = 1$

15. $25 + 2(14 - 9) = 25 + 2(5)$
    $\quad = 25 + 10$
    $\quad = 35$

17. $12 + 8\sqrt{25} \div 4 = 12 + 8 \cdot 5 \div 4$
    $\quad = 12 + 40 \div 4$
    $\quad = 12 + 10$
    $\quad = 22$

19. $7\sqrt{64} - 40 \div 5 \cdot 2 + 9 = 7 \cdot 8 - 40 \div 5 \cdot 2 + 9$
    $\quad = 56 - 8 \cdot 2 + 9$
    $\quad = 56 - 16 + 9$
    $\quad = 40 + 9$
    $\quad = 49$

21. $2^6 - 18 \div 3 \cdot 5 - \sqrt{100} = 64 - 18 \div 3 \cdot 5 - 10$
    $\quad = 64 - 6 \cdot 5 - 10$
    $\quad = 64 - 30 - 10$
    $\quad = 34 - 10$
    $\quad = 24$

23. $3^2 \cdot 4 \div 6 - 5 + \sqrt{16} = 9 \cdot 4 \div 6 - 5 + 4$
    $\quad = 36 \div 6 - 5 + 4$
    $\quad = 6 - 5 + 4$
    $\quad = 1 + 4$
    $\quad = 5$

25. $2(14 + 3) - 8 \cdot 2 + (9 - 2) = 2 \cdot 17 - 8 \cdot 2 + 7$
    $\quad = 34 - 16 + 7$
    $\quad = 18 + 7$
    $\quad = 25$

27. $(2 + 9)(19 - 16)^2 = (11)(3)^2 = 11 \cdot 9 = 99$

29. $2^5 \div (14 - 6) + 7(23 - 12) = 2^5 \div 8 + 7 \cdot 11$
    $\quad = 32 \div 8 + 7 \cdot 11$
    $\quad = 4 + 77$
    $\quad = 81$

31. $(13 - 8)3^2 - 48 \div (15 - 9) = 5 \cdot 3^2 - 48 \div 6$
    $\quad = 5 \cdot 9 - 48 \div 6$
    $\quad = 45 - 8$
    $\quad = 37$

33. $58 \div \left(2\sqrt{49} - 6 \cdot 2\right) = 58 \div (2 \cdot 7 - 6 \cdot 2)$
    $\quad = 58 \div (14 - 12)$
    $\quad = 58 \div 2$
    $\quad = 29$

35. $(3 + 4)\sqrt{121} + 3(6 - 4) = 7\sqrt{121} + 3(2)$
    $\quad = 7 \cdot 11 + 3 \cdot 2$
    $\quad = 77 + 6$
    $\quad = 83$

37. $[14 - (3 + 2)] \div 3 + 4(17 - 6)$
$= [14 - 5] \div 3 + 4(17 - 6)$
$= 9 \div 3 + 4 \cdot 11$
$= 3 + 44$
$= 47$

39. $\{18 - 4[21 \div (3 + 4)]\} + 3\sqrt{16} \cdot 4$
$= \{18 - 4[21 \div 7]\} + 3\sqrt{16} \cdot 4$
$= \{18 - 4 \cdot 3\} + 3\sqrt{16} \cdot 4$
$= \{18 - 12\} + 3\sqrt{16} \cdot 4$
$= 6 + 3\sqrt{16} \cdot 4$
$= 6 + 3\sqrt{64}$
$= 6 + 3 \cdot 8$
$= 6 + 24$
$= 30$

41. $4^3 - 5 \ (28 - 4) \div 2^3 \ + \sqrt{100 \div 25}$
$= 4^3 - 5 \ 24 \div 2^3 \ + \sqrt{4}$
$= 64 - 5[24 \div 8] + 2$
$= 64 - 5 \cdot 3 + 2$
$= 64 - 15 + 2$
$= 49 + 2$
$= 51$

43. $12 \div 4 \cdot 3 + 1^5 - \sqrt{17 - 1}$
$= 12 \div 4 \cdot 3 + 1 - \sqrt{16}$
$= 12 \div 4 \cdot 3 + 1 - 4$
$= 3 \cdot 3 + 1 - 4$
$= 9 + 1 - 4$
$= 10 - 4$
$= 6$

45. $\dfrac{9^2 - 21}{56 - 2(4 + 1)^2}$
$= \dfrac{81 - 21}{56 - 2 \cdot 25}$
$= \dfrac{60}{56 - 50}$
$= \dfrac{60}{6}$
$= 10$

47. $\dfrac{38 - 4(15 - 12)}{(3 + 5)^2 - 2(39 - 8)}$
$= \dfrac{38 - 4 \cdot 3}{8^2 - 2 \cdot 31}$
$= \dfrac{38 - 4 \cdot 3}{64 - 2 \cdot 31}$
$= \dfrac{38 - 12}{64 - 62}$
$= \dfrac{26}{2}$
$= 13$

49. $[485 - (68 + 39)] + 4^5 - 24 \cdot 16 \div 8$
$= [485 - 107] + 4^5 - 24 \cdot 16 \div 8$
$= 378 + 4^5 - 24 \cdot 16 \div 8$
$= 378 + 1024 - 24 \cdot 16 \div 8$
$= 378 + 1024 - 384 \div 8$
$= 378 + 1024 - 48$
$= 1354$

51. Mistake: subtracted 6 from 48 instead of multiplying 5 by 6
Correct: Multiplication must be done before subtraction.
$48 - 6(9 - 4) = 48 - 6 \cdot 5$
$= 48 - 30$
$= 18$

53. Mistake: squared 3 and 5 instead of squaring their sum
Correct: Perform operations inside parenthesis before raising to a power.
$(3 + 5)^2 - 2\sqrt{49} = 8^2 - 2\sqrt{49}$
$= 64 - 2 \cdot 7$
$= 64 - 14$
$= 50$

55. Mean: Divide the sum of scores by the number of scores.
$\dfrac{22 + 35 + 45 + 46}{4} = \dfrac{148}{4} = 37$

Median: Since the scores are in order from least to greatest, find the arithmetic mean of the middle two scores, which are 35 and 45.
$\dfrac{35 + 45}{2} = \dfrac{80}{2} = 40$

57. Mean: Divide the sum of scores by the number of scores.
$\dfrac{90 + 37 + 68 + 54 + 76 + 47}{6} = \dfrac{372}{6} = 62$

Median: Arrange the scores in order from least to greatest: 37, 47, 54, 68, 76, 90

Then find the arithmetic mean of the middle two scores, which are 54 and 68.
$$\frac{54+68}{2}=\frac{122}{2}=61$$

59. Mean: Divide the sum of scores by the number of scores.
$$\frac{77+76+74+73}{4}=\frac{300}{4}=75 \text{ in.}$$

Median: Arrange the scores in order from least to greatest: 73, 74, 76, 77

Then find the arithmetic mean of the middle two scores, which are 74 and 76.
$$\frac{74+76}{2}=\frac{150}{2}=75 \text{ in.}$$

61. Mean: Divide the sum of scores by the number of scores.
$$\begin{array}{r} 230,600+156,200+124,000+92,100 \\ +88,700+86,500+77,900+64,100 \\ +57,000+56,700 \\ \hline 10 \end{array}$$
$$=\frac{1,034,600}{10}$$
$$=103,460$$

Since the salary is in thousands we must add 3 zeros to our answer: $103,460,000

Median: Arrange the scores in order from least to greatest: 56,700, 57,000, 64,100, 77,900, 86,500, 88,700, 92,100, 124,000, 156,200, 230,600

Then find the arithmetic mean of the middle two scores, which are 86,500 and 88,700.
$$\frac{86,500+88,700}{2}=\frac{175,200}{2}=87,600$$

Since the salary is in thousands we must add 3 zeros to our answer: $87,600,000

Review Exercises

1.
$$\begin{array}{r} 40,000 \\ +30,000 \\ \hline 70,000 \end{array}$$

2.
$$\begin{array}{r} 498,503 \\ \times 209 \\ \hline 4486527 \\ 000000 \\ 997006 \\ \hline 104,187,127 \end{array}$$

3.
$$17\overline{)21253} \quad \text{Answer: 1250 r3}$$
$$\begin{array}{r} 1250 \\ \underline{17} \\ 42 \\ \underline{34} \\ 85 \\ \underline{85} \\ 03 \end{array}$$

4. To determine the amount of molding, we must find the perimeter of the ceiling.
$$P=2l+2w$$
$$P=2\cdot12+2\cdot16$$
$$P=24+32$$
$$P=56 \text{ ft.}$$

56 feet of molding will be needed.

To determine the cost, multiply the number of feet needed by the cost per foot. $56\times2=\$112$. It will cost $112 to install the ceiling molding.

5. To install flooring tiles, we cover the entire area.
$$A=l\times w$$
$$A=12\cdot16$$
$$A=192 \text{ ft.}^2$$

Since the area is 192 square feet, 192 tiles will be needed.

Exercise Set 1.6

1. Replace the variables with the corresponding known values, then solve for the unknown variable.

3. The total number of cubic units that completely fills an object.

5. According to the problem-solving outline, understand the problem.

7. Add the areas of the two smaller shapes.

9. $P=2l+2w$
$P=2(29)+2(18)$
$P=58+36$
$P=94 \text{ cm}$

11. $P=2l+2w$
$P=2(32)+2(32)$
$P=64+64$
$P=128 \text{ in.}$

13. $P=2l+2w$
$P=2(24)+2(8)$
$P=48+16$
$P=64 \text{ km}$

15. $A=bh$
$A=19\cdot12$
$A=228 \text{ m}^2$

17. $A = bh$
$A = 12 \cdot 12$
$A = 144 \text{ in.}^2$

19. $A = bh$
$A = 16 \cdot 7$
$A = 112 \text{ km}^2$

21. $V = lwh$
$V = 7 \cdot 4 \cdot 10$
$V = 280 \text{ ft.}^3$

23. $V = lwh$
$V = 6 \cdot 6 \cdot 6$
$V = 216 \text{ in.}^3$

25. $V = lwh$
$V = 14 \cdot 3 \cdot 5$
$V = 210 \text{ km}^3$

27. $A = bh$
$144 = 16h$
$144 \div 16 = h$
$9 \text{m} = h$

29. $A = bh$
$1922 = b \cdot 31$
$1922 \div 31 = b$
$62 \text{ in.} = b$

31. $A = lw$
$180 = l \cdot 9$
$180 \div 9 = l$
$20 \text{ in.} = l$

33. $V = lwh$
$90 = 9 \cdot 5 \cdot h$
$90 = 45 \cdot h$
$90 \div 45 = h$
$2 \text{ cm} = h$

35. $V = lwh$
$22,320 = 31 \cdot w \cdot 40$
$22,320 = 1240 \cdot w$
$22,320 \div 1240 = w$
$18 \text{ in.} = w$

37. $V = lwh$
$24 = 3 \cdot 2 \cdot h$
$24 = 6 \cdot h$
$24 \div 6 = h$
$4 \text{ ft.} = h$

39. **Understand:** We need to find the width of the Antonov AN 124 cargo aircraft compartment.

Plan: Find the width of the rectangular solid.

Evaluate: $V = lwh$
$35,280 = 120 \cdot w \cdot 14$
$35,280 = 1680w$
$35,280 \div 1680 = w$
$21 \text{ft.} = w$

Answer: The width of Antonov AN 124 compartment is 21 ft.

Check: Verify that a box with a length of 120 feet, width of 21 feet, and height of 14 feet has a volume of 35,280 cubic feet.

$V = 120 \cdot 21 \cdot 14$

$V = 35,280$ It checks.

41. **Understand:** We must find the total cost to install crown molding.

Plan: To get the total amount of crown molding we must find the perimeter and then multiply by the cost per foot to find the total cost.

Execute: $P = 2l + 2w$
$P = 2 \cdot 20 + 2 \cdot 16$
$P = 40 + 32$
$P = 72 \text{ ft.}$
To find the total cost, multiply. $72 \times 4 = \$288$

Answer: The cost to install crown molding is \$288.

Check: Estimate the perimeter by rounding so that we have only one digit other than zero.

$P = 2 \cdot 20 + 2 \cdot 20$
$P = 40 + 40$
$P = 80 \text{ ft.}$
Estimate the cost $80 \times 4 = \$320$. The actual answer is reasonable.

43. **Understand:** We must find the total cost to install a baseboard.

Plan: We must find the perimeter of the room and subtract the door length. We do not need to subtract the window since it is above the baseboard. Considering the baseboard sections are sold in 8-foot lengths, we must determine how many pieces are necessary. Then we will calculate the cost.

Execute: $P = 2l + 2w$
$P = 2 \cdot 13 + 2 \cdot 14$
$P = 26 + 28$
$P = 54 \text{ ft.}$
Subtract the doorway: $54 - 4 = 50 \text{ ft.}$

Determine the number of sections using division:

$50 \div 8 = 6\text{r}2$. We will need 7 sections since you can not buy part of a section.

Determine the cost: $7 \times 4 = \$28$.

Answer: The cost of the baseboard is \$28.

Check: Let's work backwards. If we buy 7 sections and each section covers 8 feet, the total coverage is $7 \times 8 = 56 \text{ ft.}$ Since the perimeter is 50 ft., this checks.

45. **Understand:** We must find the total cost to include the fence and gate.

Plan: We must find the perimeter of the lot and subtract the length of the gate. Then find the total cost of the fencing material and add the cost of the gate for the total amount.

Execute: Perimeter of the lot is

120 + 380 + 200 + 400 = 1100 ft.

Subtracting the length of the gate, Mr. Williams needs 1100 – 10 = 1090 ft. of fencing material.

The cost of the fencing material is

$1090 \times 5 = \$5450$.

Adding the cost of the gate, the total cost is

5450 + 75 = $5525.

Answer: The cost to install a fence and a 10-foot gate is $5,525.

Check: Estimate the perimeter by rounding so that we only have one digit other than zero.

P = 400 + 200 + 400 + 100 = 1,100 ft. Since the total perimeter and estimated perimeter are close, the answer is reasonable.

47. **Understand:** We must determine the a) number of wood posts needed and cost, b) length of barb wire needed and cost, and c) the number of U-nails needed, how many boxes, and the cost.

Plan: a) On the 380 ft. side, we will need 49 posts, on the 200 ft. side, we will need 25 posts, on the 400 ft. side, we will need 50 posts, and on the 120 ft. side (subtracting out the 10 ft. fence) we will need 14 posts. b) Determine the length of barb wire needed by finding the perimeter. c) Multiply the number of posts by 4 to determine the total number of nails needed.

Execute: a) The total number of posts needed = 49 + 25 + 50 + 14 = 138 posts. The cost is 2(138) = $276.

b) P = 120 + 380 + 200 + 400 = 1100 ft. Subtract the gate. 1100 – 10 = 1090 ft. Since there are four rows, we will multiply the number of feet by 4. 4(1090) = 4360 feet. To find out how many rolls of barbed wire to buy, we must divide the total number of feet needed by the amount on each roll. 4360 ÷ 1320 = 3 r400. He will need 4 rolls of barbed wire. The cost is 4(45) = $180.

c) We must multiply the number of posts by 4 nails for each post. 4(138) = 552 total nails. Mr. Williams will need to buy 540 ÷ 100 = 6 boxes of nails. The cost will be 6(3) = $18.

d) The total cost of materials is

276 + 180 + 18 + 75 = $549

This is considerably less than the company will charge him.

e) He must consider the time it will require, the price of the tools, and whether building a fence is something he knows how to do.

Answer: He will need 138 posts costing $276, 4 rolls of barb wire costing $180, and 6 boxes of nails costing $18 for a total of $549, including the gate.

49. **Understand:** We must find the area of the floor and the number of plywood sheets needed to cover it. We must also find the total cost.

Plan: We must find the area of the composite figure. Then divide by the area of each sheet of plywood to determine the number needed to cover the floor. Finally, we must determine the total cost.

Execute: Begin by separating the figure into recognizable rectangles.

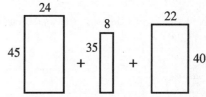

Find the area of each of the rectangles as shown in the figure. Then find the total area.

$= 45 \cdot 24 + 35 \cdot 8 + 40 \cdot 22$

$= 1080 + 280 + 880 = 2240 \text{ ft.}^2$

We must find the area of each sheet of plywood.

$4 \times 8 = 32 \text{ ft.}^2$

To find the number of sheets needed, find the total area divided by the area of each sheet of plywood. 2240 ÷ 32 = 70 sheets are needed.

Multiply the number of sheets needed by the cost per sheet. The total cost is $70 \times 15 = \$1050$.

Answer: The floor plan will need 70 sheets of plywood for a total cost of $1050.

Check: Let's find the area of the composite figure differently to see if we find the same total area.

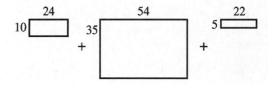

Find the area of each of the rectangles as shown in the figure. Find the total area.

$10 \cdot 24 + 35 \cdot 54 + 5 \cdot 22$

$= 240 + 1890 + 110$

$= 2240 \text{ ft.}^2$

The total areas are the same, so it checks.

51. **Understand:** We must find the total area needed to be painted and determine the number of gallons of paint needed.

Plan: We must find the area of each wall subtracting out entryways and windows. Add all of the areas to get a total. Then we must determine the number of gallons by division.

Execute: Two walls are 24 feet long and two walls are 15 feet long. All four walls are 9 feet tall.

Begin by finding the area of each wall.

Wall 1: $A = bh$
$A = 24 \cdot 9$
$A = 216$ ft.^{2}

Since there are 2 walls this size, the total area is $2 \times 216 = 432$ ft.^{2}.

Wall 2: $A = bh$
$A = 15 \cdot 9$
$A = 135$ ft.^{2}

and since there are 2 walls this size, the total area is $2 \times 135 = 270$ sq. ft.

Now adding the wall space, we get the total area is $432 + 270 = 702$ ft.^{2}.

But we need to take out the area for the entryway and windows.

Entryway: $A = bh$
$A = 6 \cdot 7$
$A = 42$ ft.^{2}

Window: $A = bh$
$A = 3 \cdot 2$
$A = 6$ ft.^{2}

Since there are 2 windows, the total window area is $2 \times 6 = 12$ ft.^{2}.

The total area we need to subtract for the wall space is $42 + 12 = 54$ ft.^{2}.

Finally, the difference of wall space and entryway/window space to find the area to be painted:

$702 - 54 = 648$ ft.^{2}

Considering each gallon of paint covers 400 ft.^{2}, we must divide the total area to be painted by the area covered in a gallon.

$648 \div 400 = 1 r 248$

Answer: We must paint 648 ft.^{2} and purchase 2 gallons of paint.

Check: Let's use estimation to see if our answer is reasonable. To do a quick calculation, let's round the ceiling height to 10 ft. Now find the area of the walls.

Wall 1: $A = bh$ $240 \times 2 = 480$ ft.^{2}
$A = 24 \cdot 10$
$A = 240$ ft.^{2}

Wall 2: $A = bh$ $150 \times 2 = 300$ ft.^{2}
$A = 15 \cdot 10$
$A = 150$ ft.^{2}

Adding the total space for the four walls, $480 + 300$ gives us 780 ft.^{2}. Since the actual wall space was 702 ft.^{2}, our answer is reasonable.

53. **Understand:** We must find the area to be paved.

Plan: We must find the area of the building and subtract it from the surrounding area.

Execute: Find the area of the rectangles.

$A = lw$ $A = lw$
$A = 72 \cdot 84$ $A = 245 \cdot 170$
$A = 6048$ ft.^{2} $A = 41,650$ ft.^{2}

Subtract to find the remaining area.
$41,650 - 6048 = 35,602$ ft.^{2}

Answer: The area to be paved is 35,602 ft.^{2}.

Check: Reverse the process. We start with 35,602 ft.^{2} and see if we can work our way back. The last step in the process was $41,650 - 6048 = 35,602$, so we can check by adding. $35,602 + 6048 = 41,650$. This is true. Now let's check the 6048 and 41,650. The 6048 ft^2 came from multiplying 84×72 so we can reverse this step and divide $6048 \div 84$ to see if we get 72. $6048 \div 84 = 72$. It checks. Now, check the 41,650. Since we multiplied 245×170, we can reverse this step and divide $41,650 \div 245$ to see if we get 170. $41,650 \div 650 = 245$. It checks.

55. **Understand:** We must find the total volume of the building.

Plan: We must find the sum of the volumes of the two wings.

Execute: Find the volume of each wing.

$V = lwh$ $V = lwh$
$V = 90 \cdot 40 \cdot 35$ $V = 120 \cdot 60 \cdot 50$
$V = 126,000$ ft.^{3} $V = 360,000$ ft.^{3}
Add to find the total volume.
$126,000 + 360,000 = 486,000$ ft.^{3}

Answer: The volume of the building is 486,000 ft.^{3}

Check: Let's reverse the process. We start with 486,000 cubic feet and work our way back to the original dimensions. The last step in the preceding process was 126,000 + 360,000 = 486,000, so we can check by subtracting.
486,000 − 360,000 = 126,000
This is true.
Now let's check 360,000 cubic feet. Since we multiplied (120)(60)(50), we can check by dividing $360,000 \div 50 \div 60$ to see if we get 120.

$360,000 \div 50 \div 60$

$= 7200 \div 60$

$= 120$ It checks.

Now let's check 126,000 cubic feet. Since we multiplied (90)(45)(35), we can check by dividing $126,000 \div 90 \div 40$ to see if we get 35.

$126,000 \div 90 \div 40$

$= 1400 \div 40$

$= 35$ It checks.

57. **Understand:** We must find the volume of the padding needed to form the chair.

Plan: We must find the difference of the volumes of the two rectangular solids, the larger one and the one needed to be removed from the solid to form the seat of the chair.

Execute: Find the volume of each piece.

$V = lwh$ $V = lwh$

$V = 4 \cdot 3 \cdot 3$ $V = 2 \cdot 2 \cdot 2$

$V = 36$ ft.^{3} $V = 8$ ft.^{3}
Subtract to find the total volume.
$36 - 8 = 28$ ft.^{3}

Answer: The volume of padding needed to dorm the chair is 28 ft.^{3}

Check: Let's reverse the process. We start with 28 cubic feet and work our way back to the original dimensions. The last step in the preceding process was 38 − 8 = 28, so we can check by adding.
28 + 8 = 36
This is true.
Now let's check 8 cubic feet. Since we multiplied (2)(2)(2), we can check by dividing $8 \div 2 \div 2$ to see if we get 2.

$8 \div 2 \div 2$

$= 4 \div 2$

$= 2$ It checks.

Now let's check 36 cubic feet. Since we multiplied (4)(3)(3), we can check by dividing

$36 \div 4 \div 3$ to see if we get 3.

$36 \div 4 \div 3$

$= 9 \div 3$

$= 3$ It checks.

Review Exercises

1. $x + 175 = 2104$ Check: $1929 + 175 = 2104$
     $x = 2104 - 175$ $2104 = 2104$
     $x = 1929$

2. $2,408,073 = 2 \times 1,000,000 + 4 \times 100,000$
                $+ 8 \times 1000 + 7 \times 10 + 3 \times 1$
             $= 2 \times 10^6 + 4 \times 10^5 + 8 \times 10^3 + 7 \times 10^1$
               $+ 3 \times 1$

3. Estimate $500 \times 71,200 = 35,600,000$.
   Actual: 71,203
            ×452
            142406
         356015
       284812
        32,183,756

4. $17 \cdot y = 3451$ Check: $17 \cdot 203 = 3451$
     $y = 3451 \div 17$ $3451 = 3451$
     $y = 203$

5. $5^3 - [(9+3)^2 - 2^6] + \sqrt{100 - 36}$
   $= 5^3 - [12^2 - 2^6] + \sqrt{64}$
   $= 5^3 - [144 - 64] + \sqrt{64}$
   $= 5^3 - 80 + 8$
   $= 125 - 80 + 8$
   $= 45 + 8$
   $= 53$

Chapter 1 Review Exercises

1. false   2. true   3. false

4. true   5. true   6. false

7. Replace x with 8 and verify that the sum is 17.

8. To check, we multiply the quotient by the divisor, 0, to equal the dividend, but the product of any number and 0 is 0, not 14.

9. Mistake: 17 + 3 was added instead of multiplying 3(7).
   Correct: $17 + 3(9 - 2)$
            $= 17 + 3(7)$
            $= 17 + 21$
            $= 38$

10. The square of a number is the number multiplied by itself. The square root of a number is a number whose square is the given number.

11. 5,680,901
$= 5 \times 1,000,000 + 6 \times 100,000 + 8 \times 10,000 + 9 \times 100 + 1 \times 1$
$= 5 \times 10^6 + 6 \times 10^5 + 8 \times 10^4 + 9 \times 10^2 + 1 \times 1$

12. $42,519 = 4 \times 10,000 + 2 \times 1000 + 5 \times 100 + 1 \times 10 + 9 \times 1$
$= 4 \times 10^4 + 2 \times 10^3 + 5 \times 10^2 + 1 \times 10 + 9 \times 1$

13. 98,274

14. 8,020,096

15. 700,928,006

16. forty-seven million, six hundred nine thousand, two hundred four

17. nine thousand, four hundred twenty-one

18. one hundred twenty-three million, four hundred five thousand, six hundred

19. 14 < 19, because 14 is less than 19.

20. 2930 > 2899, because 2930 is greater than 2899.

21.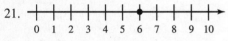

22.

23. 3,380,000

24. 1949–1950 to 1969–1970

25. 5,690,000: The 8 is in the ten thousands place. Look at the 9 on the right. Since 9 is greater than 5, we round the 8 up.

26. 3,000,000: The 2 is in the millions place. Look at the 5 on the right. Since 5 is 5 or greater, we round the 2 up.

27. 46,000,000: The 5 is in the millions place. Look at the 5 on the right. Since 5 is 5 or greater, we round the 5 up.

28. 2,600,000: The 5 is in the millions place. Look at the 4 on the right. Since 4 is less than 5, we round the 5 down.

29. Since the numbers are already in order from least to greatest we choose the number in the middle, 78.

30. Arrange the numbers in order from least to greatest. 150, 180, 192, 210, 214, 250, 340
The median is 210, the number in the middle.

31. Estimate: 46,000   Actual: 45,902
 +7,000   +6,819
 53,000   52,721

32. Estimate: 550   Actual: 545
 9,090   9,091
 30   28
 +30,010   +30,009
 39,680   39,673

33. Estimate: 540,000   Actual: 541,908
 −60,000   −56,192
 480,000   485,716

34. Estimate: 8000   Actual: 8002
 − 300   − 295
 7700   7707

35. $29 + x = 54$ Check: $29 + 25 = 54$
 $x = 54 - 29$ $54 = 54$
 $x = 25$

36. $y + 14 = 203$ Check: $189 + 14 = 203$
 $y = 203 - 14$ $203 = 203$
 $y = 189$

37. To her balance , we must add any deposits and subtract any payments.

   beginning balance + deposits – payments = ending balance

   $349 + 429 - 72 = 778 - 72 = \$706$

   Her new balance is $706.

38. $46,857,000$
 $-6,018,000$
 $40,839,000$

39. Estimate: $70 \times 30 = 2100$.

   Actual: 72
 $\times\ 25$
 360
 144
 1800

40. Estimate: $600 \times 300 = 180,000$. Think 6×3 and add 4 zeros.

   Actual: 591
 $\times307$
 4137
 000
 1773
 181,437

41. $2^7 = 2 \cdot 2 \cdot 2 \cdot 2 \cdot 2 \cdot 2 \cdot 2 = 128$

42. $5^3 = 5 \cdot 5 \cdot 5 = 125$

43. 10^5 ; since there are 5 tens, the exponent is 5.

44. 7^8 ; since there are 8 sevens, the exponent is 8.

45. There are 6 bits (slots) with a 0 or 1 choice. In other words 6 slots each containing 2 choices. Multiply $2 \times 2 \times 2 \times 2 \times 2 \times 2 = 2^6 = 64$ to get the total number of codes.

46. We must find the area of the room.

$A = lw$
$A = 18 \cdot 16$
$A = 288$ ft.^{2}

We will need 288 ft.^{2} of carpet.

47.
$$
\begin{array}{r}
4127 \\
19 \overline{)78413} \\
\underline{76} \\
24 \\
\underline{19} \\
51 \\
\underline{38} \\
133 \\
\underline{133} \\
0
\end{array}
$$

48.
$$
\begin{array}{r}
3209 \\
26 \overline{)83451} \quad \text{Answer: } 3{,}209 \text{ r}17 \\
\underline{78} \\
54 \\
\underline{52} \\
251 \\
\underline{234} \\
17
\end{array}
$$

49. $19 \cdot b = 456$ Check: $19 \cdot 24 = 456$
     $b = 456 \div 19$ $456 = 456$
     $b = 24$

50. $8 \cdot k = 2448$ Check: $8 \cdot 306 = 2448$
     $k = 2448 \div 8$ $2448 = 2448$
     $k = 306$

51. There are 60 minutes in an hour. We must divide the total Pitocin by the number of minutes. Jana should administer $120 \div 60 = 2$ ml of Pitocin each minute.

52. There are 12 months in a year. We must divide the total salary by the number of months. Andre's gross monthly salary is $29{,}256 \div 12 = \$2{,}438$.

53. 14   54. 15

55. $24 \div 8 \cdot 6 + \sqrt{121} - 5^2$
$= 24 \div 8 \cdot 6 + 11 - 25$
$= 3 \cdot 6 + 11 - 25$
$= 18 + 11 - 25$
$= 29 - 25$
$= 4$

56. $64 - [(28 - 4) \div 2^2] + \sqrt{100 \div 25}$
$= 64 - [24 \div 2^2] + \sqrt{4}$
$= 64 - [24 \div 4] + 2$
$= 64 - 6 + 2$
$= 58 + 2$
$= 60$

57. $3\sqrt{25 - 16} + [25 + 2(14 - 9)^2]$
$= 3\sqrt{9} + [25 + 2 \cdot 5^2]$
$= 3 \cdot 3 + [25 + 2 \cdot 25]$
$= 3 \cdot 3 + [25 + 50]$
$= 3 \cdot 3 + 75$
$= 9 + 75$
$= 84$

58. $\dfrac{9^2 - 21}{56 - 2(4 + 1)^2}$

$= \dfrac{9^2 - 21}{56 - 2 \cdot 5^2}$

$= \dfrac{81 - 21}{56 - 2 \cdot 25}$

$= \dfrac{60}{56 - 50}$

$= \dfrac{60}{6}$

$= 10$

59. We divide the sum of the memberships by the number of memberships.
$$\frac{\begin{pmatrix} 27 + 21 + 14 + 9 + 17 + 7 \\ +5 + 8 + 15 + 6 + 4 + 11 \end{pmatrix}}{12} = \frac{144}{12} = 12$$

60. Arrange the numbers in order from least to greatest.
4, 5, 6, 7, 8, 9, 11, 14, 15, 17, 21, 27
Since there are an even number of memberships, we find the arithmetic mean of the middle two numbers, which are 9 and 11.
$$\frac{9 + 11}{2} = \frac{20}{2} = 10$$

61. $A = bh$ 62. $V = lwh$
     $A = 14 \cdot 12$ $V = 9 \cdot 7 \cdot 6$
     $A = 168$ m^2 $V = 378$ ft.^{3}

63. $A = lw$
        $288 = 18 \cdot w$
    $288 \div 18 = w$
    $16 \text{ ft.} = w$

64. We must find the number that when squared results in 225. $\sqrt{225} = 15$. The patio dimensions are 15 ft. × 15 ft.

65. To determine the amount of trim needed to be placed along the edge of the desk, we must find the perimeter of the desk. To find the perimeter, add all of the sides. You will note that two of the sides do not have lengths labeled. We must first find them.

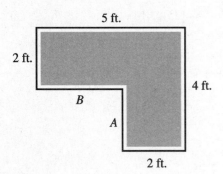

Side A: $4 - 2 = 2$ ft.

Side B: $5 - 2 = 3$ ft.

Now find the perimeter by adding all of the sides.
$2 + 5 + 4 + 2 + 2 + 3 = 18$ ft.

Now multiply the perimeter for 1 desk by the total number of desks.

$18 \cdot 500 = 9000$ ft.

There must be 9000 feet of trim purchased.

66. To find the area of the parking lot, we must find the area of the total lot and the area of the department store. We will then subtract the department store area from the total area and that will be the area of the parking lot.

Area of the total lot: $A = lw$
                          $A = 220 \cdot 450$
                          $A = 99,000$

Area of the store: $A = lw$
                      $A = 95 \cdot 100$
                      $A = 9500$

The remaining area of the parking lot is $99,000 - 9500 = 89,500$ ft.^{2}.

Chapter 1 Practice Test

1. 5 is in the ten thousands place.

2. 48,210,907
   $= 4 \times 10,000,000 + 8 \times 1,000,000 + 2 \times 100,000 + 1 \times 10,000 + 9 \times 100 + 7 \times 1$
   $= 4 \times 10^7 + 8 \times 10^6 + 2 \times 10^5 + 1 \times 10^4 + 9 \times 10^2 + 7 \times 1$

3. $19,304 > 19,204$, because 19,304 is greater than 19,204.

4. sixty-seven million, one hundred ninety-four thousand, two hundred ten

5. $\begin{array}{r} 2,300,000 \\ + \ 500,000 \\ \hline 2,800,000 \end{array}$ 6. $\begin{array}{r} 63,209 \\ +4,981 \\ \hline 68,190 \end{array}$ 7. $\begin{array}{r} 480,091 \\ -54,382 \\ \hline 425,709 \end{array}$

8. $27 + x = 35$ Check: $27 + 8 = 35$
         $x = 35 - 27$ $35 = 35$
         $x = 8$

9. $\begin{array}{r} 6910 \\ \times 357 \\ \hline 48370 \\ 34550 \\ 20730 \\ \hline 2,466,870 \end{array}$ 10. $4^3 = 4 \times 4 \times 4 = 64$

11. $\begin{array}{r} 1607 \\ 28 \overline{)45019} \\ \underline{28} \\ 170 \\ \underline{168} \\ 219 \\ \underline{196} \\ 23 \end{array}$ Answer: 1607 r23

12. $7 \cdot y = 126$ Check: $7 \cdot 18 = 126$
         $y = 126 \div 7$ $126 = 126$
         $y = 18$

13. 14

14. $14 + 2^5 - 8 \cdot 3$
    $= 14 + 32 - 8 \cdot 3$
    $= 14 + 32 - 24$
    $= 46 - 24$
    $= 22$

15. $20 - 6(3 + 2) \div 15$
    $= 20 - 6 \cdot 5 \div 15$
    $= 20 - 30 \div 15$
    $= 20 - 2$
    $= 18$

16. $\dfrac{(14+26)}{8} - \sqrt{25-9}$ 17. $\dfrac{5^2 + 47}{(4+2)^2}$

$= \dfrac{40}{8} - \sqrt{16}$ $= \dfrac{5^2 + 47}{6^2}$

$= \dfrac{40}{8} - 4$ $= \dfrac{25 + 47}{36}$

$= 5 - 4$ $= \dfrac{72}{36}$

$= 1$ $= 2$

18. a. 23,000,000: The 2 is in the millions place. Look at the number to the right of the 2 which is 7. Since 7 is greater than 5 we round the 2 up.

    b. Arrange the number in order from least to greatest. 22,751, 27,377, 27,380, 29,316, 30,135 Choose the number in the middle which is 27,380. The median is 27,380,000 because the visitors are in thousands.

19. $29,316,000$
    $\underline{-27,377,000}$
      $1,939,000$

20. We must divide the total number of students by the number allowed in a single class. $2487 \div 30 = 82 \text{ r}27$. Since there cannot be more than 30 students in a class, we must round up to 83 classes to accommodate the remaining 27 students.

21. a. We divide the sum of the scores by the number of scores.
    $$\frac{(84 + 98 + 86 + 100 + 97 + 93)}{6} = \frac{558}{6} = 93$$

    b. Arrange the numbers in order from least to greatest.
    84, 86, 93, 97, 98, 100
    Since there are an even number of scores, we find the arithmetic mean of the middle two scores, which are 93 and 97.
    $$\frac{93 + 97}{2} = \frac{190}{2} = 95$$

22. $A = bh$
    $A = 18 \cdot 9$
    $A = 162 \text{ ft.}^2$

23. a. Find the perimeter of the playground.

    $P = 2l + 2w$
    $P = 2 \cdot 80 + 2 \cdot 65$
    $P = 160 + 130$
    $P = 290 \text{ ft.}$

    b. Multiply the total perimeter by the cost per foot.

    $290 \times 6 = \$1740$

It will cost $1740 to put a fence around the playground.

24. $V = lwh$
    $V = 8 \cdot 6 \cdot 7$
    $V = 336 \text{ ft.}^3$

25. Find the area of the wall.

    $A = bh$
    $A = 16 \cdot 9$
    $A = 144 \text{ ft.}^2$

    Find the area of the door.

    $A = bh$
    $A = 3 \cdot 7$
    $A = 21 \text{ ft.}^2$

    Subtract the area of the door from the area of the wall. Total area to be painted is $144 - 21 = 123$ ft.^{2}.

Chapter 2 Integers

Exercise Set 2.1

1. ...,−3, −2, −1, 0, 1, 2, 3, ...   3. positive

5. zero   7. positive   9. +450

11. −35   13. +29,028 15. −13,200

17. +100   19. −20   21. −8°F

23. −20°F

25. The average verbal score of students whose parents have no high school diploma is 93 points below the overall average verbal score.

27. The average math score of students whose parents have a high school diploma is 44 points below the overall average math score.

29. The average verbal score of students whose parents have a bachelor's degree is 15 points above the overall average verbal score.

31. The average math score of students whose parents have a bachelor's degree is 13 points above the overall average math score.

33.
−1 0 1 2 3 4 5 6 7 8 9

35.
−2 −1 0 1 2 3 4 5 6 7 8

37.
−10 −9 −8 −7 −6 −5 −4 −3 −2 −1 0

39.
−7 −6 −5 −4 −3 −2 −1 0 1 2 3

41.
−5 −4 −3 −2 −1 0 1 2 3 4 5

43. −20 < 17: The number 17 is farther to the right on a number line, so it is larger.

45. 20 > 16: The number 20 is farther to the right on a number line, so it is larger.

47. −30 < −16: The number −16 is farther to the right on a number line, so it is larger.

49. −9 > −12: The number −9 is farther to the right on a number line, so it is larger.

51. 0 > −16: The number 0 is farther to the right on a number line, so it is larger.

53. −19 < −17: The number −17 is farther to the right on a number line, so it is larger.

55. 26   57. 21   59. 18   61. 10   63. 0

65. 14   67. 18   69. 2004 71. 8   73. 0

75. 47

77. 18: The additive inverse of −18 is 18.

79. −61: The additive inverse of 61 is −61.

81. 0: The additive inverse of 0 is 0.

83. −8: The additive inverse of the additive inverse of −8 is −8.

85. 43: The additive inverse of the additive inverse of 43 is 43.

87. −6: The additive inverse of 6 is −6.

89. 14: The additive inverse of −14 is 14.

91. 0: The additive inverse of 0 is 0.

93. −63: The additive inverse of 63 is −63.

95. 4: The additive inverse of −4 is 4. Or an even number of minus signs will have a positive outcome.

97. −12: The additive inverse of the additive inverse of −12 is −12. Or an odd number of minus signs will have a negative outcome.

99. 87: The additive inverse of the additive inverse of the additive inverse −87 is 87. Or an even number of minus signs will have a positive outcome.

101. The absolute value of negative four is 4.

103. The additive inverse of negative eight is 8.

105. The additive inverse of the additive inverse of fourteen is 14.

107. The additive inverse of the absolute value of negative five is −5.

109. The additive inverse of the absolute value of the additive inverse negative four is −4.

Review Exercises

1. 84,759
    +9,506
    ‾‾‾‾‾
    94,265

2. 900,406
    −35,918
    ‾‾‾‾‾
    864,488

3. $\quad\quad\quad 457$
    $\quad\quad\underline{\times 609}$
    $\quad\quad\;4113$
    $\quad\quad\;0000$
    $\quad\quad\underline{2742\quad}$
    $\quad\;278,313$

4. $42\overline{)85314}$ Ans: 2031 r12
    $\quad\quad\;\;2031$
    $\quad\quad\underline{84\;}$
    $\quad\quad\;131$
    $\quad\quad\underline{126}$
    $\quad\quad\;\;54$
    $\quad\quad\;\;\underline{42}$
    $\quad\quad\;\;12$

5. ending balance = beginning balance + total deposits – total withdraws

    $452 + 45 + 98 + 88 + 54 + (-220) + (-25) + (-10)$
    $= \$482$

    The ending balance is $482.

Exercise Set 2.2

1. Adding a debt of $6 to a debt of $24 increases the debt to $30, so the result is –30.

3. add, keep the same sign

5. $14 + 9 = 23$ 7. $-15 + (-8) = -23$

9. $-9 + (-6) = -15$

11. $12 + 18 + 16 = 30 + 16$
    $\quad\quad\quad\quad\quad\quad = 46$

13. $-38 + (-17) + (-21) = -55 + (-21)$
    $\quad\quad\quad\quad\quad\quad\quad\quad\;\; = -76$

15. $28 + (-18) = 10$ 17. $-34 + 20 = -14$

19. $-21 + 35 = 14$ 21. $35 + (-53) = -18$

23. $-24 + 80 = 56$

25. Two credits, so we add and the result is a credit/positive; 138.

27. Two debts, so we add and the result is a debt/negative; –72.

29. A credit with a debt, so we subtract and since the credit 84 is more than the debt, the result is a credit/positive; 61.

31. A debt with a credit, so we subtract and since the credit 42 is more than the debt, the result is a credit/positive; 27.

33. A debt with a credit, so we subtract and since the debt, –81 is more than the credit, the result is a debt/negative; –21.

35. A credit with a debt, so we subtract and since the debt, –58 is more than the credit, the result is a debt/negative; –21.

37. Paying 45 on a 45 debt means the balance is 0.

39. The total credits are 64 and the total debts are 30. Since the credits outweigh the debts, we end up with a credit of 34.

41. The total credits are 94 and the total debts are 62. Since the credits outweigh the debts, we end up with a credit of 32.

43. The total credits are 50 and the total debts are 50. Paying 50 on a 50 debt means the balance is 0.

45. The total credits are 47 and the total debts are 106. Since the debts outweigh the credits, we end up with a debt of 59; –59.

47. We must find Jason's new checking account balance. Write an addition statement adding all assets and debts.

    $24 + (-40) + (-17)$
    $= 24 + (-57)$
    $= -\$33$

    Since the total debt –57 is a number with higher absolute value than the total assets 24, Jason's balance is a debt/negative amount of –$33.

49. Net worth is the sum of all assets and debts. Write an addition statement adding all assets and debts.

    $1498 + 2148 + 18,901 + 3845$
    $+ (-1841) + (-74,614) + (-5488)$
    $= 26,392 + (-81,943)$
    $= -55,551$

    Since the total debt –81,943 is a number with larger absolute value than the total assets 26,392, the Smith's net worth was a debt/negative amount of –$55,551.

51. We must find Lea's new credit card balance. Write an addition statement adding the closing balance, all charges, and the payment.

    $-125 + (-12) + (-8) + (-120) + (-22) + (-9)$
    $+ 325$
    $= \$29$

    Lea has a credit of $29.

53. We must calculate the expected temperature. Since the summer temperature is –21° F and it is to drop another 57° F, we can add to get the expected temperature.

    Expected temperature: $-21 + (-57) = -78$

    The expected night temperature is –78°F.

55. We must find the depth of the submarine. Since the –147 ft. is its starting position and it is to be

raised 69 ft., we can add to get the current elevation.

Current elevation: $-147 + 69 = -78$ ft.

The current elevation of the submarine is –78 ft. i.e. 78 ft. below the surface.

57. Draw a diagram to get a sense of the direction of the forces. Since weight is a force due to gravity pulling downward on objects, the force has a negative value. The wires are pulling upward against gravity so they have positive values.

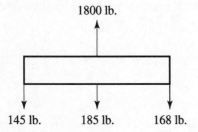

1800 lb.

145 lb.   185 lb.   168 lb.

To find the resultant force, we compute the sum of the forces. Resultant force:
$$-745 + (-300) + (-145) + (-185) + (-168) + 1800$$
$$= -1543 + 1800$$
$$= 257 \text{ lb.}$$

The resultant force is 257 lb. Since the resultant force is positive, it means the elevator is traveling upward.

59. a.

	No H. S. diploma	H. S. diploma	Associate degree
Expression	$508 + (-93)$	$508 - (39)$	$508 + (-22)$
Sum	415	469	486

	Bachelor's degree	Graduate degree
Expression	$508 + 15$	$508 + 50$
Sum	523	558

b. Students whose parents achieved higher levels of education tend to perform better on the SAT.

Review Exercises

1. 42,561,009
$$= 4 \times 10,000,000 + 2 \times 1,000,000 + 5 \times 100,000$$
$$+ 6 \times 10,000 + 1 \times 1000 + 9 \times 1$$
$$= 4 \times 10^7 + 2 \times 10^6 + 5 \times 10^5 + 6 \times 10^4$$
$$+ 1 \times 10^3 + 9 \times 1$$

2. two million, four hundred seven thousand, six

3. $\begin{array}{r} 60,041 \\ -4,596 \\ \hline 55,445 \end{array}$

4. $15 + n = 28$ Check: $15 + 13 = 28$
     $n = 28 - 15$ $28 = 28$
     $\quad n = 13$

5.   Find the total perimeter and subtract the two gates. $P = 98 + 345 + 212 + 380 = 1035$ ft. Now, subtract the two gates; $1035 - 20 = 1015$ ft. To find the cost of the fencing multiply: $1015 \times 7 = \$7105$ and find the cost of the gates $2 \times 80 = \$160$. The total cost is the fence plus the gates: $7105 + 160 = \$7265$.

Exercise Set 2.3

1. Change the operation sign from – to + and change the subtrahend (second number) to its additive inverse.

3. Answers will vary but will have the form of a negative number subtracted from positive number, as in $8 - (-2)$.

5. $18 - 25 = 18 + (-25)$
     $\quad\quad = -7$

7. $-15 - 18 = -15 + (-18)$
     $\quad\quad\quad = -33$

9. $20 - (-8) = 20 + 8$
     $\quad\quad\quad = 28$

11. $-14 - (-18) = -14 + 18$
     $\quad\quad\quad\quad = 4$

13. $-15 - (-8) = -15 + 8$
     $\quad\quad\quad\quad = -7$

15. $0 - (-5) = 0 + 5$
     $\quad\quad\quad = 5$

17. $-21 - 19 = -21 + (-19)$
     $\quad\quad\quad\quad = -40$

19. $-4 - (-19) = -4 + 19$
     $\quad\quad\quad\quad = 15$

21. $31 - 44 = 31 + (-44)$
     $\quad\quad\quad = -13$

23. $35 - (-10) = 35 + 10$
     $\quad\quad\quad\quad = 45$

25. $-28 - (-16) = -28 + 16$
     $\quad\quad\quad\quad = -12$

27. $0 - 18 = 0 + (-18)$
$\qquad = -18$

29. $-27 - (-16) = -27 + 16$
$\qquad\qquad = -11$

31. $18 + t = 12$ Check: $18 + (-6) = 12$
$\quad\ t = 12 - 18$ $12 = 12$
$\quad\ t = 12 + (-18)$
$\quad\ t = -6$

33. $d + (-6) = 9$ Check: $15 + (-6) = 9$
$\quad\ d = 9 - (-6)$ $9 = 9$
$\quad\ d = 9 + 6$
$\quad\ d = 15$

35. $-28 + u = -15$ Check: $-28 + 13 = -15$
$\quad\ u = -15 - (-28)$ $-15 = -15$
$\quad\ u = -15 + 28$
$\quad\ u = 13$

37. $-14 + m = 0$ Check: $-14 + 14 = 0$
$\quad\ m = 0 - (-14)$ $0 = 0$
$\quad\ m = 0 + 14$
$\quad\ m = 14$

39. $h + (-35) = -40$ Check: $-5 + (-35) = -40$
$\quad\ h = -40 - (-35)$ $-40 = -40$
$\quad\ h = -40 + 35$
$\quad\ h = -5$

41. $-13 + k = -25$ Check: $-13 + (-12) = -25$
$\quad\ k = -25 - (-13)$ $-25 = -25$
$\quad\ k = -25 + 13$
$\quad\ k = -12$

43. We must find Belinda's bank account balance.
    a) A check is a deduction, so we subtract it from
       the initial balance. $126 - 245$
$$= 126 + (-245)$$
$$= -\$119$$
    Belinda's balance after the check clears is
    $-\$119$.
    b) Both the check and service charge are
       deductions, so we subtract them from the
       initial balance.
$$(-119) + (-20) = -\$139$$
      Belinda's balance after the bank assesses her
      account a service charge is $-\$139$.

45. We must find the net given the revenue and cost.
    The formula for net is $N = R - C$.
    $N = R - C$
    $N = 14,363 - 12,137$
    $N = \$2,226$ million

The net for 2005 was $2,226 million, which means the business had a profit of $2,226,000,000.

47. We must find the net given the revenue and the cost. The formula for net is $N = R - C$.
    $N = R - C$
    $N = 4500 - 26,458$
    $N = -\$21,958$

The net for the car is $-\$21,958$ which means Florence had a loss of $21,958.

49. The year 1980 had the greatest net loss. The formula for net is $N = $ Income $-$ Disbursements.
    $N = Income - Disbursements$
    $N = 100,051,000,000 - 103,228,000,000$
    $N = -\$3,177,000,000$

51. Since we are asked to find the amount of decrease, we must calculate a difference. Find the difference between 19 and -27.

    temperature difference: $19 - (-27) = 19 + 27$
$$= 46°F$$

From 19° F to -27° F is a decrease of 46°F.

53. "How much colder" indicates a difference.

    temperature difference:
    $-218 - (-273) = -218 + 273$
$$= 55°C$$

It will have to get 55°C colder.

55. Find the minimum deposit to avoid further service charges. Write a missing addend equation, then solve.

    $-37 + x = 30$
       $x = 30 - (-37)$
       $x = 30 + 37$
       $x = \$67$

Derrick must deposit $67 into his account to avoid further charges.

57 Find the distance from the lowest level of the Sears Towers to the sky deck. Write a missing addend equation, then solve.

    $-43 + x = 1353$
       $x = 1353 - (-43)$
       $x = 1353 + 43$
       $x = 1396$ ft.

A person will rise 1396 feet going from the lowest level to the sky deck in the Sears Towers.

59. a.

	Barrow, Alaska	Duluth, Minnesota	Nashville, Tennessee
Expression	$-8-(-20)$	$18-(-1)$	$46-28$
Difference	12	19	18

    b. Duluth, Minnesota

Review Exercises

1.   145
    $\times 209$
    1305
    0000
    29000
    30,305

2. $3^5 = 3 \cdot 3 \cdot 3 \cdot 3 \cdot 3 = 243$

3. $3 \cdot y = 24$
    $y = 24 \div 3$
    $y = 8$

4. $\sqrt{225} = 15$

5. Find the area of the rectangular strip. $A = lw$
    $A = (8)(2)$
    $A = 16$

Subtract the amount of the strip from the total Jaqueline owns. $20 - 16 = 4$ mi.^{2}. She will have 4 mi.^{2} left.

Exercise Set 2.4

1. positive.
3. positive
5. two

7. Find the additive inverse of the principal square root of n, which is the negative square root of n.

9. $-2 \cdot 16 = -32$

11. $-7 \cdot 9 = -63$

13. $4 \cdot (-2) = -8$

15. $13 \cdot (-5) = -65$

17. $0 \cdot (-9) = 0$

19. $(-1) \cdot (-32) = 32$

21. $-13 \cdot 0 = 0$

23. $15 \cdot (-1) = -15$

25. $-21 \cdot (-8) = 168$

27. $19(-20) = -380$

29. $(-1)(-5)(-7) = 5(-7)$
                     $= -35$

31. $(-1)(-1)(-6)(-9) = 1(-6)(-9)$
                            $= -6(-9)$
                            $= 54$

33. $(-20)(-2)(-1)(3)(-1) = 40(-1)(3)(-1)$
                                   $= -40(3)(-1)$
                                   $= -120(-1)$
                                   $= 120$

35. $(-1)^2 = (-1)(-1)$
          $= 1$

37. $(-3)^2 = (-3)(-3)$
          $= 9$

39. $(-7)^2 = (-7)(-7)$
          $= 49$

41. $(-4)^3 = (-4)(-4)(-4)$
          $= 16(-4)$
          $= -64$

43. $(-3)^4 = (-3)(-3)(-3)(-3)$
          $= 9(-3)(-3)$
          $= -27(-3)$
          $= 81$

45. $(-2)^6 = (-2)(-2)(-2)(-2)(-2)(-2)$
          $= 4(-2)(-2)(-2)(-2)$
          $= -8(-2)(-2)(-2)$
          $= 16(-2)(-2)$
          $= -32(-2)$
          $= 64$

47. $-8^2 = -(8 \cdot 8)$
       $= -64$

49. $-10^6 = -(10 \cdot 10 \cdot 10 \cdot 10 \cdot 10 \cdot 10)$
         $= -(100 \cdot 10 \cdot 10 \cdot 10 \cdot 10)$
         $= -(1000 \cdot 10 \cdot 10 \cdot 10)$
         $= -(10,000 \cdot 10 \cdot 10)$
         $= -(100,000 \cdot 10)$
         $= -(1,000,000)$
         $= -1,000,000$

51. $-1^2 = -(1 \cdot 1)$
       $= -(1)$
       $= -1$

53. $-3^2 = -(3 \cdot 3)$
       $= -(9)$
       $= -9$

55. $36 \div (-3) = -12$

57. $-81 \div 27 = -3$

59. $-32 \div (-4) = 8$

61. $0 \div (-2) = 0$

63. $31 \div (-1) = -31$

65. $\dfrac{65}{-13} = -5$

67. $\dfrac{-41}{0}$ is undefined.

69. $\dfrac{-124}{-4} = 31$

71. $-\dfrac{28}{4} = -7$

73. $4x = 12$

   $x = \dfrac{12}{4}$

   $x = 3$

   Check: $4(3) = 12$

   $12 = 12$

75. $9x = -18$

   $x = \dfrac{-18}{9}$

   $x = -2$

   Check: $9(-2) = -18$

   $-18 = -18$

77. $-12t = 48$

   $t = \dfrac{48}{-12}$

   $t = -4$

   Check: $-12(-4) = 48$

   $48 = 48$

79. $-6m = -54$

   $m = \dfrac{-54}{-6}$

   $m = 9$

   Check: $-6(9) = -54$

   $-54 = -54$

81. $-2a = -24$

   $a = \dfrac{-24}{-2}$

   $a = 12$

   Check: $-2(12) = -24$

   $-24 = -24$

83. $0v = -14$

   $v = \dfrac{-14}{0}$

   no solution

85. $-1c = 17$

   $c = \dfrac{17}{-1}$

   $c = -17$

   Check: $-1(-17) = 17$

   $17 = 17$

87. $-18m = 0$

   $m = \dfrac{0}{-18}$

   $m = 0$

   Check: $-18(0) = 0$

   $0 = 0$

89. $-2(-5)d = -50$

   $10d = -50$

   $d = \dfrac{-50}{10}$

   $d = -5$

   Check: $-2(-5)(-5) = -50$

   $10(-5) = -50$

   $-50 = -50$

91. $-1(-1)(-7)g = -63$

   $1(-7)g = -63$

   $-7g = -63$

   $g = \dfrac{-63}{-7}$

   $g = 9$

   Check: $-1(-1)(-7)(9) = -63$

   $1(-7)(9) = -63$

   $-7(9) = -63$

   $-63 = -63$

93. $\pm\sqrt{16} = \pm 4$ because $4^2 = 16$ and $(-4)^2 = 16$.

95. $\pm\sqrt{36} = \pm 6$ because $6^2 = 36$ and $(-6)^2 = 36$.

97. $\pm\sqrt{100} = \pm 10$ because $10^2 = 100$ and $(-10)^2 = 100$.

99. There is no integer solution because there is no integer that can be squared to equal -81.

101. $\sqrt{81} = 9$ because $9^2 = 81$

103. $\sqrt{64} = 8$ because $8^2 = 64$

105. $\sqrt{49} = 7$ because $7^2 = 49$

107. There is no integer solution because there is no integer that can be squared to equal -169.

109. $-\sqrt{121} = -11$ We are being asked to find the additive inverse of the positive square root.

111. $\sqrt{0} = 0$ because $0^2 = 0$

113. Since her balance tripled, we are repeatedly adding the same amount three times. Since multiplication is repeated addition, we can simply multiply to calculate the total.
   $3(-214) = -\$642$

   Rohini has a total balance of $-\$642$.

115. Since they were assessed charges on 7 occasions, we are repeatedly adding the same amount seven times. $7(-17) = -\$119$

   The Morrison's paid $\$119$ in insufficient funds charges.

117. Since she agreed to equal monthly payments, we must take the total and divide it by the number of months. Since 3 years equals 36 months, the total must be divided by 36.

   $1656 \div 36 = \$46$

Each payment should be $46.

Review Exercises

1. Find the perimeter of the room. Add all the sides. $P = 14 + 14 + 16 + 16 = 60$ ft.

2. $19 - 6 \cdot 3 + 2^3 = 19 - 6 \cdot 3 + 8$
    $$= 19 - 18 + 8$$
    $$= 1 + 8$$
    $$= 9$$

3. $7\sqrt{64} - 2(5 + 3) = 7\sqrt{64} - 2(8)$
    $$= 7 \cdot 8 - 2(8)$$
    $$= 56 - 16$$
    $$= 40$$

4. Find the area of the wall and the area of the window. We must subtract the area of the window from the area of the wall to find the area to be painted.

    area of wall: $A = lw$
    $$A = 15 \cdot 10$$
    $$A = 150 \text{ ft.}^2$$

    area of window: $A = lw$
    $$A = 4 \cdot 3$$
    $$A = 12 \text{ ft.}^2$$

    total area to be painted: $150 - 12 = 138 \text{ ft.}^2$

5. We must find the volume of two boxes.

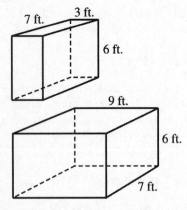

    box 1: $V = lwh$
    $$V = 9 \cdot 7 \cdot 6$$
    $$V = 378 \text{ ft.}^3$$

    box 2: $V = lwh$
    $$V = 3 \cdot 7 \cdot 6$$
    $$V = 126 \text{ ft.}^3$$

    The total volume is $378 + 126 = 504 \text{ ft.}^3$

Exercise Set 2.5

1. 1. Grouping symbols

    2. Exponents or roots from left to right

    3. Multiplication or division from left to right

    4. Addition or Subtraction from left to right

3. Subtract 8 from 3 within the parentheses. We work within grouping symbols before performing any other operations.

5. $3 - 10 \div 2 = 3 - 5$
    $$= -2$$

7. $9 + 4 \cdot (-6) = 9 + (-24)$
    $$= -15$$

9. $3 - 2 \cdot 4 + 11 = 3 - 8 + 11$
    $$= -5 + 11$$
    $$= 6$$

11. $9 + 20 \div (-4) \cdot 3 = 9 + (-5) \cdot 3$
    $$= 9 + (-15)$$
    $$= -6$$

13. $3 - (-2)^2 = 3 - 4$
    $$= -1$$

15. $-5 + (-4)^2 - 1 = -5 + 16 - 1$
    $$= 11 - 1$$
    $$= 10$$

17. $9^2 - 6 \cdot (-4) \div 8 = 81 - 6 \cdot (-4) \div 8$
    $$= 81 - (-24) \div 8$$
    $$= 81 - (-3)$$
    $$= 81 + 3$$
    $$= 84$$

19. $3 + (4 - 6^2) = 3 + (4 - 36)$
    $$= 3 + (-32)$$
    $$= -29$$

21. $3 + (4 - 6)^2 = 3 + (-2)^2$
    $$= 3 + 4$$
    $$= 7$$

23. $\sqrt{25 - 9} = \sqrt{16}$
    $$= 4$$

25. $\sqrt{25} - \sqrt{9} = 5 - 3$
    $$= 2$$

27. $(-2)^4 = (-2)(-2)(-2)(-2)$
    $= 16$

29. $-2^5 = -(2 \cdot 2 \cdot 2 \cdot 2 \cdot 2)$
    $= -32$

31. $-(-3)^2 = -\left[(-3) \cdot (-3)\right]$
    $= -9$

33. $\left[-(-3)\right]^2 = \left[3\right]^2$
    $= 9$

35. $9 \div 3 - (-2)^2 = 9 \div 3 - 4$
    $= 3 - 4$
    $= -1$

37. $3 - (-2)(-3)^2 = 3 - (-2)(9)$
    $= 3 - (-18)$
    $= 3 + 18$
    $= 21$

39. $\left|18 - 5(-4)\right| = \left|18 - (-20)\right|$
    $= \left|18 + 20\right|$
    $= \left|38\right|$
    $= 38$

41. $-\left|-43 + 6 \cdot 4\right| = -\left|-43 + 24\right|$
    $= -\left|-19\right|$
    $= -19$

43. $-5\left|26 \div 13 - 7 \cdot 2\right| = -5\left|2 - 14\right|$
    $= -5\left|-12\right|$
    $= -5 \cdot 12$
    $= -60$

45. $(-3)^2 - 2\left[3 - 5(1+4)\right] = (-3)^2 - 2\left[3 - 5(5)\right]$
    $= (-3)^2 - 2\left[3 - 25\right]$
    $= (-3)^2 - 2\left[-22\right]$
    $= 9 - 2\left[-22\right]$
    $= 9 - \left[-44\right]$
    $= 9 + 44$
    $= 53$

47. $28 \div (-7) + \sqrt{49} + (-3)^3 = 28 \div (-7) + 7 + (-27)$
    $= -4 + 7 + (-27)$
    $= 3 + (-27)$
    $= -24$

49. $12 + 8\sqrt{81} \div (-6) = 12 + 8 \cdot 9 \div (-6)$
    $= 12 + 72 \div (-6)$
    $= 12 + (-12)$
    $= 0$

51. $4^2 - 7 \cdot 5 + \sqrt{121} - 21 \div (-7)$
    $= 16 - 7 \cdot 5 + 11 - 21 \div (-7)$
    $= 16 - 35 + 11 - (-3)$
    $= 16 - 35 + 11 + 3$
    $= -19 + 11 + 3$
    $= -8 + 3$
    $= -5$

53. $15 + (-3)^3 + (-5)\left|14 + (-2)9\right|$
    $= 15 + (-3)^3 + (-5)\left|14 + (-18)\right|$
    $= 15 + (-3)^3 + (-5)\left|-4\right|$
    $= 15 + (-3)^3 + (-5) \cdot 4$
    $= 15 + (-27) + (-5) \cdot 4$
    $= 15 + (-27) + (-20)$
    $= -12 + (-20)$
    $= -32$

55. $(-15 + 12) + 5(-6) - \left[9 - (-12)\right] \div 3$
    $= -3 + 5(-6) - \left[9 + 12\right] \div 3$
    $= -3 + 5(-6) - 21 \div 3$
    $= -3 + (-30) - 7$
    $= -33 - 7$
    $= -40$

57. $(-2) \cdot (5 - 8)^2 \div 6 + (4 - 2)$
    $= (-2) \cdot (-3)^2 \div 6 + 2$
    $= (-2) \cdot 9 \div 6 + 2$
    $= -18 \div 6 + 2$
    $= -3 + 2$
    $= -1$

59. $-\left|38 - 14 \cdot 2\right| + 44 \div (5 - (-6)) - 2^4$
    $= -\left|38 - 28\right| + 44 \div (5 + 6) - 2^4$
    $= -\left|10\right| + 44 \div 11 - 2^4$
    $= -10 + 44 \div 11 - 2^4$
    $= -10 + 44 \div 11 - 16$
    $= -10 + 4 - 16$
    $= -6 - 16$
    $= -22$

61. $\left\{19-2\left[4+(-9)\right]\right\}-18\div\sqrt{25-16}$
$=\left\{19-2\left[-5\right]\right\}-18\div\sqrt{9}$
$=\left\{19-\left[-10\right]\right\}-18\div\sqrt{9}$
$=\left\{19+10\right\}-18\div\sqrt{9}$
$=29-18\div\sqrt{9}$
$=29-18\div 3$
$=29-6$
$=23$

63. $[(-14+2)+5]\div 0+9^4\cdot 7$
$=[-12+5]\div 0+9^4\cdot 7$
$=-7\div 0+9^4\cdot 7$

Undefined because we cannot divide by zero.

65. $\left\{(-3)^2+4[(8-20)\div(-2)]\right\}+(-5)\sqrt{25\cdot 4}$
$=\left\{(-3)^2+4[-12\div(-2)]\right\}+(-5)\sqrt{100}$
$=\left\{(-3)^2+4\cdot 6\right\}+(-5)\sqrt{100}$
$=\left\{9+4\cdot 6\right\}+(-5)\sqrt{100}$
$=\left\{9+24\right\}+(-5)\sqrt{100}$
$=33+(-5)\sqrt{100}$
$=33+(-5)\cdot 10$
$=33+(-50)$
$=-17$

67. $[30-3(-4)]\div[(15-6)+(-2)]+(5-3)^4$
$=[30-(-12)]\div[9+(-2)]+2^4$
$=[30+12]\div 7+2^4$
$=42\div 7+2^4$
$=42\div 7+16$
$=6+16$
$=22$

69. $\left\{\left|-12(-5)-38\right|\div 2\right\}\div 3[-9+\sqrt{49}]^3$
$=\left\{\left|60-38\right|\div 2\right\}\div 3[-9+7]^3$
$=\left\{\left|22\right|\div 2\right\}\div 3[-2]^3$
$=\left\{22\div 2\right\}\div 3\cdot(-8)$
$=24\div 3\cdot(-8)$
$=8\cdot(-8)$
$=-64$

71. $\left\{2[14-11]+\sqrt{(4)(-9)}\right\}-\left[12-6\cdot 9\right]$
$=\{2\cdot 3+\sqrt{-36}\}-[12-54]$

Because $\sqrt{-36}$ has no integer solution, this problem cannot be simplified.

73. $\dfrac{6(-4)+16}{(4+7)-9}=\dfrac{6(-4)+16}{11-9}$
$=\dfrac{-24+16}{11-9}$
$=\dfrac{-8}{2}$
$=-4$

75. $\dfrac{3(-12)+1}{3^2-2}=\dfrac{3(-12)+1}{9-2}=$
$=\dfrac{-36+1}{9-2}$
$=\dfrac{-35}{7}$
$=-5$

77. $\dfrac{(5-19)+3^3}{(-2)(-6)-5^2}=\dfrac{-14+3^3}{(-2)(-6)-5^2}$
$=\dfrac{-14+27}{(-2)(-6)-25}$
$=\dfrac{-14+27}{12-25}$
$=\dfrac{13}{-13}$
$=-1$

79. $\dfrac{3\cdot(-6)-6^2}{-2(2-5)^2}=\dfrac{3\cdot(-6)-6^2}{-2(-3)^2}$
$=\dfrac{3\cdot(-6)-36}{-2(9)}$
$=\dfrac{-18-36}{-2(9)}$
$=\dfrac{-54}{-18}$
$=3$

81. $\dfrac{2[5(3-7)+(-3)^3]-6}{15\cdot 5-(12-7)^2}=\dfrac{2[5(-4)+(-3)^3]-6}{15\cdot 5-5^2}$
$=\dfrac{2[5(-4)+(-27)]-6}{15\cdot 5-25}$
$=\dfrac{2[-20+(-27)]-6}{75-25}$
$=\dfrac{2[-47]-6}{75-25}$
$=\dfrac{-94-6}{50}$
$=\dfrac{-100}{50}$
$=-2$

83.

$$\frac{3[14-2(9)]+(5-11)^2}{(9-3)^2-4(10-1)} = \frac{3[14-18]+(-6)^2}{6^2-4\cdot9}$$

$$= \frac{3[-4]+(-6)^2}{6^2-4\cdot9}$$

$$= \frac{3[-4]+36}{36-4\cdot9}$$

$$= \frac{-12+36}{36-36}$$

$$= \frac{-12+36}{0}$$

$$= \frac{24}{0}$$

undefined

85. $(-2)^4$ means to find the fourth power of -2, which is $(-2)(-2)(-2)(-2)=16$. -2^4 means to find the additive inverse of the fourth power of 2, which is $-(2\cdot2\cdot2\cdot2)=-16$.

87. The exponent is odd.

89. Mistake: Subtraction was performed before the multiplication.

Correct:
$$28-5(24-30) = 28-5(-6)$$
$$= 28-(-30)$$
$$= 28+30$$
$$= 58$$

91. Mistake: The exponential form was simplified before performing the operations within the parenthesis.

Correct:
$$4-(9-4)^2 = 4-5^2$$
$$= 4-25$$
$$= -21$$

93. Mistake: The brackets were eliminated before performing all the operations inside them.

Correct:
$$34-[3\cdot5-(14+8)] = 34-[3\cdot5-22]$$
$$= 34-[15-22]$$
$$= 34-[-7]$$
$$= 34+7$$
$$= 41$$

95. Mistake: The square roots were found before subtracting in the radical.

Correct:
$$\sqrt{169-25} = \sqrt{144}$$
$$= 12$$

Review Exercises

1. Subtract the check amounts from her opening balance. $185-45-68-95=-23$ Felicia has a $-\$23$ balance. She is overdrawn.

2.
$$N = R-C$$
$$N = 32-245$$
$$N = -\$213$$

Tina's net is $-\$213$. It is a loss.

3. This is a missing addend situation.
$$-98+x = -25$$
$$x = -25-(-98)$$
$$x = -25+98$$
$$x = 73$$

The submarine ascended 73 ft.

4. $3(-235)=-705$ Jeff's new balance is $-\$705$.

Exercise Set 2.6

1. N represents the net, R represents revenue, and C represents cost

3. V represents voltage, I represents current, and r represents resistance.

5. **Understand:** We must calculate a net amount. To calculate the net we need total revenue and total cost. The formula for net is $N = R-C$. Karen's total revenue is the amount that she sold the paintings for, which was given to be $1500. Her total cost is the sum of all the money she spent for the paintings.

Plan: Calculate total cost, then subtract the cost from the total revenue to get the net.

Execute: total cost = cost for three paintings + cost for two paintings

$$\text{cost} = 3(250)+2(400)$$
$$\text{cost} = 750+800$$
$$\text{cost} = \$1550$$

Now replace R with 1500 and C with 1550 in $N = R-C$ to get the net.

$$N = R-C$$
$$N = 1500-1550$$
$$N = -\$50$$

Answer: Karen's net is $-\$50$, which is a loss of $50.

Check: Reverse the process. The net added to cost should produce the revenue.

$$-50+1550 = 1500$$
$$1500 = 1500$$

Subtracting the costs for the two paintings from the total cost should produce the cost for the three paintings.

$$1550 - 2(400) = 3(250)$$
$$1550 - 800 = 750$$
$$750 = 750$$

7. **Understand:** We must calculate a net amount. To calculate the net we need total revenue and total cost. The formula for net is $N = R - C$. Darwin's total revenue is the amount that he sold the car for, which was given to be $4000. His total cost is the sum of all the money he spent on the car.

Plan: Calculate total cost, then subtract the cost from the total revenue to get the net.

Execute: cost = amount down + total of all payments + maintenance costs

cost = 800+60(288)+950
cost = $800 + 17,280 + 950$
cost = $19,030

Now replace R with 4000 and C with 19,030 in $N = R - C$ to get the net.

$N = R - C$
$N = 4000 - 19,030$
$N = -\$15,030$

Answer: Darwin's net is -$15,030, which is a loss of $15,030.

Check: Reverse the process. The net added to cost should produce the revenue.

$-15,030 + 19,030 = 4000$
$4000 = 4000$

Subtracting the maintenance costs and total of all payments from the total cost should produce the down payment.

$19,030 - 950 - 17,280 = 800$
$800 = 800$

Dividing the total of all the payments by 60 should produce the amount of each payment.

$17,280 \div 60 = 288$
$288 = 288$

9. **Understand:** We must calculate a net amount. To calculate the net we need total revenue and total cost. The formula for net is $N = R - C$. Lynn's total revenue is the amount that she sold the house for, which was given to be $80,560. Her total cost is the sum of all the money she spent on the house.

Plan: Calculate total cost, then subtract the cost from the total revenue to get the net.

Execute: cost = loan payoff + repairs

cost = $71,484 + 4500$
cost = $75,984

Now replace R with 80,560 and C with 75,984 in $N = R - C$ to get the net.

$N = R - C$
$N = 80,560 - 75,984$
$N = \$4576$

Answer: Lynn's net is $4576, which is a profit.

Check: Reverse the process. The net added to cost should produce the revenue.

$4576 + 75,984 = 80,560$
$80,560 = 80,560$

Subtract the loan payoff from the cost should produce the loan payoff.

$75,984 - 4500 = 71,484$
$71,484 = 71,484$

11. **Understand:** We must calculate a net amount. To calculate the net we need total revenue and total cost. The formula for net is $N = R - C$. Greg's total revenue is the amount that he sold the stock for. His total cost is the sum of the amounts for the share purchases.

Plan: Calculate total cost and revenue, then subtract the cost from the total revenue to get the net.

Execute: cost = (number of shares purchased)(amount per share)

cost = $(40)(54)$
cost = $2160

revenue = (number of shares sold)(amount per share)

revenue = $(25)(56) + (40 - 25)(50)$
revenue = $1400 + (15)(50)$
revenue = $1400 + 750$
revenue = $2150

Now replace R with 2150 and C with 2160 in $N = R - C$ to get the net.

$N = R - C$
$N = 2150 - 2160$
$N = -\$10$

Answer: Greg's net is –$10, which is a loss of $10.

Check: Reverse the process. The net added to cost should produce the revenue.

$$-10 + 2160 = 2150$$
$$2150 = 2150$$

Subtracting the cost of one stock trade from the revenue produce the cost of the second stock trade.

$$2150 - 1400 = 750$$
$$750 = 750$$

Dividing the cost by the number of shares purchased should produce the amount paid per share.

$$2160 \div 40 = 54$$
$$54 = 54$$

13. **Understand:** We must calculate voltage given resistance and current. The formula that relates voltage, current, and resistance is $V = ir$.

Plan: Substitute –9 for i and 7 for r, then multiply.

Execute: $V = ir$
$$V = (-9)(7)$$
$$V = -63 \text{ volts}$$

Answer: The voltage is –63 volts.

Check: Voltage divided by current should produce the resistance. $-63 \div \left(-9\right) = 7$
$$7 = 7$$

15. **Understand:** We must calculate current given voltage and resistance. We use the formula $V = ir$.

Plan: Substitute –220 for V and 5 for r, then solve.

Execute: $V = ir$
$$-220 = i \cdot 5$$
$$-220 \div 5 = i$$
$$-44 = i$$

Answer: The current is –44 A.

Check: Reverse the process. $-44 \cdot 5 = -220$
$$-220 = -220$$

17. **Understand:** We must calculate resistance given voltage and current. We use the formula $V = ir$.

Plan: Substitute –120 for V and –4 for i, then solve.

Execute: $V = ir$
$$-120 = -4 \cdot r$$
$$-120 \div 4 = i$$
$$-30 = i$$

Answer: The resistance is –30Ω not 40Ω. The technician was told the incorrect resistance.

Check: Reverse the process. $-30 \cdot 4 = -120$
$$-120 = -120$$

19. **Understand:** To find average rate we must consider the total distance of the trip and the total time. The formula that relates distance, rate, and time is $d = rt$.

Plan: Use the formula $d = rt$ and solve for r.

Execute: $d = rt$
$$968 = r \cdot 44$$
$$968 \div 44 = r$$
$$22 = r$$

Answer: The average rate was 22 ft./sec.

Check: Reverse the process. If the elevator traveled 22 ft/sec., in 44 seconds would it then travel 968 ft.? $d = rt$
$$d = (22)(44)$$
$$d = 968$$

21. **Understand:** The rate was given to be 17,060 mph. The time was given to be 3 hours. The formula that relates distance, rate and time is $d = rt$.

Plan: Replace r with 17,060 and t with 3, then solve for d.

Execute: $d = rt$
$$d = (17,060)(3)$$
$$d = 51,180 \text{ mi.}$$

Answer: The space shuttle travels 51,180 miles.

Check: Reverse the process. A space shuttle covering 51,180 miles in 3 hours should have an average velocity of 17,060 mph.
$$(17,060)(3) = 51,180$$
$$51,180 = 51,180$$

23. **Understand:** To find average speed we must consider the total distance of the trip and the total time. The trip was broken into three parts. We were given the distances for all three parts. We were also given a departure time and arrival time, but there were two 30-minute stops we must take

into account. The formula that relates distance, rate, and time is $d = rt$.

Plan: Find the total distance of the trip and total time of the trip, then use the formula $d = rt$ to solve for r.

Execute: total distance: $105 + 140 + 80 = 325$ mi.

total time: From 11 A.M. to 5 P.M. is 6 hours. However, Devin stopped for a total of 1 hours. So the actual travel time was 5 hours.

Because we now have the total distance of 325 mi. and total time of 5 hr., we can use $d = rt$.

$$d = rt$$
$$325 = r \cdot 5$$
$$325 \div 5 = r$$
$$65 = r$$

Answer: The average speed was 65 mph.

Check: Reverse the process. If Devin traveled 65 mph for 5 hours, would he travel 325 miles? $d = rt$
$$d = (65)(5)$$
$$d = 325$$

Review Exercises

1.

2. -5 3. 15

4. $k + 76 = -34$ Check: $-110 + 76 = -34$
     $k = -34 - 76$ $-34 = -34$
     $k = -110$

5. $-14m = 56$ Check: $-14(-4) = 56$
     $m = \dfrac{56}{-14}$ $56 = 56$
     $m = -4$

Chapter 2 Review Exercises

1. true   2. false   3. false

4. true   5. false   6. true

7. subtract their absolute values and keep the sign of the number with the greater absolute value

8. change the operation symbol from $-$ to $+$ and change the subtrahend to its additive inverse

9. positive   10. negative

11. $-13,000$ 12. $+212$

13. a)

b)

14. a) $-15 < 0$; The number 0 is farther to the right on a number line, so it is larger.

b) $-26 > -35$; The number -26 is farther to the right on a number line, so it is larger.

c) $12 > -41$; The number 12 is farther to the right on a number line, so it is larger.

15. 41   16. -27 17. 16

18. 17   19. -26 20. -12

21. $-16 + 25 = 9$ 22. $-12 + (-14) = -26$

23. $18 + (-30) = -12$ 24. $24 - 31 = -7$

25. $-20 - 17 = -20 + (-17)$
$$= -37$$

26. $-13 - (-19) = -13 + 19$
$$= 6$$

27. $-27 - (-22) = -27 + 22$
$$= -5$$

28. $6 \cdot (-9) = -54$ 29. $-5(-12) = 60$

30. $(-1)(-3)(-6)(-2) = 3(-6)(-2)$
$$= -18(-2)$$
$$= 36$$

31. $(-5)(-2)(-3) = 10(-3)$
$$= -30$$

32. $\dfrac{-48}{8} = -6$ 33. $\dfrac{-63}{-9} = 7$

34. $\dfrac{-12}{0}$ is undefined

35. $14 + x = -27$ Check: $14 + (-41) = -27$
     $x = -27 - 14$ $-27 = -27$
     $x = -27 + (-14)$
     $x = -41$

36. $n + (-12) = -7$ Check: $5 + (-12) = -7$
     $n = -7 - (-12)$ $-7 = -7$
     $n = -7 + 12$
     $n = 5$

37. $7k = -63$ Check: $7(-9) = -63$
$$k = \frac{-63}{7}$$
$$-63 = -63$$
$$k = -9$$

38. $(-10)h = -80$ Check: $(-10)(8) = -80$
$$-10h = -80$$
$$-80 = -80$$
$$h = \frac{-80}{-10}$$
$$h = 8$$

39. $(-3)^4 = (-3)(-3)(-3)(-3) = 81$

40. $(-10)^3 = (-10)(-10)(-10) = -1000$

41. $-2^4 = -(2)(2)(2)(2) = -16$

42. $-3^3 = -(3)(3)(3) = -27$

43. The square roots of 64 are 8 and -8 because $8 \cdot 8 = 64$ and $(-8)(-8) = 64$.

44. $\sqrt{100} = 10$

45. $-\sqrt{121} = -11$

46. There is no integer solution because there is no integer that can be squared to equal -36.

47. $-16 - 28 \div (-7) = -16 - (-4)$
$$= -16 + 4$$
$$= -12$$

48. $-|26 - 3(-2)| = -|26 - (-6)|$
$$= -|26 + 6|$$
$$= -|32|$$
$$= -32$$

49. $13 + 4(6 - 15) + 2^3 = 13 + 4(-9) + 2^3$
$$= 13 + 4(-9) + 8$$
$$= 13 + (-36) + 8$$
$$= -23 + 8$$
$$= -15$$

50. $-3\sqrt{49} + 4(2 - 6)^2 = -3 \cdot 7 + 4(-4)^2$
$$= -3 \cdot 7 + 4 \cdot 16$$
$$= -21 + 64$$
$$= 43$$

51. $[-12 + 4(15 - 10)] + \sqrt{100 - 36}$
$$= [-12 + 4 \cdot 5] + \sqrt{64}$$
$$= [-12 + 20] + 8$$
$$= 8 + 8$$
$$= 16$$

52. $\dfrac{4 + (5 + 3)^2}{3(-13) + 5} = \dfrac{4 + 8^2}{-39 + 5}$
$$= \frac{4 + 64}{-34}$$
$$= \frac{68}{-34}$$
$$= -2$$

53. **Understand:** We must find William's new credit card balance.

Plan: The closing balance is the sum of the previous balance, all charges, and the payment.

Execute:
$$-1100 + (-23) + (-14) + (-9) + (-15) + 900$$
$$= -1161 + 900$$
$$= -\$261$$

Answer: William has a balance of $-\$261$.

Check: Instead of adding the total charges and payments we can compute from left to right.
$$-1100 + (-23) + (-14) + (-9) + (-15) + 900$$
$$= -1123 + (-14) + (-9) + (-15) + 900$$
$$= -1137 + (-9) + (-15) + 900$$
$$= -1146 + (-15) + 900$$
$$= -1161 + 900$$
$$= -\$261$$

54. **Understand:** We must find the total force on the concrete block.

Plan: The total force is the sum of all forces acting on the block. The weight of the concrete block is a downward force, so we assign it a negative value, -600 pounds. The two cables will be assigned positive values, 300 lb. each, since they are directed upward.

Execute:
$$-600 + 300 + 300$$
$$= -600 + 600$$
$$= 0 \text{ lb.}$$

Answer: The resultant force is 0 lb. The concrete block is not moving in either direction.

Check: Instead of adding the total charges and payments we can compute from left to right.

$$-600 + 300 + 300$$
$$= -300 + 300$$
$$= 0$$

55. **Understand:** We must calculate the net of the business given the total revenue and the total cost. The formula for net is $N = R - C$.

Plan: Replace R with 1,648,200 and C with 928,600 in $N = R - C$, then subtract.

Execute:

$$N = R - C$$
$$N = 1,648,200 - 928,600$$
$$N = \$719,600$$

Answer: The business's net was $719,600. This was a profit of $719,600.

Check: Reverse the process. Adding the cost, $928,600, to the net, $719,600, should produce the revenue, $1,648,200.

$$928,600 + 719,600 \overset{?}{=} 1,648,200$$
$$1,648,200 = 1,648,200$$

56. **Understand:** We must calculate the decrease of temperature between sunset and midnight. Because we are asked to find the amount of decrease, we mist calculate the difference.

Plan: Find the difference between 12 and –19.

Execute:

temperature decrease: $12 - (-19)$
$$= 12 + 19$$
$$= 31°F$$

Answer: The temperature decreased 31°F.

Check: Verify that if the temperature of 12°F decreases by 31°F it will result in –19°F.

$$12 - 31 \overset{?}{=} -19$$
$$-19 = -19$$

57. **Understand:** The words deposit indicates that Arturo must add an amount to –$45 to reach a value of $25 so this is a missing addend situation.

Plan: Write a missing addend equation, then solve.

Execute:

$$-45 + x = 25$$
$$x = 25 - (-45)$$
$$x = 25 + 45$$
$$x = \$70$$

Answer: Arturo must deposit $70 to avoid further charges.

Check: Verify that if Arturo deposits $70 into his account with –$45, he will have a balance of $25.

$$-45 + 70 \overset{?}{=} 25$$
$$25 = 25$$

58. **Understand:** Since the $4272 is a debt, we write it as –$4272. Making one payment each month for two years means that he makes $2(12) = 24$ payments over the two years.

Plan: Since he is splitting the amount into equal payments, we divide.

Execute: $-4272 \div 24 = -\$178$

Answer: Branford will pay $178 each month.

Check: Reverse the process. Verify that if Branford makes 24 payments of $178, he will pay a total of $4272.

$$178 \cdot 24 \overset{?}{=} 4272$$
$$4272 = 4272$$

59. **Understand:** We must calculate voltage given resistance and current. The formula that relates voltage, resistance, and current is $V = ir$.

Plan: Replace i with –3 and r with 12 in $V = ir$, then multiply.

Execute:

$$V = ir$$
$$V = (-3)(12)$$
$$V = -36 \text{ volts}$$

Answer: The voltage is –36 volts.

Check: Reverse the process. Divide voltage, –36 V, by current, 12 A, to see if you get resistance, –3 ohms.

$$(-36) \div (12) \overset{?}{=} -3$$
$$-3 = -3$$

60. **Understand:** We must calculate current given resistance and voltage. The formula that relates voltage, resistance, and current is $V = ir$.

Plan: Replace V with –110 and r with 10 in $V = ir$, then solve for i.

Execute:

$$V = ir$$
$$-110 = i(10)$$
$$-110 \div 10 = i$$
$$-11A = i$$

Answer: The current is –11A.

Check: Reverse the process. Multiply current, –11, by resistance, 10, to see if you get voltage, –110.

$$11(10) \overset{?}{=} 110$$
$$110 = 110$$

61. **Understand:** We are asked to find the distance traveled given the average rate and time. The formula which related average rate, time, and distance is $d = rt$.

Plan: Replace r with 65 and t with 3 in $d = rt$, then multiply.

Execute:

$$d = rt$$
$$d = 65(3)$$
$$d = 195 \text{ mi.}$$

Answer: Jacquelyn travels 195 mi.

Check: Reverse the process. Divide distance, 195 mi., by time, 3 hr., to see if you get rate, 65 mph.

$$195 \div 3 \overset{?}{=} 65$$
$$65 = 65$$

62. **Understand:** To find average rate we must consider the total distance of the trip and the total time. The trip was broken into three parts. We were given the distance for all three parts. We are also given the departure time and the arrival time, but there were two 30-minute stops that we must take into account. The formula which related average rate, time, and distance is $d = rt$.

Plan: Find the total distance of the trip and total time of the trip, then use the formula $d = rt$ to solve for r.

Execute: The total distance is $150 + 110 + 40 = 300$ miles.

The total time from 7 A.M. to 1 P.M is 6 hours. However, Steve had two 30-minute breaks which total 1 hour so the actual travel time of 5 hours.

We can now replace d with 300 and t with 5 in $d = rt$.

$$d = rt$$
$$300 = r \cdot 5$$
$$300 \div 5 = r$$
$$60 = r$$

Answer: Steve's average speed was 60 mph.

Check: Reverse the process. Verify that if Steve traveled for 5 hr. at the speed of 60 mph, he would cover 300 mi.

$$300 \overset{?}{=} 60 \cdot 5$$
$$300 = 300$$

Chapter 2 Practice Test

1.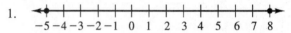

2. 26

3. 18

4. $17 + (-29) = -12$

5. $-31 + (-14) = -45$

6. $20 - 34 = 20 + (-34)$
   $\qquad = -14$

7. $-16 - 19 = -16 + (-19)$
   $\qquad = -35$

8. $-30 - (-14) = -30 + 14$
   $\qquad = -16$

9. $9(-12) = -108$

10. $(-2)(-6)(-7) = 12(-7)$
    $\qquad = -84$

11. $-30 \div 6 = -5$ 12. $\dfrac{-48}{-12} = 4$

13. $-19 + k = 25$ Check: $-19 + 44 = 25$
    $k = 25 - (-19)$ $25 = 25$
    $k = 25 + 19$
    $k = 44$

14. $6n = -54$ Check: $6(-9) = -54$
    $n = \dfrac{-54}{6}$ $-54 = -54$
    $n = -9$

15. $(-4)^3 = (-4)(-4)(-4)$
    $= 16(-4)$
    $= -64$

16. $-2^2 = -(2)(2) = -4$

17. The square roots of 81 are 9 and -9
    because $9 \cdot 9 = 81$ and $(-9)(-9) = 81$.

18. $\sqrt{144} = 12$ because $12^2 = 144$.

19. There is not an integer solution.

20. $28 - 6 \cdot 5 = 28 - 30$
    $= -2$

21. $19 - 4(7 + 2) + 2^4 = 19 - 4 \cdot 9 + 2^4$
    $= 19 - 4 \cdot 9 + 16$
    $= 19 - 36 + 16$
    $= -17 + 16$
    $= -1$

22. $|28 \div (2 - 6)| + 2(4 - 5) = |28 \div (-4)| + 2(-1)$
    $= |-7| + (-2)$
    $= 7 - 2$
    $= 5$

23. $[18 \div 2 + (4 - 6)] - \sqrt{49}$
    $= [18 \div 2 + (-2)] - \sqrt{49}$
    $= [9 + (-2)] - \sqrt{49}$
    $= 7 - \sqrt{49}$
    $= 7 - 7$
    $= 0$

24. $\dfrac{5^2 + 3(-12)}{36 \div 2 - 18} = \dfrac{25 + 3(-12)}{36 \div 2 - 18}$
    $= \dfrac{25 + (-36)}{18 - 18}$
    $= \dfrac{-11}{0}$
    undefined

25. $\dfrac{(-7)^2 + 11}{(-2)(5) + \sqrt{64}} = \dfrac{49 + 11}{(-2)(5) + 8}$
    $= \dfrac{49 + 11}{-10 + 8}$
    $= \dfrac{60}{-2}$
    $= -30$

26. $-487 + (-76) + 125 + (-14) = -577 + 125$
    $= -452$

    Allison's new balance is $-\$452$.

27. **Understand:** Since Natasha is being charged the same amount each time she overdrafts five times, we are repeatedly adding the same amount. Repeated addition means that we can multiply.

    Plan: Multiply the overdraft amount by 5.

    Execute: $5(-14) = -\$70$

    Answer: Natasha's total overdraft charges were $-\$70$ for the month.

    Check: Reverse the process. Total overdraft charges, $-\$70$, divided by the number of overdraft charges, 5, should produce the amount per charge, $-\$14$.

    $$\dfrac{-70}{5} \overset{?}{=} -14$$
    $$-14 = -14$$

28. **Understand:** We must calculate the net. We need the total cost and the total revenue. The formula for net is $N = R - C$.

    Howard's total revenue is the amount that he sold the car for, which was given to be $15,500.

    Howard's total cost is the sum of all the money he spent on the car.

    Plan: Calculate the total cost, then subtract the cost from the total revenue to get the net.

    Execute: cost = amount down + total of all payments + maintenance costs cost $= 12{,}000 + 55(89) + 347$

cost $= 12{,}000 + 4895 + 347$

cost $= \$17{,}242$

Now replace R with 15,500 and C with 17,242 in $N = R - C$ to get the net.

$N = R - C$

$N = 15{,}500 - 17{,}242$

$N = -\$1742$

Answer: Howard's net is –$1742, which is a loss of $1742.

Check: Reverse the process. The net, –$1742, added to cost, $17,242, should produce the revenue, $15,500.

$$-1742 + 17{,}242 \overset{?}{=} 15{,}500$$
$$15{,}500 = 15{,}500$$

Subtracting maintenance cost, $347, and total payments, $4895, from the total cost, $17,242 should produce the down payment, $12,000.

$$17{,}242 - 347 - 4895 \overset{?}{=} 12{,}000$$
$$12{,}000 = 12{,}000$$

Dividing the total of all payments, $4895, by the number of payments, 55, should produce the amount of each payment, $89.

$$\frac{4895}{55} \overset{?}{=} 89$$
$$89 = 89$$

29. **Understand:** We must calculate voltage given resistance and current. The formula that relates voltage, resistance, and current is $V = ir$.

Plan: Replace i with –8 and r with 12 in $V = ir$, then multiply.

Execute: $V = ir$

$\quad\quad\quad V = -8(12)$

$\quad\quad\quad V = -96$ volts

Answer: The voltage is –96V.

Check: Reverse the process. Divide voltage, –96 V, by current, –8A, to see if you get resistance, 12 ohms.

$$-96 \div (-8) \overset{?}{=} 12$$
$$12 = 12$$

30. **Understand:** We are asked to find the distance traveled given the average rate and time. The formula which related average rate, time, and distance is $d = rt$.

Plan: Replace r with 62 and t with 2 in $d = rt$, then multiply.

Execute:

$$d = rt$$
$$d = 62(2)$$
$$d = 124 \text{ mi.}$$

Answer: Lashanda travels 124 mi.

Check: Reverse the process. Divide distance, 124 mi., by time, 2 hr., to see if you get rate, 62 mph.

$$124 \div 2 \overset{?}{=} 62$$
$$62 = 62$$

Chapters 1 & 2 Cumulative Review Exercises

1. true   2. false   3. false

4. true   5. false   6. True

7. $5 \times 1{,}000{,}000 + 6 \times 100{,}000 + 8 \times 10{,}000$
$\quad + 9 \times 100 + 1 \times 1$

8. 50,836,009

9. four hundred nine million, two hundred fifty-four thousand, six

10. 23,000,000: The 3 is in the millions place. Look at the 4 to its right. Since 4 is less than 5 we round the 3 down.

11. 50,000
$\quad\underline{+7{,}000}$
$\quad 57{,}000$

12. Addends can exchange places: $2 + 3 = 3 + 2$.

13. associative, multiplication

14. 1. Grouping symbols
$\quad$ 2. Exponents or roots from left to right
$\quad$ 3. Multiplication or division from left to right
$\quad$ 4. Addition or subtraction from left to right

15. The mistake is calculating the square roots of 16 and 9 when they should have first been added. Correct: 33.

16.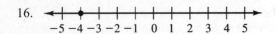
$$-5\ -4\ -3\ -2\ -1\quad 0\quad 1\quad 2\quad 3\quad 4\quad 5$$

17. $135 > -450$ 18. $-930 > -932$

19. 16 20. 5 21. -8 22. -4

23. $287 + 48 + (-160) + (-82) = 335 + (-242)$
$$= 93$$

24. $-19 - 24 = -19 + (-24)$
$$= -43$$

25. $-64 - (-14) = -64 + 14$
$$= -50$$

26. $-14(6) = -84$

27. $-12(-8) = 96$

28. $-1(12)(-6)(-2) = -12(-6)(-2)$
$$= 72(-2)$$
$$= -144$$

29. $3^5 = 3 \cdot 3 \cdot 3 \cdot 3 \cdot 3 = 243$

30. $(-2)^6 = (-2)(-2)(-2)(-2)(-2)(-2) = 64$

31. $-2^6 = -(2 \cdot 2 \cdot 2 \cdot 2 \cdot 2 \cdot 2) = -64$

32. $\begin{array}{r} 201 \\ 6\overline{)1208} \\ \underline{12} \\ 008 \\ \underline{6} \\ 2 \end{array}$ Answer: 201 r2

33. $-105 \div 7 = -15$

34. $\sqrt{121} = 11$

35. $-15 - 2[36 \div (3 + 15)] = -15 - 2[36 \div 18]$
$$= -15 - 2 \cdot 2$$
$$= -15 - 4$$
$$= -19$$

36. $-4(2)^3 + [18 - 12(2)] - \sqrt{25 \cdot 4}$
$$= -4(2)^3 + [18 - 24] - \sqrt{100}$$
$$= -4(2)^3 + (-6) - \sqrt{100}$$
$$= -4 \cdot 8 + (-6) - 10$$
$$= -32 + (-6) + (-10)$$
$$= -38 + (-10)$$
$$= -48$$

37. $29 + x = 54$ Check: $29 + 25 = 54$
$$x = 54 - 29 \qquad\qquad\qquad 54 = 54$$
$$x = 25$$

38. $y + 60 = 23$ Check: $-37 + 60 = 23$
$$y = 23 - 60 \qquad\qquad\qquad 23 = 23$$
$$y = -37$$

39. $26a = 2652$ Check: $26 \cdot 102 = 2652$
$$a = \frac{2652}{26} \qquad\qquad\qquad 2652 = 2652$$
$$a = 102$$

40. $-18c = 126$ Check: $-18(-7) = 126$
$$c = \frac{126}{-18} \qquad\qquad\qquad 126 = 126$$
$$c = -7$$

41. perimeter $= 17 + 20 + 17 + 20 = 74$ m

area: $A = bh$
$$A = 20 \cdot 15$$
$$A = 300 \text{ m}^2$$

42. $V = lwh$
$$V = 6 \cdot 3 \cdot 2$$
$$V = 18 \cdot 2$$
$$V = 36 \text{ in.}^3$$

43. **Understand:** We must find the new balance. The new balance is the sum of the old balance, all charges, and deposits.

Plan: Add the old balance, $45, the $62 deposit, and all charges, –$23 and –$57.

Execute:

$45 + (-23) + 62 + (-57)$

$= 22 + 62 + (-57)$

$= 84 + (-57)$

$= \$27$

Answer: Shanisse's balance is $27.

Check: Instead of computing from left to right we can add the total charges and deposits.

$45 + (-23) + 62 + (-57)$

$= 107 + (-80)$

$= 27$

44. **Understand:** We must find the total cost of shares given the number of shares and the price of each share.

Plan: Multiply the number of shares by the price of each share.

Execute: cost = number of shares · price of each share

cost = 350 · 15

cost = $5250

Answer: The shares will cost $5250.

Check: Reverse the process. Dividing the total cost, $5250, by the number of shares, 350, should produce the price of each share, $15.

$$\frac{5250}{350} \overset{?}{=} 15$$
$$15 = 15$$

45. **Understand:** We must find the amount of Heparin per minute given the total amount Heparin and the total time.

Plan: The word *each* indicates that we must divide. Divide the total number of milliliters of Heparin, 300 ml, by the total number of minute, 60 minutes. There are 60 minute in an hour.

Execute: $300 \div 60 = 5$

Answer: The patient should receive 5 ml of Heparin each minute.

Check: Reverse the process. Verify that if the patient receives 5 ml of Heparin each minute for 60 minute, he will receive a total of 300 ml of Heparin.

$$5 \cdot 60 \overset{?}{=} 300$$
$$300 = 300$$

46. The mean is found by dividing the sum of the scores by the number of scores.
$$\frac{\left(\begin{array}{c}85+94+75+50+68+91\\+80+76+72+95+100+74\end{array}\right)}{12} = \frac{9600}{12} = 80$$
The mean is 80.

The median is found by arranging the scores from least to greatest. Since there are an even number of scores, we find the mean of the middle two scores which are 76 and 80.
50, 68, 72, 74, 75, 76, 80, 85, 91, 94, 95, 100

$$\frac{76+80}{2} = \frac{156}{2} = 78$$
The median is 78.

47. **Understand:** We must find the area of the yard.

Plan: We must subtract the area of the house from the area of the lot.

Execute: Find the area of the lot.

$$A = lw$$
$$A = 280 \cdot 85$$
$$A = 23,800$$

Now, find the area of the house.

$$A = lw$$
$$A = 45 \cdot 55$$
$$A = 2475$$

Subtract to find the area of yard.

$$23,800 - 2475 = 21,325 \text{ ft.}^2$$

Answer: The area of the yard is 21,325 ft.^{2}.

Check: Reverse the process. Verify that the sum of the areas of the house, 2475 ft.^{2}, and the yard, 21,325 ft.^{2}, is the total area of the lot, 23,800 ft.^{2}.

$$2475 + 21,325 \overset{?}{=} 23,800$$
$$23,800 = 23,800$$

Divide the area of the house, 2475 ft.^{2}, by the length of the house, 45 ft., to check that you get its width, 55 ft.

$$\frac{2475}{45} \overset{?}{=} 55$$
$$55 = 55$$

Divide the area of the lot, 23,800 ft.^{2}, by the length of the lot, 280 ft., to check that you get its width, 85 ft.

$$\frac{23,800}{280} \overset{?}{=} 85$$
$$85 = 85$$

48. a. $-40 - (-76) = -40 + 76 = 36°\text{F}$

    b. $-18 - (-76) = -18 + 76 = 58°\text{F}$

49. **Understand:** We must calculate the net. We need the total cost and the total revenue. The formula for net is N = R – C.

Sonya's total revenue is the amount that he sold the car for, which was given to be $4300.

Sonya's total cost is the sum of all the money he spent on the car.

Plan: Calculate the total cost, then subtract the cost from the total revenue to get the net.

Execute: cost = amount down + total of all payments + maintenance costs

cost = 1500 + 60(276) + 1450

cost = 1500 + 16,560 + 1450

cost = 19,510

Now replace R with 4300 and C with 19,510 in $N = R - C$ to get the net.

$N = R - C$

$N = 4300 - 19,510$

$N = -\$15,210$

Answer: Sonya's net is –$15,210, which is a loss of $15,210.

Check: Reverse the process. The net, –$15,210, added to cost, $19,510, should produce the revenue, $4300.

$$-15,210 + 19,510 \overset{?}{=} 4300$$
$$4300 = 4300$$

Subtracting maintenance cost, $1450, and total payments, $16,560, from the total cost, $19,510 should produce the down payment, $1500.

$$19,510 - 1450 - 16,560 \overset{?}{=} 1500$$
$$1500 = 1500$$

Dividing the total of all payments, $16,560, by the number of payments, 60, should produce the amount of each payment, $276.

$$\frac{16,560}{60} \overset{?}{=} 276$$
$$276 = 276$$

Check: Reverse the process. Verify that an eider duck flying for 8 hr. at 45 mph will cover 360 mi.

$$8 \cdot 45 \overset{?}{=} 360$$
$$360 = 360$$

50. **Understand:** We must find time give the rate and distance. The formula that relates distance, time, and rate is $d = rt$.

Plan: Replace d with 360 and r with 45 in $d = rt$, then solve for t.

Execute:

$$d = rt$$
$$360 = 45t$$
$$\frac{360}{45} = t$$
$$8 = t$$

Answer: The eider duck flies for 8 hours.

Chapter 3 Expressions and Polynomials

Exercise Set 3.1

1. An equation has an equal sign, whereas an expression does not.

3. the divisor is 0 with a nonzero dividend

5. equation 　　　　　7. expression

9. expression 　　　　11. equation

13. equation 　　　15. $4(4) - 9 = 16 - 9$
$$= 7$$

17. $3(-5) + 8 = -15 + 8$
$$= -7$$

19. $-2(4) + 7 = -8 + 7$
$$= -1$$

21. $2(3) - 4(-2) = 6 - (-8)$
$$= 6 + 8$$
$$= 14$$

23. $3(4) - 5(4 + 2) = 3(4) - 5(6)$
$$= 12 - 30$$
$$= -18$$

25. $3^2 - 2(3) + 4 = 9 - 2(3) + 4$
$$= 9 - 6 + 4$$
$$= 3 + 4$$
$$= 7$$

27. $(-2)^2 + 9(-2) - 2 = 4 + 9(-2) - 2$
$$= 4 - 18 - 2$$
$$= -14 - 2$$
$$= -16$$

29. $(-3)^2 + 5(1) - 3 = 9 + 5(1) - 3$
$$= 9 + 5 - 3$$
$$= 14 - 3$$
$$= 11$$

31. $3(-2)^2 - 4(4) + 1 = 3(4) - 4(4) + 1$
$$= 12 - 16 + 1$$
$$= -4 + 1$$
$$= -3$$

33. $(-3)^3 - 4(-3)(-1) = -27 - 4(-3)(-1)$
$$= -27 - (-12)(-1)$$
$$= -27 - 12$$
$$= -39$$

35. $(-4)^2 - 4(-3)(2) = 16 - 4(-3)(2)$
$$= 16 - (-12)(2)$$
$$= 16 - (-24)$$
$$= 16 + 24$$
$$= 40$$

37. $\left| 2^2 + (-10) \right| = \left| 4 + (-10) \right|$
$$= \left| 4 - 10 \right|$$
$$= \left| -6 \right|$$
$$= 6$$

39. $-\left| 4(-5) + 7(2) \right| = -\left| -20 + 14 \right|$
$$= -\left| -6 \right|$$
$$= -6$$

41. $-6(-3) - \left| (-2)^3 \right| = -6(-3) - \left| -8 \right|$
$$= -6(-3) - 8$$
$$= 18 - 8$$
$$= 10$$

43. $\sqrt{144} + \sqrt{25} = 12 + 5$
$$= 17$$

45. $\sqrt{144 + 25} = \sqrt{169} = 13$

47. $\sqrt{(-3)^2 + 4^2} = \sqrt{9 + 16}$
$$= \sqrt{25}$$
$$= 5$$

49. $\dfrac{3(10) + 5}{7 - 2(1)} = \dfrac{30 + 5}{7 - 2}$
$$= \dfrac{35}{5}$$
$$= 7$$

51. $\dfrac{-7(-5) + (1)^2}{4(-5) + 2(1)} = \dfrac{-7(-5) + 1}{4(-5) + 2(1)}$
$$= \dfrac{35 + 1}{-20 + 2}$$
$$= \dfrac{36}{-18}$$
$$= -2$$

53.

x	$3x-1$
-4	$3(-4)-1=-12-1=-13$
-2	$3(-2)-1=-6-1=-7$
0	$3(0)-1=0-1=-1$
2	$3(2)-1=6-1=5$
4	$3(4)-1=12-1=11$

55.

m	m^2-2m+4
-1	$(-1)^2-2(-1)+4$ $=1+2+4$ $=7$
0	$(0)^2-2(0)+4$ $=0-0+4$ $=4$
1	$(1)^2-2(1)+4$ $=1-2+4$ $=3$
2	$(2)^2-2(2)+4$ $=4-4+4$ $=4$
3	$(3)^2-2(3)+4$ $=9-6+4$ $=7$

57.

y	$|y+2|$
-4	$|(-4)+2|=|-2|=2$
-3	$|(-3)+2|=|-1|=1$
-2	$|(-2)+2|=|0|=0$
-1	$|(-1)+2|=|1|=1$
0	$|0+2|=|2|=2$

59. $x=10$ **61.** $y=0$

63. $a=-4$ or $a=7$

Review Exercises

1. 2

2. $279=2\times100+7\times10+9\times1$

3. 1, because 7 raised to the first power is 7.

4. $2^6=2\cdot2\cdot2\cdot2\cdot2\cdot2=64$

5. $3\cdot10^5=3\cdot100,000$
   $=300,000$

6. $3+(-7)+(-9)$

Exercise Set 3.2

1. A monomial is an expression that is a constant, or a product of a constant and variable that are raised to whole number powers. Examples may vary, but three possibilities are 12, $3x$, and $-7y^3$.

3. The degree of a monomial is the sum of the exponents of all its variables.

5. A polynomial is an expression that is a sum of monomials. Examples will vary; but one example is $3x^2+x-9$.

7. The degree of a polynomial is the greatest degree of all its terms.

9. $5x^3$ is a monomial because it is a product of a constant, 5 and a variable, x, that has an exponent of 3, which is a whole number.

11. $8m+5$ is not a monomial because it is not a product of a constant with variables.

13. $-x$ is a monomial because it is a product of a constant, -1 and a variable x with a whole number exponent.

15. -7 is a monomial because it is a constant.

17. $\dfrac{4x^2}{5y}$ is not a monomial because it is not a product of a constant with variables.

19. $-9x^2y$ is a monomial because it is a product of a constant, -9 and two variables with whole number exponents.

21. $3x^8$; coefficient: 3; degree: 8

23. $-9x$; coefficient: -9; degree: 1

25. 8; coefficient: 8; degree: 0

27. xy^2; coefficient: 1; degree: $1+2=3$

29. $-2t^3u^4$; coefficient: -2; degree: $3+4=7$

31. -1; coefficient: -1; degree: 0

33. $8x$ and $-5x$ are like terms because they both have the variable x, raised to the same exponent.

35. $-5m$ and $-7n$ are not like terms because they do not have the same variables.

37. $4n^2$ and $7n^2$ are like terms because they both have the variable n^2.

39. $-9K$ and $-10K$ are like terms because the variable is exactly the same.

41. $5n$ and $9N$ are not like terms because the variables are not exactly the same. One letter is in lowercase and the other is uppercase.

43. 14 and -6 are like terms because they are both constants.

45. first term: $5x^2$ coefficient: 5
     second term: $8x$ coefficient: 8
     third term: -7 coefficient: -7

47. first term: $6t$ coefficient: 6
     second term: -1 coefficient: -1

49. first term: $-6x^3$ coefficient: -6
     second term: x^2 coefficient: 1
     third term: $-9x$ coefficient: -9
     fourth term: 4   coefficient: 4

51. Binomial, because it contains 2 terms.

53. Monomial, because it contains 1 term.

55. no special name

57. Trinomial, because it contains 3 terms.

59. Monomial, because it contains 1 term.

61. Binomial, because it contains 2 terms.

63. degree: 3   65. degree: 6

67. degree: 12

69. $14t^6 - 8t^4 + 9t^3 + 5t^2 - 1$

71. $y^5 - 18y^3 - 10y^2 + 12y + 9$

73. $9a^5 + 7a^3 + 2a^2 - a - 6$

Review Exercises

1. $-7 + 5 = -2$ 2. $-9 - 7 = -9 + (-7)$
                            $= -16$

3. $8 - 15 = 8 + (-15)$
         $= -7$

4. $-15 + 8 - 21 + 2 = -15 + 8 + (-21) + 2$
                     $= -7 + (-21) + 2$
                     $= -28 + 2$
                     $= -26$

5. No, because subtraction is not commutative.

6. Yes, because addition is commutative.

7. Yes, because multiplication is commutative.

8. $P = 2l + 2w$
    $P = 2 \cdot 10 + 2 \cdot 4$
    $P = 20 + 8$
    $P = 28$ m

Exercise Set 3.3

1. Add or subtract the coefficients and keep the variables and their exponents the same.

3. Combine like terms.

5. $3x + 9x = 12x$ 7. $3y + y = 3y + 1y$
                                     $= 4y$

9. $-4u - 8u = -12u$ 11. $-15a - a = -15a - 1a$
                                     $= -16a$

13. $-6x + 8x = 2x$ 15. $-10m + 3m = -7m$

17. $12m - 7m = 5m$ 19. $9t - 12t = -3t$

21. $8a^2 + 5a^2 = 13a^2$ 23. $-4m^5 - 4m^5 = -8m^5$

25. $7x^3 - 2x^3 = 5x^3$ 27. $2y^2 - 9y^2 = -7y^2$

29. $-10t^2 + 9t^2 = -1t^2 = -t^2$

31. $2b^2 - 2b^2 = 0b^2 = 0$

33. $9x^2 + 5x + 2x^2 - 3x + 4 = 9x^2 + 2x^2 + 5x - 3x + 4$
                                    $= 11x^2 + 2x + 4$

35. $7a - 8a^3 - a + 4 - 5a^3$
     $= -8a^3 - 5a^3 + 7a - 1a + 4$
     $= -13a^3 + 6a + 4$

37. $-9c^2 + c + 4c^2 - 5c = -9c^2 + 4c^2 + 1c - 5c$
                             $= -5c^2 - 4c$

39. $-3n^2 - 2n + 7 - 4n^2 = -3n^2 - 4n^2 - 2n + 7$
                           $= -7n^2 - 2n + 7$

41. $12m^2 + 5m^3 - 11m^2 + 3 + 2m^3 - 7$
     $= 5m^3 + 2m^3 + 12m^2 - 11m^2 + 3 - 7$
     $= 7m^3 + m^2 - 4$

43. $-7x + 5x^4 - 2 + 3x^4 + 7x + 10 - 3x^2$
     $= 5x^4 + 3x^4 - 3x^2 - 7x + 7x - 2 + 10$
     $= 8x^4 - 3x^2 + 8$

45. $(2x + 5) + (3x + 1) = 2x + 3x + 5 + 1$
                          $= 5x + 6$

47. $(7y-4)+(2y-1)=7y+2y-4-1$
    $\qquad\qquad\qquad\quad=9y-5$

49. $(10x+7)+(3x-9)=10x+3x+7-9$
    $\qquad\qquad\qquad\qquad=13x-2$

51. $(4t-11)+(t+13)=4t+1t-11+13$
    $\qquad\qquad\qquad\qquad=5t+2$

53. $(2x^2+x+3)+(5x+7)=2x^2+1x+5x+3+7$
    $\qquad\qquad\qquad\qquad\qquad=2x^2+6x+10$

55. $(9n^2-14n+7)+(6n^2+2n-4)$
    $=9n^2+6n^2-14n+2n+7-4$
    $=15n^2-12n+3$

57. $(4p^2-7p-9)+(3p^2+4p+5)$
    $=4p^2+3p^2-7p+4p-9+5$
    $=7p^2-3p-4$

59. $(-3b^3+2b^2-9b)+(-5b^3-3b^2+11)$
    $=-3b^3-5b^3+2b^2-3b^2-9b+11$
    $=-8b^3-b^2-9b+11$

61. $(3a^3-a^2+10a-4)+(6a^2-9a+2)$
    $=3a^3-1a^2+6a^2+10a-9a-4+2$
    $=3a^3+5a^2+a-2$

63. $P=(a-2)+(3a+4)+(a-2)+(3a+4)$
    $P=1a+3a+1a+3a-2+4-2+4$
    $P=8a+4$

65. $P=(x+3)+(x+3)+(2x+1)$
    $P=1x+1x+2x+3+3+1$
    $P=4x+7$

67. Mistake: combined $4y^2-2y^2$ to get $-2y^2$
    Correct: $9y^3+1y^3+4y^2-2y^2-6y$
    $\qquad\qquad=10y^3+2y^2-6y$

69. Mistake: combined unlike terms
    Correct: $8x^6+1x^5-9x^4+15x^3-8x^3-1-3$
    $\qquad\qquad=8x^6+x^5-9x^4+7x^3-4$

71. $(5x+9)-(3x+2)=(5x+9)+(-3x-2)$
    $\qquad\qquad\qquad\qquad=2x+7$

73. $(8t+3)-(5t+3)=(8t+3)+(-5t-3)$
    $\qquad\qquad\qquad\qquad=3t$

75. $(10n+1)-(2n+7)=(10n+1)+(-2n-7)$
    $\qquad\qquad\qquad\qquad=8n-6$

77. $(5x-6)-(2x+1)=(5x-6)+(-2x-1)$
    $\qquad\qquad\qquad\qquad=3x-7$

79. $(6x^2+8x-9)-(9x+2)$
    $=(6x^2+8x-9)+(-9x-2)$
    $=6x^2-x-11$

81. $(8a^2-10a+2)-(-4a^2+a+8)$
    $=(8a^2-10a+2)+(4a^2-1a-8)$
    $=12a^2-11a-6$

83. $(7x^2+5x-8)-(7x^2+9x+2)$
    $=(7x^2+5x-8)+(-7x^2-9x-2)$
    $=-4x-10$

85. $(-8m^3+9m^2-17m-2)-(-8m^3+5m^2+m-6)$
    $=(-8m^3+9m^2-17m-2)+(8m^3-5m^2-m+6)$
    $=4m^2-18m+4$

87. Mistake: did not change all the signs in the subtrahend
    Correct: $2x^2-2x$

89. Mistake: added $-9t^2$ to $-9t^2$ to get 0.
    Correct: $-5t^3-18t^2+4t+5$

91.

	Expression 1	Expression 2
x	$5x+4$	$3x-7$
4	$5(4)+4$ $=20+4$ $=24$	$3(4)-7$ $=12-7$ $=5$
-2	$5(-2)+4$ $=-10+4$ $=-6$	$3(-2)-7$ $=-6-7$ $=-13$

	Sum	Difference
x	$(5x+4)+(3x-7)$ $=5x+3x+4-7$ $=8x-3$	$(5x+4)-(3x-7)$ $=5x+4-3x+7$ $=5x-3x+4+7$ $=2x+11$
4	$8(4)-3$ $=32-3$ $=29$	$2(4)+11$ $=8+11$ $=19$
-2	$8(-2)-3$ $=-16-3$ $=-19$	$2(-2)+11$ $=-4+11$ $=7$

	Expression 1	Expression 2
n	$n^2 + 3n - 1$	$4n^2 - n - 7$
2	$(2)^2 + 3(2) - 1$ $= 4 + 6 - 1$ $= 9$	$4(2)^2 - (2) - 7$ $= 16 - 2 - 7$ $= 7$
-3	$(-3)^2 + 3(-3) - 1$ $= 9 - 9 - 1$ $= -1$	$4(-3)^2 - (-3) - 7$ $= 36 + 3 - 7$ $= 32$

93.

	Sum
n	$(n^2 + 3n - 1) + (4n^2 - n - 7)$ $= n^2 + 4n^2 + 3n - n - 1 - 7$ $= 5n^2 + 2n - 8$
2	$5(2)^2 + 2(2) - 8$ $= 20 + 4 - 8$ $= 16$
-3	$5(-3)^2 + 2(-3) - 8$ $= 45 - 6 - 8$ $= 31$

	Difference
n	$(n^2 + 3n - 1) - (4n^2 - n - 7)$ $= n^2 - 4n^2 + 3n + n - 1 + 7$ $= -3n^2 + 4n + 6$
2	$-3(2)^2 + 4(2) + 6$ $= -12 + 8 + 6$ $= 2$
-3	$-3(-3)^2 + 4(-3) + 6$ $= -27 + 12 + 6$ $= -33$

Review Exercises

1. $2 \cdot (-341) = -682$

2. $(-3)^4 = -3 \cdot -3 \cdot -3 - 3 = 81$

3. $(-2)^5 = -2 \cdot -2 \cdot -2 \cdot -2 \cdot -2 = -32$

4. $2^3 \cdot 2^4 = 8 \cdot 16$
    $= 128$

5. $A = lw$
    $A = 25 \cdot 15$
    $A = 375 \text{ m}^2$

6. $5 \cdot 4 = 20$ choices

Exercise Set 3.4

1. add   3. multiply

5. Use the distributive property.

7. Conjugates are binomials that differ only in the sign separating the terms. Examples will vary, but one pair is $x + 5$ and $x - 5$

9. $x^3 \cdot x^4 = x^{3+4} = x^7$ 11. $t \cdot t^9 = t^{1+9} = t^{10}$

13. $7a^2 \cdot 3a = 7 \cdot 3a^{2+1}$ 15. $2n^3 \cdot 5n^4 = 2 \cdot 5n^{3+4}$
     $= 21a^3$ $= 10n^7$

17. $-8u^4 \cdot 3u^2 = -8 \cdot 3u^{4+2}$
     $= -24u^6$

19. $-y^5(-8y^2) = -1 \cdot (-8)u^{5+2}$
     $= 8y^7$

21. $5x(9x^2)(2x^4) = 5 \cdot 9 \cdot 2x^{1+2+4}$
     $= 90x^7$

23. $3m(-3m^2)(4m) = 3 \cdot (-3) \cdot 4m^{1+2+1}$
     $= -36m^4$

25. $3y^4(-5y^3)(-4y^2) = 3 \cdot -5 \cdot -4y^{4+3+2}$
     $= 60y^9$

27. Mistake: multiplied the exponents
     Correct: $9 \cdot 5x^{3+4} = 45x^7$

29. $(2x^3)^2 = 2^2 x^{3 \cdot 2}$ 31. $(-7m^5)^2 = (-7)^2 m^{5 \cdot 2}$
     $= 4x^6$ $= 49m^{10}$

33. $(-2y^6)^5 = (-2)^5 y^{6 \cdot 5}$ 35. $(-4x^5)^4 = (-4)^4 x^{5 \cdot 4}$
     $= -32y^{30}$ $= 256x^{20}$

37. $(5y^6)^2 = 5^2 y^{6 \cdot 2}$ 39. $(-5v^6)^3 = (-5)^3 v^{6 \cdot 3}$
     $= 25y^{12}$ $= -125v^{18}$

41. $(3x)(2x^3)^4 = (3x) \cdot 2^4 x^{3 \cdot 4}$
     $= (3x) \cdot 16x^{12}$
     $= 3 \cdot 16x^{1+12}$
     $= 48x^{13}$

43. Mistake: did not raise 3 to the 4th power, added exponents of x
     Correct: $3^4 x^{16} = 81x^{16}$

45. $4(2x+3) = 4 \cdot 2x + 4 \cdot 3$
$= 8x + 12$

47. $5(3x-7) = 5 \cdot 3x + 5(-7)$
$= 15x - 35$

49. $-3(5t+2) = -3 \cdot 5t - 3 \cdot 2$
$= -15t - 6$

51. $-4(7a-9) = -4 \cdot 7a - 4(-9)$
$= -28a + 36$

53. $6u(3u+4) = 6u \cdot 3u + 6u \cdot 4$
$= 18u^{1+1} + 24u$
$= 18u^2 + 24u$

55. $-2a(3a+1) = -2a \cdot 3a - 2a \cdot 1$
$= -6a^{1+1} - 2a$
$= -6a^2 - 2a$

57. $-2x^3(5x-8) = -2x^3 \cdot 5x - 2x^3(-8)$
$= -10x^{3+1} + 16x^3$
$= -10x^4 + 16x^3$

59. $8x(2x^2+3x-4) = 8x \cdot 2x^2 + 8x \cdot 3x + 8x(-4)$
$= 16x^{1+2} + 24x^{1+1} - 32x$
$= 16x^3 + 24x^2 - 32x$

61. $-x^2(5x^2-6x+9) = -x^2 \cdot 5x^2 - x^2(-6x) - x^2 \cdot 9$
$= -5x^{2+2} + 6x^{2+1} - 9x^2$
$= -5x^4 + 6x^3 - 9x^2$

63. $-2p^2(3p^2+4p-5)$
$= -2p^2 \cdot 3p^2 - 2p^2 \cdot 4p - 2p^2(-5)$
$= -6p^{2+2} - 8p^{2+1} + 10p^2$
$= -6p^4 - 8p^3 + 10p^2$

65. $(x+4)(x+2) = x \cdot x + x \cdot 2 + 4 \cdot x + 4 \cdot 2$
$= x^2 + 2x + 4x + 8$
$= x^2 + 6x + 8$

67. $(m-3)(m+5) = m \cdot m + m \cdot 5 - 3 \cdot m - 3 \cdot 5$
$= m^2 + 5m - 3m - 15$
$= m^2 + 2m - 15$

69. $(y-8)(y+1) = y \cdot y + y \cdot 1 - 8 \cdot y - 8 \cdot 1$
$= y^2 + y - 8y - 8$
$= y^2 - 7y - 8$

71. $(3x+2)(4x-5)$
$= 3x \cdot 4x + 3x \cdot (-5) + 2 \cdot 4x + 2 \cdot (-5)$
$= 12x^2 - 15x + 8x - 10$
$= 12x^2 - 7x - 10$

73. $(4x-1)(3x-5)$
$= 4x \cdot 3x + 4x \cdot (-5) - 1 \cdot 3x - 1 \cdot (-5)$
$= 12x^2 - 20x - 3x + 5$
$= 12x^2 - 23x + 5$

75. $(3t-5)(4t+1)$
$= 3t \cdot 4t + 3t \cdot 1 - 5 \cdot 4t - 5 \cdot 1$
$= 12t^2 + 3t - 20t - 5$
$= 12t^2 - 17t - 5$

77. $(a-7)(2a-5)$
$= a \cdot 2a + a \cdot (-5) - 7 \cdot 2a - 7 \cdot (-5)$
$= 2a^2 - 5a - 14a + 35$
$= 2a^2 - 19a + 35$

79. $x+7$ 81. $2x-5$

83. $-b-2$ 85. $-6x+9$

87. $(x+3)(x-3) = x^2 - 3^2 = x^2 - 9$

89. $(5t+6)(5t-6) = (5t)^2 - (6)^2 = 25t^2 - 36$

91. $(-6x-1)(-6x+1) = (-6x)^2 - (1)^2 = 36x^2 - 1$

93. Mistake: did not square the initial x and the 5 should be negative
Correct: $6x^2 + 2x - 15x - 5 = 6x^2 - 13x - 5$

95. $(x+5)(2x^2-3x+1)$
$= x \cdot 2x^2 + x \cdot (-3x) + x \cdot 1 + 5 \cdot 2x^2 + 5 \cdot (-3x) + 5 \cdot 1$
$= 2x^3 - 3x^2 + x + 10x^2 - 15x + 5$
$= 2x^3 + 7x^2 - 14x + 5$

97. $(2y-3)(4y^2+y-6)$
$= 2y \cdot 4y^2 + 2y \cdot y + 2y(-6) - 3 \cdot 4y^2 - 3 \cdot y - 3(-6)$
$= 8y^3 + 2y^2 - 12y - 12y^2 - 3y + 18$
$= 8y^3 - 10y^2 - 15y + 18$

99. $2a(a-1)(a+2)$
$= 2a[a \cdot a + a \cdot 2 - 1 \cdot a - 1 \cdot 2]$
$= 2a[a^2 + 2a - 1a - 2]$
$= 2a[a^2 + a - 2]$
$= 2a \cdot a^2 + 2a \cdot a + 2a(-2)$
$= 2a^3 + 2a^2 - 4a$

101. $-x^2(2x+3)(x+1)$

$=-x^2\left[2x\cdot x+2x\cdot 1+3\cdot x+3\cdot 1\right]$

$=-x^2\left[2x^2+2x+3x+3\right]$

$=-x^2\left[2x^2+5x+3\right]$

$=-x^2\cdot 2x^2-x^2\cdot 5x-x^2\cdot 3$

$=-2x^4-5x^3-3x^2$

103. $-3q^2(3q-1)(3q+1)$

$=-3q^2\left(9q^2-1\right)$

$=-3q^2\cdot 9q^2-3q^2(-1)$

$=-27q^4+3q^2$

105. $A=lw$

$A=(3y+1)(8y)$

$A=3y\cdot 8y+1\cdot 8y$

$A=24y^2+8y$

107. $V=lwh$

$V=(x+2)(x-1)(3x)$

$V=\left[x\cdot x+x\cdot(-1)+2\cdot x+2\cdot(-1)\right](3x)$

$V=\left(x^2-x+2x-2\right)(3x)$

$V=\left(x^2+x-2\right)(3x)$

$V=x^2\cdot 3x+x\cdot 3x-2\cdot 3x$

$V=3x^3+3x^2-6x$

109.

	Expression 1	Expression 2	Product
x	$5x$	$-3x$	$(5x)(-3x)$ $=-15x^{1+1}$ $=-15x^2$
2	$5(2)=10$	$-3(2)=-6$	$10(-6)=-60$
-3	$5(-3)=-15$	$-3(-3)=9$	$-15(9)=-135$

111.

	Expression 1	Expression 2
n	$n+3$	$n-7$
1	$1+3=4$	$1-7=-6$
-3	$-3+3=0$	$-3-7=-10$

Product
$(n+3)(n-7)$ $=n\cdot n+n\cdot(-7)+3\cdot n+3\cdot(-7)$ $=n^2-7n+3n-21$ $=n^2-4n-21$
$4(-6)=-24$
$0(-10)=0$

Review Exercises

1. $2\overline{)846}$

$\underline{8}$

04

$\underline{4}$

06

$\underline{6}$

0

quotient 423

2. $-6\cdot(?)=24$

$(?)=24\div-6$

$(?)=-4$

3. $A=lw$

$132=12\cdot w$

$\dfrac{132}{12}=w$

$11\text{ ft.}=w$

4. $V=lwh$

$210=7\cdot w\cdot 6$

$210=42\cdot w$

$\dfrac{210}{42}=w$

$5\text{ ft.}=w$

5. $3x-7x=-4x$

Exercise Set 3.5

1. A prime number is a natural number that has exactly two different factors, 1 and the number itself. Examples will vary, but three possibilities are 2, 3, and 5.

3. The GCF of a set of numbers is the greatest number that divides all the given numbers with no remainder.

5. composite   7. prime

9. neither   11. prime

13. composite   15. composite

17. prime   19. prime

21. No, 2 is an even prime number.

23. $2\cdot 2\cdot 2\cdot 2\cdot 5=2^4\cdot 5$

25. $2\cdot 2\cdot 3\cdot 13=2^2\cdot 3\cdot 13$

27. $2 \cdot 2 \cdot 67 = 2^2 \cdot 67$

29. $2 \cdot 2 \cdot 2 \cdot 5 \cdot 5 = 2^3 \cdot 5^2$

31. $7 \cdot 7 \cdot 7 = 7^3$

33. $3 \cdot 5 \cdot 5 \cdot 13 = 3 \cdot 5^2 \cdot 13$

35. $2 \cdot 3 \cdot 3 \cdot 3 \cdot 7 = 2 \cdot 3^3 \cdot 7$

37. $2 \cdot 2 \cdot 2 \cdot 7 \cdot 17 = 2^3 \cdot 7 \cdot 17$

39. $1, 2, 3, 5, 6, 9, 10, 15, 18, 30, 45, 90$

41. $1, 3, 9, 27, 81$

43. $1, 2, 3, 4, 5, 6, 8, 10, 12, 15, 20, 24, 30, 40, 60, 120$

45. $24: 1, 2, 3, 4, 6, 8, 12, 24$
    $60: 1, 2, 3, 4, 5, 6, 10, 12, 15, 20, 30, 60$
    GCF = 12

47. $81: 1, 3, 9, 27, 81$
    $65: 1, 5, 13, 65$
    GCF = 1

49. $72: 1, 2, 3, 4, 6, 8, 9, 12, 18, 24, 36, 72$
    $120: 1, 2, 3, 4, 5, 6, 8, 10, 12, 15, 20, 24, 30, 40, 60, 120$
    GCF = 24

51. $140 = 2^2 \cdot 5 \cdot 7$
    $196 = 2^2 \cdot 7^2$
    GCF $= 2^2 \cdot 7 = 4 \cdot 7 = 28$

53. $130 = 2 \cdot 5 \cdot 13$
    $78 = 2 \cdot 3 \cdot 13$
    GCF $= 2 \cdot 13 = 26$

55. $336 = 2^4 \cdot 3 \cdot 7$
    $504 = 2^3 \cdot 3^2 \cdot 7$
    GCF $= 2^3 \cdot 3 \cdot 7 = 168$

57. $99 = 3^2 \cdot 11$
    $140 = 2^2 \cdot 5 \cdot 7$
    GCF = 1

59. $60 = 2^2 \cdot 3 \cdot 5$
    $120 = 2^3 \cdot 3 \cdot 5$
    $140 = 2^2 \cdot 5 \cdot 7$
    GCF $= 2^2 \cdot 5 = 20$

61. $64 = 2^6$
    $160 = 2^5 \cdot 5$
    $224 = 2^5 \cdot 7$
    GCF $= 2^5 = 32$

63. We must find the GCF of 60 and 70 to find the largest size square to be cut.
    $60: 1, 2, 3, 4, 5, 6, 10, 12, 15, 20, 30, 60$
    $70: 1, 2, 5, 7, 10, 14, 35, 70$
    GCF = 10 in.
    The square should be 10 in. by 10 in.

65. We must find the GCF of 40, 32, and 24 to find the longest length of PVC pipe he can use and not require further cutting. $40: 1, 2, 4, 5, 8, 10, 20, 40$
    $32: 1, 2, 4, 8, 16, 32$
    $24: 1, 2, 3, 4, 6, 8, 12, 24$
    GCF = 8 ft.

The longest piece he can use requiring no further cutting is 8 ft. He can use 5 in the 40-ft. trench, 4 in the 32-ft. trench, and 3 in the 24-ft. trench.

67. $20x = 2^2 \cdot 5 \cdot x$
    $30 = 2 \cdot 3 \cdot 5$
    GCF $= 2 \cdot 5 = 10$

69. $8x^2 = 2^3 \cdot x^2$
    $14x^5 = 2 \cdot 7 \cdot x^5$
    GCF $= 2 \cdot x^2 = 2x^2$

71. $24h^3 = 2^3 \cdot 3 \cdot h^3$
    $35h^4 = 5 \cdot 7 \cdot h^4$
    GCF $= h^3$

73. $27n^4 = 3^3 \cdot n^4$
    $49 = 7^2$
    GCF = 1

75. $56x^9 = 2^3 \cdot 7 \cdot x^9$
    $72x^7 = 2^3 \cdot 3^2 \cdot x^7$
    GCF $= 2^3 \cdot x^7 = 8x^7$

77. $18n^2 = 2 \cdot 3^2 \cdot n^2$
    $24n^3 = 2^3 \cdot 3 \cdot n^3$
    $30n^4 = 2 \cdot 3 \cdot 5 \cdot n^4$
    GCF $= 2 \cdot 3 \cdot n^2 = 6n^2$

79. $42x^4 = 2 \cdot 3 \cdot 7 \cdot x^4$
    $35x^6 = 5 \cdot 7 \cdot x^6$
    $28x^3 = 2^2 \cdot 7 \cdot x^3$
    GCF $= 7 \cdot x^3 = 7x^3$

Review Exercises

1. $5 \times 1,000,000 + 7 \times 100,000 + 8 \times 10,000 + 4 \times 1000 + 2 \times 100 + 9 \times 1$

2. $17 \overline{)34328}$ Answer: 2019 r5

   $$\begin{array}{r} 2019 \\ 17\overline{)34328} \\ \underline{34} \\ 032 \\ \underline{17} \\ 158 \\ \underline{153} \\ 5 \end{array}$$

3. $2^7 = 2 \cdot 2 \cdot 2 \cdot 2 \cdot 2 \cdot 2 \cdot 2 = 128$

4. There are two slots. The first is turned a digit between $0 - 9$ (10) and the second is filled with the letters of the alphabet (26). Multiply the 2 slots for the total combinations.

   $10 \times 26 = 260$ combinations. This is not enough combinations for 265 employees.

5. $\sqrt{36} = 6$ ft. The dimensions must be 6 ft. $\times$ 6 ft.

Exercise Set 3.6

1. subtract

3. Divide each term in the polynomial by the monomial.

5. $x^9 \div x^2 = x^{9-2}$
   $ = x^7$

7. $\dfrac{m^7}{m} = m^{7-1}$
   $\phantom{\dfrac{m^7}{m}} = m^6$

9. $u^4 \div u^4 = u^{4-4}$
   $ = u^0$
   $ = 1$

11. $12a^7 \div \left(3a^4\right) = \dfrac{12}{3}a^{7-4}$
    $ = 4a^3$

13. $\dfrac{-20t^6}{4t^2} = \dfrac{-20}{4}t^{6-2}$
    $\phantom{\dfrac{-20t^6}{4t^2}} = -5t^4$

15. $-40n^9 \div (-2n) = \dfrac{-40}{-2}n^{9-1}$
    $ = 20n^8$

17. $12x^5 \div \left(-4x\right) = \dfrac{12}{-4}x^{5-1} = -3x^4$

19. $\dfrac{-36b^7}{9b^5} = \dfrac{-36}{9}b^{7-5} = -4b^2$

21. $14b^4 \div (-b^4) = \dfrac{14}{-1}b^{4-4}$
    $ = -14b^0$
    $ = -14$

23. $(9a+6) \div 3 = 9a \div 3 + 6 \div 3$
    $ = 3a + 2$

25. $\dfrac{14c-8}{-2} = \dfrac{14c}{-2} + \dfrac{-8}{-2}$
    $\phantom{\dfrac{14c-8}{-2}} = -7c + 4$

27. $\left(12x^2+8x\right) \div \left(4x\right) = 12x^2 \div (4x) + 8x \div (4x)$
    $ = 12 \div 4 \cdot x^{2-1} + 8 \div 4 \cdot x^{1-1}$
    $ = 3x + 2$

29. $\dfrac{-63d^2+49d}{7d} = \dfrac{-63d^2}{7d} + \dfrac{49d}{7d}$
    $\phantom{\dfrac{-63d^2+49d}{7d}} = \dfrac{-63}{7}d^{2-1} + \dfrac{49}{7}d^{1-1}$
    $\phantom{\dfrac{-63d^2+49d}{7d}} = -9d + 7$

31. $\left(16a^5+24a^3\right) \div \left(4a^2\right)$
    $= 16a^5 \div \left(4a^2\right) + 24a^3 \div \left(4a^2\right)$
    $= 16 \div 4 \cdot a^{5-2} + 24 \div 4 \cdot a^{3-2}$
    $= 4a^3 + 6a$

33. $\dfrac{16x^4-8x^3+4x^2}{4x} = \dfrac{16x^4}{4x} - \dfrac{8x^3}{4x} + \dfrac{4x^2}{4x}$
    $\phantom{\dfrac{16x^4-8x^3+4x^2}{4x}} = 4x^{4-1} - 2x^{3-1} + x^{2-1}$
    $\phantom{\dfrac{16x^4-8x^3+4x^2}{4x}} = 4x^3 - 2x^2 + x$

35. $\left(6a^5-9a^4+12a^3\right) \div (-3a^2)$
    $= 6a^5 \div (-3a^2) - 9a^4 \div (-3a^2) + 12a^3 \div (-3a^2)$
    $= -2a^{5-2} + 3a^{4-2} - 4a^{3-2}$
    $= -2a^3 + 3a^2 - 4a$

37. $\dfrac{30a^4-24a^3}{6a^3} = \dfrac{30a^4}{6a^3} - \dfrac{24a^3}{6a^3}$
    $\phantom{\dfrac{30a^4-24a^3}{6a^3}} = 5a^{4-3} - 4a^{3-3}$
    $\phantom{\dfrac{30a^4-24a^3}{6a^3}} = 5a - 4$

39. $\dfrac{32c^9+16c^5-40c^3}{-4c^2} = \dfrac{32c^9}{-4c^2} + \dfrac{16c^5}{-4c^2} - \dfrac{40c^3}{-4c^2}$
    $\phantom{\dfrac{32c^9+16c^5-40c^3}{-4c^2}} = -8c^{9-2} - 4c^{5-2} + 10c^{3-2}$
    $\phantom{\dfrac{32c^9+16c^5-40c^3}{-4c^2}} = -8c^7 - 4c^3 + 10c$

41. $5a \cdot (?) = 10a^3$
    $(?) = \dfrac{10a^3}{5a}$
    $(?) = 2a^{3-1}$
    $(?) = 2a^2$

43. $(?) \cdot \left(-6x^3\right) = 42x^7$
    $(?) = \dfrac{42x^7}{-6x^3}$
    $(?) = -7x^{7-3}$
    $(?) = -7x^4$

45. $-36u^5 = (?) \cdot \left(-9u^4\right)$
    $\dfrac{-36u^5}{-9u^4} = (?)$
    $4u^{5-4} = (?)$
    $4u = (?)$

47. $8a^9 = 8a^4(?)$

$$\frac{8a^9}{8a^4} = (?)$$

$$a^{9-4} = (?)$$

$$a^5 = (?)$$

49. $8x + 12 = 4 \cdot (?)$

$$\frac{8x+12}{4} = (?)$$

$$\frac{8x}{4} + \frac{12}{4} = (?)$$

$$2x + 3 = (?)$$

51. $5a^2 - a = a(?)$

$$\frac{5a^2 - a}{a} = (?)$$

$$\frac{5a^2}{a} - \frac{a}{a} = (?)$$

$$5a - 1 = (?)$$

53. $18x^3 - 30x^4 = 6x^3(?)$

$$\frac{18x^3 - 30x^4}{6x^3} = (?)$$

$$\frac{18x^3}{6x^3} - \frac{30x^4}{6x^3} = (?)$$

$$3x^{3-3} - 5x^{4-3} = (?)$$

$$3 - 5x = (?)$$

55. $6t^5 - 9t^4 + 12t^2 = 3t^2(?)$

$$\frac{6t^5 - 9t^4 + 12t^2}{3t^2} = (?)$$

$$\frac{6t^5}{3t^2} - \frac{9t^4}{3t^2} + \frac{12t^2}{3t^2} = (?)$$

$$2t^{5-2} - 3t^{4-2} + 4t^{2-2} = (?)$$

$$2t^3 - 3t^2 + 4 = (?)$$

57. $45n^4 - 36n^3 - 18n^2 = 9n^2(?)$

$$\frac{45n^4 - 36n^3 - 18n^2}{9n^2} = (?)$$

$$\frac{45n^4}{9n^2} - \frac{36n^3}{9n^2} - \frac{18n^2}{9n^2} = (?)$$

$$5n^{4-2} - 4n^{3-2} - 2n^{2-2} = (?)$$

$$5n^2 - 4n - 2 = (?)$$

59. $A = lw$

$$63m^2 = 7m(?)$$

$$\frac{63m^2}{7m} = (?)$$

$$9m = (?)$$

61. $V = lwh$

$$60t^5 = 3t^2 \cdot 5t \cdot (?)$$

$$60t^5 = 15t^3(?)$$

$$\frac{60t^5}{15t^3} = (?)$$

$$4t^2 = (?)$$

63. $8x - 4 = 4\left(\dfrac{8x-4}{4}\right)$

$$= 4\left(\frac{8x}{4} - \frac{4}{4}\right)$$

$$= 4(2x - 1)$$

65. $10a + 20 = 10\left(\dfrac{10a+20}{10}\right)$

$$= 10\left(\frac{10a}{10} + \frac{20}{10}\right)$$

$$= 10(a + 2)$$

67. $2n^2 + 6n = 2n\left(\dfrac{2n^2+6n}{2n}\right)$

$$= 2n\left(\frac{2n^2}{2n} + \frac{6n}{2n}\right)$$

$$= 2n(n + 3)$$

69. $7x^3 - 3x^2 = x^2\left(\dfrac{7x^3-3x^2}{x^2}\right)$

$$= x^2\left(\frac{7x^3}{x^2} - \frac{3x^2}{x^2}\right)$$

$$= x^2(7x - 3)$$

71. $20r^5 - 24r^3 = 4r^3\left(\dfrac{20r^5-24r^3}{4r^3}\right)$

$$= 4r^3\left(\frac{20r^5}{4r^3} - \frac{24r^3}{4r^3}\right)$$

$$= 4r^3(5r^2 - 6)$$

73. $6y^3 + 3y^2 = 3y^2 \left(\dfrac{6y^3 + 3y^2}{3y^2} \right)$

$= 3y^2 \left(\dfrac{6y^3}{3y^2} + \dfrac{3y^2}{3y^2} \right)$

$= 3y^2 \left(2y + 1 \right)$

75. $12x^3 + 20x^2 + 32x = 4x \left(\dfrac{12x^3 + 20x^2 + 32x}{4x} \right)$

$= 4x \left(\dfrac{12x^3}{4x} + \dfrac{20x^2}{4x} + \dfrac{32x}{4x} \right)$

$= 4x \left(3x^2 + 5x + 8 \right)$

77. $9a^7 - 12a^5 + 18a^3 = 3a^3 \left(\dfrac{9a^7 - 12a^5 + 18a^3}{3a^3} \right)$

$= 3a^3 \left(\dfrac{9a^7}{3a^3} - \dfrac{12a^5}{3a^3} + \dfrac{18a^3}{3a^3} \right)$

$= 3a^3 \left(3a^4 - 4a^2 + 6 \right)$

79. $14m^8 + 28m^6 + 7m^5$

$= 7m^5 \left(\dfrac{14m^8 + 28m^6 + 7m^5}{7m^5} \right)$

$= 7m^5 \left(\dfrac{14m^8}{7m^5} + \dfrac{28m^6}{7m^5} + \dfrac{7m^5}{7m^5} \right)$

$= 7m^5 \left(2m^3 + 4m + 1 \right)$

81. $10x^9 - 20x^5 - 40x^3$

$= 10x^3 \left(\dfrac{10x^9 - 20x^5 - 40x^3}{10x^3} \right)$

$= 10x^3 \left(\dfrac{10x^9}{10x^3} - \dfrac{20x^5}{10x^3} - \dfrac{40x^3}{10x^3} \right)$

$= 10x^3 \left(x^6 - 2x^2 - 4 \right)$

Review Exercises

1. $P = 4 + 5 + 3 + 7 + 1 = 20$ ft.

2. $A = bh$
   $A = 18 \cdot 8$
   $A = 144$ m^2

3. $V = lwh$
   $V = 9 \cdot 3 \cdot 4$
   $V = 108$ in.^{3}

4. $\left(6x^3 - 9x^2 + x - 12 \right) - \left(4x^3 + x + 5 \right)$

$= \left(6x^3 - 9x^2 + x - 12 \right) + \left(-4x^3 - x - 5 \right)$

$= 2x^3 - 9x^2 - 17$

5. $(x + 4)(x - 6) = x \cdot x + x \cdot (-6) + 4 \cdot x + 4 \cdot (-6)$

$= x^2 - 6x + 4x - 24$

$= x^2 - 2x - 24$

Exercise Set 3.7

1. Surface measure is the total number of square units that completely cover the outer shell of an object.

3. The formula is used to calculate the height of a falling object, where h represents the height in feet after falling for t seconds from an initial height of h_0 feet.

5. **Understand:** To find the perimeter around the shape, we must add the lengths of all the sides. Because those sides are expressions, we must add the expression.

   Plan: Add the expressions to get an expression for the perimeter.

   Execute: Perimeter $= x + (x + 7) + x + (x + 7)$

   $= x + x + 7 + x + x + 7$

   $= 4x + 14$

   Answer: The expression for the perimeter is $4x + 14$.

7. **Understand:** To find the perimeter around the shape, we must add the lengths of all the sides. Because those sides are expressions, we must add the expression.

   Plan: Add the expressions to get an expression for the perimeter.

   Execute: Perimeter $= (n - 4) + n + (n - 2)$

   $= n - 4 + n + n - 2$

   $= 3n - 6$

   Answer: The expression for the perimeter is $3n - 6$.

9. **Understand:** To find the perimeter around the shape, we must add the lengths of all the sides. Because those sides are expressions, we must add the expression.

   Plan: Add the expressions to get an expression for the perimeter.

   Execute:
   Perimeter $= 2b + (b - 3) + (b - 2) + (b - 2) + (b - 3) + 2b$

   $= 2b + b - 3 + b - 2 + b - 2 + b - 3 + 2b$

   $= 8b - 10$

   Answer: The expression for the perimeter is $8b - 10$.

11. a) Evaluate $4x + 14$ using $x = 19$.
$4 \cdot 19 + 14 = 76 + 14$
$\qquad = 90$ in.

b) Evaluate $4x + 14$ using $x = 27$.
$4 \cdot 27 + 14 = 108 + 14$
$\qquad = 122$ cm

13. a) Evaluate $3n - 6$ using $n = 12$.
$3 \cdot 12 - 6 = 36 - 6$
$\qquad = 30$ ft.

b) Evaluate $3n - 6$ using $n = 8$.
$3 \cdot 8 - 6 = 24 - 6$
$\qquad = 18$ yd.

15. a) Evaluate $8b - 10$ using $b = 10$.
$8 \cdot 10 - 10 = 80 - 10$
$\qquad = 70$ in.

b) Evaluate $8b - 10$ using $b = 20$.
$8 \cdot 20 - 10 = 160 - 10$
$\qquad = 150$ ft.

17. **Understand:** To find the area of the parallelogram, we must use $A = bh$. Because those sides are expressions, we must multiply the expressions.

Plan: Multiply the expressions to get an expression for the area.

Execute: Area $= 2n(n + 4)$
$\qquad = 2n^2 + 8n$

Answer: The expression for the area is $2n^2 + 8n$.

19. a) Evaluate $2n^2 + 8n$ using $n = 6$.
$2 \cdot 6^2 + 8 \cdot 6 = 2 \cdot 36 + 8 \cdot 6$
$\qquad = 72 + 48$
$\qquad = 120$ km^2

b) Evaluate $2n^2 + 8n$ using $n = 12$.
$2 \cdot 12^2 + 8 \cdot 12 = 2 \cdot 144 + 8 \cdot 12$
$\qquad = 288 + 96$
$\qquad = 384$ ft.^{2}

21. **Understand:** To find the volume of the box, we must use $V = lwh$. Because those sides are expressions, we must multiply the expressions.

Plan: Multiply the expressions to get an expression for the volume.

Execute: Volume $= (d + 6)(d - 1)d$
$\qquad = (d^2 + 5d - 6)d$
$\qquad = d^3 + 5d^2 - 6d$

Answer: The expression for the volume is $d^3 + 5d^2 - 6d$.

23. a) Evaluate $d^3 + 5d^2 - 6d$ using $d = 3$.
$3^3 + 5 \cdot 3^2 - 6 \cdot 3 = 27 + 5 \cdot 9 - 6 \cdot 3$
$\qquad = 27 + 45 - 18$
$\qquad = 54$ in.^{3}

b) Evaluate $d^3 + 5d^2 - 6d$ using $d = 5$.
$5^3 + 5 \cdot 5^2 - 6 \cdot 5 = 125 + 5 \cdot 25 - 6 \cdot 5$
$\qquad = 125 + 125 - 30$
$\qquad = 220$ ft.^{3}

25. **Understand:** We must calculate how many square feet of cardboard are needed to produce a box. Because the cardboard will become the shell or skin of the box, we are dealing with surface area. The formula for surface area of a box is $SA = 2(lw + lh + wh)$.

Plan: Replace l with 2, w with 2, and h with 3, in the formula $SA = 2(lw + lh + wh)$.

Execute: $SA = 2(lw + lh + wh)$
$SA = 2(2 \cdot 2 + 2 \cdot 3 + 2 \cdot 3)$
$SA = 2(4 + 6 + 6)$
$SA = 2(16)$
$SA = 32$ ft.^{2}

Answer: Each box will require 32 ft.^{2} of cardboard.

27. **Understand:** We must calculate how many square feet of metal are needed to make a cube. The formula for surface area of a cube is $SA = 2(lw + lh + wh)$.

Plan: Replace l with 16, w with 16, and h with 16, in the formula $SA = 2(lw + lh + wh)$.

Execute: $SA = 2(lw + lh + wh)$
$SA = 2(16 \cdot 16 + 16 \cdot 16 + 16 \cdot 16)$
$SA = 2(256 + 256 + 256)$
$SA = 2(768)$
$SA = 1536$ ft.^{2}

Answer: The cube will require 1536 panels.

29. **Understand:** We are given the initial altitude, or height, of the cannonball and the time for the free fall. To find the height after (a) 2 sec. and (b) 3 sec. of free fall, we can use the formula $h = -16t^2 + h_0$.

Plan: Replace t with (a) 2 and (b) 3 and replace h_0 with 180, then calculate.

Execute: (a) $h = -16t^2 + h_0$

$h = -16 \cdot 2^2 + 180$

$h = -16 \cdot 4 + 180$

$h = -64 + 180$

$h = 116$ ft.

(b) $h = -16t^2 + h_0$

$h = -16 \cdot 3^2 + 180$

$h = -16 \cdot 9 + 180$

$h = -144 + 180$

$h = 36$ ft.

Answer: (a) The cannonball is at an altitude of 116 ft. after 2 seconds. (b) The cannonball is at an altitude of 36 ft. after 3 seconds.

31. **Understand:** We are given the initial altitude, or height, of the skydiver and the time for the free fall. To find the height after 22 seconds of free fall, we can use the formula $h = -16t^2 + h_0$.

Plan: Replace t with 22 and h_0 with 12,000, then calculate.

Execute: $h = -16t^2 + h_0$

$h = -16 \cdot 22^2 + 12{,}000$

$h = -16 \cdot 484 + 12{,}000$

$h = -7744 + 12{,}000$

$h = 4256$

Answer: The skydiver is at an altitude of 4256 ft. after 22 seconds.

33. **Understand:** We are to find net, given expressions for revenue and cost. The formula for net is $N = R - C$.

Plan: a) Replace R with $145r + 215s + 100$ and C with $110r + 140s + 345$, then subtract.
b) Replace r with 120 and s with 106 in the net formula, then calculate.

Execute:
$N = R - C$

$N = \left(145r + 215s + 100\right) - \left(110r + 140s + 345\right)$

$N = \left(145r + 215s + 100\right) + \left(-110r - 140s - 345\right)$

$N = 35r + 75s - 245$

Answer: a) The expression that describes net is $35r + 75s - 245$.

b) $N = 35r + 75s - 245$

$N = 35 \cdot 120 + 75 \cdot 106 - 245$

$N = 4200 + 7950 - 245$

$N = \$11{,}905$

The profit would be \$11,905.

Check: We can reverse the process. If we add the expression for net to the expression for cost, we should get the expression for revenue.

$\left(35r + 75s - 245\right) + \left(110r + 140s + 345\right)$

$= 145r + 215s + 100$

35. **Understand:** We are to find the net, given the number of small, medium, and large baskets. Suppose s represents the number of small baskets, m represents the number of medium baskets, and l represents the number of large baskets. We must find (a) the polynomial describing revenue, (b) describing cost, (c) an expression in simplest for net, and (d) the net profit or loss.

Plan and Execute:
(a) The revenue is the total earned for all of the baskets. $R = 5s + 9m + 15l$
(b) The cost is the total to make all of the baskets. $C = 2s + 4m + 7l$
(c) The expression to represent net would be
$N = R - C$

$N = (5s + 9m + 15l) - (2s + 4m + 7l)$

$N = (5s + 9m + 15l) + (-2s - 4m - 7l)$

$N = 3s + 5m + 8l$

(d) To find the net profit or loss for one day, we will substitute 6 for s, 9 for m, and 3 for l.
$N = 3s + 5m + 8l$

$N = 3 \cdot 6 + 5 \cdot 9 + 8 \cdot 3$

$N = 18 + 45 + 24$

$N = \$87$

Answer: Candice's net profit for one day is \$87.

Review Exercises

1. $\sqrt{2 \cdot 21 + 7} - (-2)^3 = \sqrt{42 + 7} - \left(-2\right)^3$

$= \sqrt{49} - (-2)^3$

$= 7 - (-8)$

$= 7 + 8$

$= 15$

2. The coefficient is -1.

3. The degree is 0.

4. $8x^3 - 9x^2 + 11 - 12x^2 + 2x - 9x^3 + 3 - 2x$

$= 8x^3 - 9x^3 - 9x^2 - 12x^2 + 2x - 2x + 11 + 3$

$= -x^3 - 21x^2 + 14$

5. $-4(3a+7) = -4 \cdot 3a - 4 \cdot 7$
    $= -12a - 28$

6. $6^0 = 1$

Chapter 3 Review Exercises

1. false   2. true   3. true

4. false   5. false   6. false

7. coefficients; variables

8. add; base

9. subtract; divisor's; dividend's

10. multiply

11. $5(-2)^2 - 9(-2) + 2 = 5 \cdot 4 - 9(-2) + 2$
    $= 20 - (-18) + 2$
    $= 20 + 18 + 2$
    $= 40$

12. $\sqrt{3 \cdot 7 + 4} - 2 \cdot 9 = \sqrt{21 + 4} - 2 \cdot 9$
    $= \sqrt{25} - 2 \cdot 9$
    $= 5 - 18$
    $= -13$

13. $4\left|3^2 - 11\right| = 4\left|9 - 11\right|$
    $= 4\left|-2\right|$
    $= 4 \cdot 2$
    $= 8$

14. $\dfrac{-3 \cdot 2 \cdot -4 + 4}{2 - (-4)^2} = \dfrac{-6 \cdot -4 + 4}{2 - 16}$
    $= \dfrac{24 + 4}{-14}$
    $= \dfrac{28}{-14}$
    $= -2$

15. Yes, it is a product of a constant and a variable whole number power.

16. No, because it is not a product.

17. coefficient: 18; degree: 1

18. coefficient: 1; degree: 3

19. coefficient: -9; degree: 0

20. coefficient: -3; degree: $5+1 = 6$

21. They have the same variables raised to the same exponents.

22. The same variables are raised to different exponents.

23. binomial   24. no special name

25. monomial   26. trinomial

27. The degree is 6.

28. $5a^6 + 7a^4 - a^2 - 9a + 13$

29. $3a^2 + 2a - 4a^2 - 1$
    $= 3a^2 - 4a^2 + 2a - 1$
    $= -a^2 + 2a - 1$

30. $3x^2 - 5x + x^7 + 13 + 5x - 6x^7$
    $= x^7 - 6x^7 + 3x^2 - 5x + 5x + 13$
    $= -5x^7 + 3x^2 + 13$

31. $(x-4) + (2x+9) = x + 2x - 4 + 9$
    $= 3x + 5$

32. $(y^4 + 2y^3 - 8y + 5) + (3y^4 - 2y - 9)$
    $= y^4 + 3y^4 + 2y^3 - 8y - 2y + 5 - 9$
    $= 4y^4 + 2y^3 - 10y - 4$

33. $\left(a^2 - a\right) - \left(-2a^2 + 4a\right)$
    $= \left(a^2 - a\right) + \left(2a^2 - 4a\right)$
    $= a^2 + 2a^2 - a - 4a$
    $= 3a^2 - 5a$

34. $(19h^3 - 4h^2 + 2h - 1) - (6h^3 + h^2 + 7h + 2)$
    $= (19h^3 - 4h^2 + 2h - 1) + (-6h^3 - h^2 - 7h - 2)$
    $= 19h^3 - 6h^3 - 4h^2 - h^2 + 2h - 7h - 1 - 2$
    $= 13h^3 - 5h^2 - 5h - 3$

35. $m^3 \cdot m^4 = m^{3+4} = m^7$

36. $2x \cdot (-5x^4) = 2 \cdot (-5)x^{1+4}$
    $= -10x^5$

37. $-4y \cdot 3y^4 = -4 \cdot 3y^{1+4}$
    $= -12y^5$

38. $-6t^3 \cdot (-t^5) = -6 \cdot (-1)t^{3+5}$
    $= 6t^8$

39. $(5x^4)^3 = 5^3 x^{4 \cdot 3}$
    $= 125x^{12}$

40. $\left(-2y^3\right)^2 = (-2)^2 y^{3 \cdot 2}$
    $= 4y^6$

41. $\left(4t^3\right)^3 = 4^3 t^{3\cdot 3}$

       $= 64t^9$

42. $(-2a^4)^3 = (-2)^3 a^{4\cdot 3}$

       $= -8a^{12}$

43. $2(3x-4) = 2\cdot 3x + 2\cdot(-4)$

         $= 6x - 8$

44. $-4y(4y-5) = -4y\cdot 4y - 4y\cdot(-5)$

           $= -16y^2 + 20y$

45. $3n(5n^2 - n + 7) = 3n\cdot 5n^2 + 3n\cdot(-n) + 3n\cdot 7$

                $= 15n^3 - 3n^2 + 21n$

46. $\left(a+5\right)\left(a-7\right) = a\cdot a + a\cdot -7 + 5\cdot a + 5\cdot(-7)$

                $= a^2 - 7a + 5a - 35$

                $= a^2 - 2a - 35$

47. $\left(2x-3\right)(x+1) = 2x\cdot x + 2x\cdot 1 - 3\cdot x - 3\cdot 1$

                $= 2x^2 + 2x - 3x - 3$

                $= 2x^2 - x - 3$

48. $\left(2y-1\right)\left(5y-8\right)$

     $= 2y\cdot 5y + 2y\cdot\left(-8\right) - 1\cdot 5y - 1\cdot(-8)$

     $= 10y^2 - 16y - 5y + 8$

     $= 10y^2 - 21y + 8$

49. $\left(3t+4\right)\left(3t-4\right)$

     $= \left(3t\right)^2 - \left(4\right)^2$

     $= 9t^2 - 16$

50. $\left(u+2\right)\left(u^2 - 5u + 3\right)$

     $= u\cdot u^2 + u\cdot(-5u) + u\cdot 3 + 2\cdot u^2 + 2\cdot(-5u) + 2\cdot 3$

     $= u^3 - 5u^2 + 3u + 2u^2 - 10u + 6$

     $= u^3 - 3u^2 - 7u + 6$

51. $3y + 4$ 52. $-7x - 2$

53. composite   54. prime

55. $360 = 2^3\cdot 3^2\cdot 5$ 56. $4200 = 2^3\cdot 3\cdot 5^2\cdot 7$

57. $140 = 2^2\cdot 5\cdot 7$

     $196 = 2^2\cdot 7^2$

     GCF $= 2^2\cdot 7 = 4\cdot 7 = 28$

58. $45 = 3^2\cdot 5$

     $28 = 2^2\cdot 7$

     GCF $= 1$

59. $48x^5 = 2^4\cdot 3\cdot x^5$

     $36x^6 y = 2^2\cdot 3^2\cdot x^6\cdot y$

     GCF $= 2^2\cdot 3\cdot x^5 = 12x^5$

60. $18a^2 = 2\cdot 3^2\cdot a^2$

     $30a^4 = 2\cdot 3\cdot 5\cdot a^4$

     $24a^3 = 2^3\cdot 3\cdot a^3$

     GCF $= 2\cdot 3\cdot a^2 = 6a^2$

61. $r^8 \div r^2 = r^{8-2}$

       $= r^6$

62. $\dfrac{20x^5}{-5x^2} = \dfrac{20}{-5}x^{5-2}$

         $= -4x^3$

63. $\left(14t^2 + 18t\right) \div \left(2t\right)$

     $= \dfrac{14t^2 + 18t}{2t}$

     $= \dfrac{14t^2}{2t} + \dfrac{18t}{2t}$

     $= \dfrac{14}{2}t^{2-1} + \dfrac{18}{2}t^{1-1}$

     $= 7t + 9$

64. $\dfrac{-30x^6 + 24x^4 - 12x^2}{-6x^2}$

     $= \dfrac{-30x^6}{-6x^2} + \dfrac{24x^4}{-6x^2} - \dfrac{12x^2}{-6x^2}$

     $= \dfrac{-30}{-6}x^{6-2} + \dfrac{24}{-6}x^{4-2} - \dfrac{12}{-6}x^{2-2}$

     $= 5x^4 - 4x^2 + 2$

65. $6m^3 n\cdot (?) = -54m^8 n$

       $(?) = \dfrac{-54m^8 n}{6m^3 n}$

       $(?) = \dfrac{-54}{6}m^{8-3}n^{1-1}$

       $(?) = -9m^5$

66. $12x^6 + 9x^4 = 3x^4(?)$

       $\dfrac{12x^6 + 9x^4}{3x^4} = (?)$

       $\dfrac{12x^6}{3x^4} + \dfrac{9x^4}{3x^4} = (?)$

     $\dfrac{12}{3}x^{6-4} + \dfrac{9}{3}x^{4-4} = (?)$

         $4x^2 + 3 = (?)$

67. $$40y^6 - 30y^4 - 20y^2 = 10y^2(?)$$

$$\frac{40y^6 - 30y^4 - 20y^2}{10y^2} = (?)$$

$$\frac{40y^6}{10y^2} - \frac{30y^4}{10y^2} - \frac{20y^2}{10y^2} = (?)$$

$$\frac{40}{10}y^{6-2} - \frac{30}{10}y^{4-2} - \frac{20}{10}y^{2-2} = (?)$$

$$4y^4 - 3y^2 - 2 = (?)$$

68. $$A = lw$$

$$42b^6 = (?)3b^5$$

$$\frac{42b^6}{3b^5} = (?)$$

$$\frac{42}{3}b^{6-5} = (?)$$

$$14b = (?)$$

69. $$30x + 12 = 6\left(\frac{30x+12}{6}\right)$$

$$= 6\left(\frac{30x}{6} + \frac{12}{6}\right)$$

$$= 6(5x + 2)$$

70. $$12y^2 + 20y = 4y\left(\frac{12y^2 + 20y}{4y}\right)$$

$$= 4y\left(\frac{12y^2}{4y} + \frac{20y}{4y}\right)$$

$$= 4y(3y + 5)$$

71. $$9n^4 - 15n^3 = 3n^3\left(\frac{9n^4 - 15n^3}{3n^3}\right)$$

$$= 3n^3\left(\frac{9n^4}{3n^3} - \frac{15n^3}{3n^3}\right)$$

$$= 3n^3(3n - 5)$$

72. $$18x^3 + 24x^2 - 36x$$

$$= 6x\left(\frac{18x^3 + 24x^2 - 36x}{6x}\right)$$

$$= 6x\left(\frac{18x^3}{6x} + \frac{24x^2}{6x} - \frac{36x}{6x}\right)$$

$$= 6x(3x^2 + 4x - 6)$$

73. a) $$P = (5x + 2) + (3x - 8) + (5x + 2) + (3x - 8)$$

$$P = 5x + 3x + 5x + 3x + 2 - 8 + 2 - 8$$

$$P = 16x - 12$$

b) $$P = 16 \cdot 9 - 12$$

$$P = 144 - 12$$

$$P = 132 \text{ m}$$

74. a) $$A = bh$$

$$A = (m + 3)(m - 5)$$

$$A = m \cdot m + m \cdot (-5) + 3 \cdot m + 3 \cdot (-5)$$

$$A = m^2 - 5m + 3m + (-15)$$

$$A = m^2 - 2m - 15$$

b) $$A = 8^2 - 2 \cdot 8 - 15$$

$$A = 64 - 2 \cdot 8 - 15$$

$$A = 64 - 16 - 15$$

$$A = 33 \text{ in.}^2$$

75. a) $$V = lwh$$

$$V = (3n - 2)(2n)(8n)$$

$$V = (3n \cdot 2n - 2 \cdot 2n)(8n)$$

$$V = (6n^2 - 4n)(8n)$$

$$V = 6n^2 \cdot 8n - 4n \cdot 8n$$

$$V = 48n^3 - 32n^2$$

b) $$V = 48 \cdot 4^3 - 32 \cdot 4^2$$

$$V = 48 \cdot 64 - 32 \cdot 16$$

$$V = 3072 - 512$$

$$V = 2560 \text{ ft.}^3$$

76. Replace l with 4 , w with 3 , and h with 2, in the formula $SA = 2(lw + lh + wh)$.

$$SA = 2(lw + lh + wh)$$

$$= 2(4 \cdot 3 + 4 \cdot 2 + 3 \cdot 2)$$

$$= 2(12 + 8 + 6)$$

$$= 2(26)$$

$$= 52 \text{ ft.}^2$$

77. We must determine the height of a coin dropped from 150 ft. after 2 seconds. $h = -16t^2 + h_0$

$$h = -16 \cdot 2^2 + 150$$

$$h = -16 \cdot 4 + 150$$

$$h = -64 + 150$$

$$h = 86 \text{ ft.}$$

78. a) $$N = R - C$$

$$N = (65n + 85b + 345) - (22n + 42b + 450)$$

$$N = (65n + 85b + 345) + (-22n - 42b - 450)$$

$$N = 43n + 43b - 105$$

b) Replace n with 487 and b with 246.

$$N = 43 \cdot 487 + 43 \cdot 246 - 105$$

$$N = 20,941 + 10,578 - 105$$

$$N = \$31,414$$

The net profit is \$31,414.

Chapter 3 Practice Test

1. $2(-2)^3 - 3 \cdot (-3)^2 = 2 \cdot (-8) - 3 \cdot 9$
$$= -16 - 27$$
$$= -43$$

2. coefficient: -1; degree: 4

3. binomial

4. degree: 4

5. $10a^4 + 3a^2 - 5a + 2a^2 + 5a - 11a^4$
$$= 10a^4 - 11a^4 + 3a^2 + 2a^2 - 5a + 5a$$
$$= -a^4 + 5a^2$$

6. $(2y^3 - 8y + 5) + (3y^3 - 2y - 9)$
$$= 2y^3 + 3y^3 - 8y - 2y + 5 - 9$$
$$= 5y^3 - 10y - 4$$

7. $(12y^5 + 5y^3 - 7y - 3) - (6y^5 + y^2 - 7y + 1)$
$$= (12y^5 + 5y^3 - 7y - 3) + (-6y^5 - y^2 + 7y - 1)$$
$$= 12y^5 - 6y^5 + 5y^3 - y^2 - 7y + 7y - 3 - 1$$
$$= 6y^5 + 5y^3 - y^2 - 4$$

8. $(-4u^5)(-5u^2) = (-4)(-5)u^{5+2}$
$$= 20u^7$$

9. $(-2a^3)^5 = (-2)^5 a^{3 \cdot 5}$
$$= -32a^{15}$$

10. $-2t(t^3 - 3t - 7)$
$$= -2t \cdot t^3 - 2t \cdot (-3t) - 2t \cdot (-7)$$
$$= -2t^{1+3} - 2 \cdot (-3)t^{1+1} - 2 \cdot (-7)t$$
$$= -2t^4 + 6t^2 + 14t$$

11. $(x - 6)(x + 6) = x^2 - 36$

12. $-2a^2(a + 1)(2a - 1)$
$$= -2a^2 \left[a \cdot 2a + a \cdot (-1) + 1 \cdot 2a + 1 \cdot (-1) \right]$$
$$= -2a^2 \left[2a^2 - a + 2a - 1 \right]$$
$$= -2a^2 \left[2a^2 + a - 1 \right]$$
$$= -2a^2 \cdot 2a^2 + (-2a^2) \cdot a + (-2a^2) \cdot (-1)$$
$$= -4a^4 - 2a^3 + 2a^2$$

13. $6x + 5$ 14. composite

15. $340 = 2^2 \cdot 5 \cdot 17$

16. $180 = 2^2 \cdot 3^2 \cdot 5$
$396 = 2^2 \cdot 3^2 \cdot 11$
$\text{GCF} = 2^2 \cdot 3^2 = 4 \cdot 9 = 36$

17. $60h^5 = 2^2 \cdot 3 \cdot 5 \cdot h^5$
$48h^7 = 2^4 \cdot 3 \cdot h^7$
$\text{GCF} = 2^2 \cdot 3h^5 = 12h^5$

18. $m^6 \div m^4 = m^{6-4} = m^2$

19. $\dfrac{16x^5 - 18x^3 + 22x^2}{-2x} = \dfrac{16x^5}{-2x} - \dfrac{18x^3}{-2x} + \dfrac{22x^2}{-2x}$
$$= \frac{16}{-2}x^{5-1} - \frac{18}{-2}x^{3-1} + \frac{22}{-2}x^{2-1}$$
$$= -8x^4 + 9x^2 - 11x$$

20. $-8x^4(?) = -40x^7$
$$(?) = \frac{-40x^7}{-8x^4}$$
$$(?) = \frac{-40}{-8}x^{7-4}$$
$$(?) = 5x^3$$

21. $12x - 20 = 4\left(\dfrac{12x - 20}{4}\right)$
$$= 4\left(\frac{12x}{4} - \frac{20}{4}\right)$$
$$= 4(3x - 5)$$

22. $10y^4 - 18y^3 + 14y^2$
$$= 2y^2\left(\frac{10y^4 - 18y^3 + 14y^2}{2y^2}\right)$$
$$= 2y^2\left(\frac{10y^4}{2y^2} - \frac{18y^3}{2y^2} + \frac{14y^2}{2y^2}\right)$$
$$= 2y^2\left(5y^2 - 9y + 7\right)$$

23. a) $A = bh$
$A = (2n + 1)(n - 4)$
$A = 2n \cdot n + 2n \cdot (-4) + 1 \cdot n + 1 \cdot (-4)$
$A = 2n^2 - 8n + n - 4$
$A = 2n^2 - 7n - 4$

b) $A = 2 \cdot 7^2 - 7 \cdot 7 - 4$
$A = 2 \cdot 49 - 7 \cdot 7 - 4$
$A = 98 - 49 - 4$
$A = 45$ cm^2

24. Replace l with 10, w with 8, and h with 4, in the formula $SA = 2(lw + lh + wh)$.

$$SA = 2\big(lw + lh + wh\big)$$
$$= 2\big(10 \cdot 8 + 10 \cdot 4 + 8 \cdot 4\big)$$
$$= 2\big(80 + 40 + 32\big)$$
$$= 2\big(152\big)$$
$$= 304 \text{ in.}^2$$

25. a) $N = R - C$
$$N = \big(145a + 176b\big) - \big(61a + 85b\big)$$
$$N = \big(145a + 176b\big) + \big(-61a - 85b\big)$$
$$N = 84a + 91b$$

   b) $N = 84 \cdot 128 + 91 \cdot 115$
$$N = 10,752 + 10,465$$
$$N = \$21,217$$

   The net profit is $21,217.

Chapter 1-3 Cumulative Review Exercises

1. false   2. true   3. false

4. false   5. true   6. true

7. two million, four hundred eighty thousand, forty-five

8. $5 \times 10,000 + 4 \times 1000 + 3 \times 100 + 7 \times 1$

9. 110,000

10. $700 \times 200 = 140,000$; think 7×2 and add 4 zeros.

11. 1. Grouping symbols
    2. Exponents or roots from left to right
    3. Multiplication or division from left to right
    4. Addition or subtraction from left to right.

12.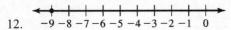
    $-9\ -8\ -7\ -6\ -5\ -4\ -3\ -2\ -1\ \ 0$

13. negative   14. negative

15. coefficient: -1 16. degree: 1

17. degree: 5

18. Multiply each term in the second polynomial by each term in the first polynomial (FOIL).

19. $|-20| = 20$

20. $-(-(-9)) = -9$

21. $-20 + \big(-8\big) = -28$

22. $-63 - (-25) = -63 + 25$
$$= -38$$

23. $-1(-9)(-3)(-4) = 9(-3)(-4)$
$$= -27(-4)$$
$$= 108$$

24. $(-5)^3 = -5 \cdot \big(-5\big) \cdot \big(-5\big) = -125$

25. $-3^4 = -(3 \cdot 3 \cdot 3 \cdot 3) = -81$

26. $42 \div (-6) = -7$

27. $\sqrt{16 \cdot 25} = \sqrt{400}$
$$= 20$$

28. $2[9 + (3 - 16)] - 35 \div 5 = 2[9 + (-13)] - 35 \div 5$
$$= 2[-4] - 35 \div 5$$
$$= -8 - 7$$
$$= -15$$

29. $4^3 - \{[6 + (2)8] - [12 + 9(6)]\}$
$$= 4^3 - \{[6 + 16] - [12 + 54]\}$$
$$= 4^3 - \{22 - 66\}$$
$$= 4^3 - (-44)$$
$$= 64 + 44$$
$$= 108$$

30. $\sqrt{100 - 36} = \sqrt{64}$
$$= 8$$

31. $5x^2 - 9x + 11x^2 - 7 - x^3 + 8x - 5$
$$= -x^3 + 5x^2 + 11x^2 - 9x + 8x - 7 - 5$$
$$= -x^3 + 16x^2 - x - 12$$

32. $(4y^3 - 6y^2 + 9y - 8) - (3y^2 - 12y - 2)$
$$= (4y^3 - 6y^2 + 9y - 8) + (-3y^2 + 12y + 2)$$
$$= 4y^3 - 3y^2 - 6y^2 + 9y + 12y - 8 + 2$$
$$= 4y^3 - 9y^2 + 21y - 6$$

33. $(9x^3)(7x) = 9 \cdot 7x^{3+1}$
$$= 63x^4$$

34. $(b - 8)(2b + 3) = b \cdot 2b + b \cdot 3 - 8 \cdot 2b - 8 \cdot 3$
$$= 2b^2 + 3b - 16b - 24$$
$$= 2b^2 - 13b - 24$$

35. $360 = 2^3 \cdot 3^2 \cdot 5$ 36. $40x^2 = 2^3 \cdot 5 \cdot x^2$
$$30x^3 y = 2 \cdot 3 \cdot 5 \cdot x^3 \cdot y$$
$$\text{GCF} = 2 \cdot 5 \cdot x^2 = 10x^2$$

37. $\dfrac{18n^5}{3n^2} = \dfrac{18}{3} n^{5-2} = 6n^3$

38. $\dfrac{4x^3 - 8x^2 + 12x}{-2x} = \dfrac{4x^3}{-2x} - \dfrac{8x^2}{-2x} + \dfrac{12x}{-2x}$

$\quad\quad = \dfrac{4}{-2}x^{3-1} - \dfrac{8}{-2}x^{2-1} + \dfrac{12}{-2}x^{1-1}$

$\quad\quad = -2x^2 + 4x - 6$

39. $12m^4 - 18m^3 + 24m^2$

$= 6m^2\left(\dfrac{12m^4 - 18m^3 + 24m^2}{6m^2}\right)$

$= 6m^2\left(\dfrac{12m^4}{6m^2} - \dfrac{18m^3}{6m^2} + \dfrac{24m^2}{6m^2}\right)$

$= 6m^2(2m^2 - 3m + 4)$

40. $x + 19 = 25$ Check: $6 + 19 = 25$

$\quad\quad x = 25 - 19$ $25 = 25$

$\quad\quad x = 6$

41. $-16x = -128$ Check: $-16 \cdot 8 = -128$

$\quad\quad x = -128 \div (-16)$ $-128 = -128$

$\quad\quad x = 8$

42. $A = bh$ 43. $P = w + 7w + w + 7w$

$\quad A = 15 \cdot 17$ $P = 16w$

$\quad A = 255 \text{ cm}^2$

44. $V = lwh$

$\quad V = 5y \cdot y \cdot (y + 1)$

$\quad V = 5y^2(y + 1)$

$\quad V = 5y^2 \cdot y + 5y^2 \cdot 1$

$\quad V = 5y^3 + 5y^2$

45. To find Andre's total mortgage payments, we must multiply. $785 \times 360 = \$282{,}600$.

46. We must divide to find the amount of solution to receive each minute. $600 \div 60 = 10$ ml

47. Mean: Divide the sum of scores by the number of scores.

$\left(\begin{array}{l} 250 + 280 + 140 + 86 + 60 + 42 \\ +40 + 36 + 46 + 98 + 160 + 226 \end{array}\right)$

over 12

$= \dfrac{1464}{12}$

$= 122$

The mean monthly cost is $122.

Median: Arrange the scores in order from least to greatest: 36, 40, 42, 46, 60, 86, 98, 140, 160, 226, 250, 280

Then find the arithmetic mean of the middle two scores, which are 86 and 98.

$\dfrac{86 + 98}{2} = \dfrac{184}{2} = 92$

The mean monthly cost is $92.

48. $-353 + 150 + (-38) + (-38) = -429 + 150$

$\quad\quad\quad\quad\quad\quad\quad\quad\quad\quad = \$-279$

Her balance will be –$279.

49. Calculate voltage given resistance and current. The formula that relates voltage, current, and resistance is $V = ir$. Substitute –8 for i and 15 for r, then multiply.

$V = ir$

$V = (-8)(15)$

$V = -120$ V

50. $SA = 2(lw + lh + wh)$

$\quad SA = 2(12 \cdot 10 + 12 \cdot 7 + 10 \cdot 7)$

$\quad SA = 2(120 + 84 + 70)$

$\quad SA = 2(274)$

$\quad SA = 548 \text{ ft.}^2$

The surface area of the box is 548 ft.^{2}.

51. We must determine the height of an item dropped from 80 ft. after 2 seconds. $h = -16t^2 + h_0$

$\quad\quad\quad h = -16 \cdot 2^2 + 80$

$\quad\quad\quad h = -16 \cdot 4 + 80$

$\quad\quad\quad 5h = -64 + 80$

$\quad\quad\quad h = 16 \text{ ft.}$

52. a)

$N = R - C$

$N = (225b + 345a + 200) - (112b + 187a + 545)$

$N = (225b + 345a + 200) + (-112b - 187a - 545)$

$N = 113b + 158a - 345$

b) $N = 113b + 158a - 345$

$\quad N = 113(88) + 158(64) - 345$

$\quad N = 9944 + 10{,}112 - 345$

$\quad N = \$19{,}711$

The net profit for one month is $19,711.

Chapter 4 Equations

Exercise Set 4.1

1. An expression has no equal sign whereas an equation does.

3. To solve an equation means to find its solution.

5. It is an equation because it contains an equal sign.

7. It is an expression because it does not contain an equal sign.

9. It is an expression because it does not contain an equal sign.

11. $x + 9 = 14$ -5 is not a solution.
$$-5 + 9 \overset{?}{=} 14$$
$$4 \neq 14$$

13. $3t = -9$ -9 is not a solution.
$$3(-9) \overset{?}{=} -9$$
$$-27 \neq -9$$

15. $3y + 1 = 13$ 3 is not a solution.
$$3 \cdot 3 + 1 \overset{?}{=} 13$$
$$9 + 1 \overset{?}{=} 13$$
$$10 \neq 13$$

17. $6x - 7 = 41$ 8 is a solution.
$$6 \cdot 8 - 7 \overset{?}{=} 41$$
$$48 - 7 \overset{?}{=} 41$$
$$41 = 41$$

19. $3m + 10 = -2m$ 5 is not a solution.
$$3 \cdot 5 + 10 \overset{?}{=} -2 \cdot 5$$
$$15 + 10 \overset{?}{=} -10$$
$$25 \neq -10$$

21. $x^2 - 15 = 2x$ -3 is a solution.
$$(-3)^2 - 15 \overset{?}{=} 2(-3)$$
$$9 - 15 \overset{?}{=} -6$$
$$-6 = -6$$

23. $b^2 = 5b + 6$ -3 is not a solution.
$$(-3)^2 \overset{?}{=} 5(-3) + 6$$
$$9 \overset{?}{=} -15 + 6$$
$$9 \neq -9$$

25. $-5y + 20 = 4y + 20$ 0 is a solution.
$$-5 \cdot 0 + 20 \overset{?}{=} 4 \cdot 0 + 20$$
$$0 + 20 \overset{?}{=} 0 + 20$$
$$20 = 20$$

27. $-2(x - 3) - x = 4(x + 5) - 21$
$$-2(3 - 3) - 3 \overset{?}{=} 4(3 + 5) - 21$$
$$-2 \cdot 0 - 3 \overset{?}{=} 4 \cdot 8 - 21$$
$$0 - 3 \overset{?}{=} 32 - 21$$
$$-3 \neq 11$$

3 is not a solution.

29. $a^3 - 5a + 6 = a^2 + 3(a - 1)$
$$(-2)^3 - 5(-2) + 6 \overset{?}{=} (-2)^2 + 3(-2 - 1)$$
$$(-2)^3 - 5(-2) + 6 \overset{?}{=} (-2)^2 + 3(-3)$$
$$-8 - 5(-2) + 6 \overset{?}{=} 4 + 3(-3)$$
$$-8 + 10 + 6 \overset{?}{=} 4 + (-9)$$
$$8 \neq -5$$

-2 is not a solution.

Review Exercises

1. $2x - 19 + 3x + 7 = 2x + 3x - 19 + 7$
$$= 5x - 12$$

2. $7y + 4 - 12 - 9y = 7y - 9y + 4 - 12$
$$= -2y - 8$$

3. $5(m + 3) = 5 \cdot m + 5 \cdot 3$
$$= 5m + 15$$

4. $-1(3m - 4) = -1 \cdot 3m - 1 \cdot (-4)$
$$= -3m + 4$$

5. $x + 8 = 15$ Check: $7 + 8 = 15$
$$x = 15 - 8 \qquad\qquad 15 = 15$$
$$x = 7$$

Exercise Set 4.2

1. A linear equation is an equation in which each variable term is a monomial of degree 1.

3. The same amount can be added to or subtracted from both sides of an equation without affecting its solution(s).

5. Add $2x$ to both sides of the equation.

7. It is linear because 1 is the highest degree involved.

9. It is not linear because there is a term with a degree higher than 1.

11. It is not linear because there is a term with a degree higher than 1.

13. It is linear because 1 is the highest degree involved.

15. It is not linear because there is a term with a degree higher than 1.

17. It is linear because 1 is the highest degree involved.

19. $n + 14 = 20$ Check: $6 + 14 = 20$
$\underline{-14 \quad -14}$ $20 = 20$
$n + 0 = 6$
$ n = 6$

21. $-7 = x - 15$ Check: $-7 = 8 - 15$
$\underline{+15 \quad +15}$ $-7 = -7$
$ 8 = x + 0$
$ 8 = x$

23. $10 = 5x - 2 - 4x$ Check: $10 = 5 \cdot 12 - 2 - 4 \cdot 12$
$10 = x - 2$ $10 = 60 - 2 - 48$
$\underline{+2 = \quad +2}$ $10 = 58 - 48$
$12 = x + 0$ $10 = 10$
$12 = x$

25. $9y - 8y + 11 = 12 - 2$
$ y + 11 = 10$
$ \underline{-11 \quad -11}$
$ y + 0 = -1$
$ y = -1$

Check: $9(-1) - 8(-1) + 11 = 12 - 2$
$ -9 - (-8) + 11 = 10$
$ -9 + 8 + 11 = 10$
$ 10 = 10$

27. $4x - 2 = 3x + 5$
$\underline{-3x \qquad -3x}$
$ x - 2 = 0 + 5$
$ x - 2 = 5$
$ \underline{+2 \quad +2}$
$ x + 0 = 7$
$ x = 7$

Check: $4 \cdot 7 - 2 = 3 \cdot 7 + 5$
$ 28 - 2 = 21 + 5$
$ 26 = 26$

29. $m - 1 = 2m - 5$ Check: $4 - 1 = 2 \cdot 4 - 5$
$\underline{-m \qquad -m}$ $3 = 8 - 5$
$0 - 1 = m - 5$ $3 = 3$
$ -1 = m - 5$
$ \underline{+5 \qquad +5}$
$ 4 = m + 0$
$ 4 = m$

31. $3n - 5 + 4n = 6n - 12$
$ 7n - 5 = 6n - 12$
$ \underline{-6n \qquad -6n}$
$ n - 5 = 0 - 12$
$ n - 5 = -12$
$ \underline{+5 \quad +5}$
$ n + 0 = -7$
$ n = -7$

Check: $3 \cdot (-7) - 5 + 4 \cdot (-7) = 6 \cdot (-7) - 12$
$ -21 - 5 - 28 = -42 - 12$
$ -26 - 28 = -54$
$ -54 = -54$

33. $7y + 1 - 3y = 2y - 10 + 3y$
$ 4y + 1 = 5y - 10$
$ \underline{-4y \qquad -4y}$
$ 0 + 1 = y - 10$
$ 1 = y - 10$
$ \underline{+10 \qquad +10}$
$ 11 = y + 0$
$ 11 = y$

Check: $7 \cdot 11 + 1 - 3 \cdot 11 = 2 \cdot 11 - 10 + 3 \cdot 11$
$ 77 + 1 - 33 = 22 - 10 + 33$
$ 78 - 33 = 12 + 33$
$ 45 = 45$

35. $3(n + 2) = 4 + 2(n - 1)$
$ 3n + 6 = 4 + 2n - 2$
$ 3n + 6 = 2 + 2n$
$ \underline{-2n \qquad -2n}$
$ n + 6 = 2 + 0$
$ n + 6 = 2$
$ \underline{-6 \quad -6}$
$ n + 0 = -4$
$ n = -4$

Check: $3(-4 + 2) = 4 + 2(-4 - 1)$
$ 3(-2) = 4 + 2(-5)$
$ -6 = 4 + (-10)$
$ -6 = -6$

37. $6t - 2(t - 5) = 9t - (6t + 2)$

    $6t - 2t + 10 = 9t - 6t - 2$

    $4t + 10 = 3t - 2$

    $\underline{-3t \qquad -3t}$

    $t + 10 = 0 - 2$

    $t + 10 = -2$

    $\underline{-10 \quad -10}$

    $t + 0 = -12$

    $t = -12$

    Check:

    $6(-12) - 2(-12 - 5) = 9(-12) - (6 \cdot (-12) + 2)$

    $6(-12) - 2(-17) = 9(-12) - (-72 + 2)$

    $-72 - (-34) = -108 - (-70)$

    $-72 + 34 = -108 + 70$

    $-38 = -38$

39. $3295 + x = 4768$

    $\underline{-3295 \qquad -3295}$

    $0 + x = 1473$

    $x = 1473$

    She needs $1473.

41. $200 + 180 + x = 450$

    $380 + x = 450$

    $\underline{-380 \qquad -380}$

    $0 + x = 70$

    $x = 70$

    The last injection must be 70 cc.

43. a)

	1999 to 2000	2000 to 2001
Equation	$1016 + x = 1019$ $\underline{-1016 \quad -1016}$ $0 + x = 3$ $x = 3$	$1019 + x = 1020$ $\underline{-1019 \quad -1019}$ $0 + x = 1$ $x = 1$
Score Increase	3	1

	2002 to 2003	1999 to 2004
Equation	$1020 + x = 1026$ $\underline{-1020 \quad -1020}$ $0 + x = 6$ $x = 6$	$1016 + x = 1026$ $\underline{-1016 \quad -1016}$ $0 + x = 10$ $x = 10$
Score Increase	6	10

b) 2003

45. $1600 + 4500 + 1935 + x = 10,500$

    $8035 + x = 10,500$

    $\underline{-8035 \qquad -8035}$

    $x = \$2465$

    He must sell $2465 to make the quota. Yes, midway through the month he has sold $8035.

47. $3 + 4 + x = 17$

    $7 + x = 17$

    $\underline{-7 \qquad -7}$

    $x = 10$ ft.

    The length of the dining area is 10 ft. Now we need to find the area. $A = lw$

    $A = 10 \cdot 8$

    $A = 80$ ft.^{2}

    This is enough room to accommodate a 4 ft. by 4 ft. square table and four chairs.

49. Perimeter is to add all of the sides. We are given the total perimeter. $P = $ side $+$ side $+$ side

    $84 = 23 + 35 + l$

    $84 = 58 + l$

    $\underline{-58 \quad -58}$

    $26 = l$

    The missing length is 26 cm.

51. $2420 + 2447 + 1200 + x = 10,300$

    $6067 + x = 10,300$

    $\underline{-6067 \qquad = -6067}$

    $0 + x = 4233$

    $x = \$4233$

    The Coleman's are lacking $4233 in itemized deductions.

Review Exercises

1. $5(-8) = -40$ 2. $24 \div (-3) = -8$

3. $12 + 3(-5 - 2) = 12 + 3(-7)$

    $= 12 + (-21)$

    $= -9$

4. $\dfrac{16 + 2(3 - 5)}{-2(5) + 4} = \dfrac{16 + 2(-2)}{-2(5) + 4}$

    $= \dfrac{16 + (-4)}{-10 + 4}$

    $= \dfrac{12}{-6}$

    $= -2$

5. $5x = 15$

$\dfrac{5x}{5} = \dfrac{15}{5}$

$1x = 3$

$x = 3$

Check: $5 \cdot 3 = 15$

$15 = 15$

Exercise Set 4.3

1. Both sides of an equation can be multiplied or divided by the same amount without affecting its solution(s).

3. $3x = 21$

$\dfrac{3x}{3} = \dfrac{21}{3}$

$1x = 7$

$x = 7$

Check: $3(7) = 21$

$21 = 21$

5. $-4a = 36$

$\dfrac{-4a}{-4} = \dfrac{36}{-4}$

$1a = -9$

$a = -9$

Check: $-4(-9) = 36$

$36 = 36$

7. $-7b = 77$

$\dfrac{-7b}{-7} = \dfrac{77}{-7}$

$1b = -11$

$b = -11$

Check: $(-7)(-11) = 77$

$77 = 77$

9. $-14a = -154$

$\dfrac{-14a}{-14} = \dfrac{-154}{-14}$

$1a = 11$

$a = 11$

Check: $(-14)(11) = -154$

$-154 = -154$

11. $-12y = -72$

$\dfrac{-12y}{-12} = \dfrac{-72}{-12}$

$1y = 6$

$y = 6$

Check: $-12 \cdot 6 = -72$

$-72 = -72$

13. $2x + 9 = 23$

$\quad \underline{-9 \quad -9}$

$2x + 0 = 14$

$2x = 14$

$\dfrac{2x}{2} = \dfrac{14}{2}$

$1x = 7$

$x = 7$

Check: $2 \cdot 7 + 9 = 23$

$14 + 9 = 23$

$23 = 23$

15. $29 = -6b - 13$

$\underline{+13 \qquad +13}$

$42 = -6b + 0$

$42 = -6b$

$\dfrac{42}{-6} = \dfrac{-6b}{-6}$

$-7 = 1b$

$-7 = b$

Check: $29 = -6(-7) - 13$

$29 = 42 - 13$

$29 = 29$

17. $9u - 17 = -53$

$\underline{+17 = +17}$

$9u + 0 = -36$

$\dfrac{9u}{9} = \dfrac{-36}{9}$

$1u = -4$

$u = -4$

Check: $9(-4) - 17 = -53$

$-36 - 17 = -53$

$-53 = -53$

19. $7x - 16 = 5x + 6$

$\underline{-5x \qquad -5x}$

$2x - 16 = 0 + 6$

$2x - 16 = 6$

$\quad \underline{+16 \quad +16}$

$2x + 0 = 22$

$2x = 22$

$\dfrac{2x}{2} = \dfrac{22}{2}$

$1x = 11$

$x = 11$

Check: $7 \cdot 11 - 16 = 5 \cdot 11 + 6$

$77 - 16 = 55 + 6$

$61 = 61$

21. $9k + 19 = -3k - 17$

$\underline{+3k \qquad +3k}$

$12k + 19 = 0 - 17$

$12k + 19 = -17$

$\quad \underline{-19 \quad -19}$

$12k + 0 = -36$

$12k = -36$

$\dfrac{12k}{12} = \dfrac{-36}{12}$

$1k = -3$

$k = -3$

Check: $9(-3) + 19 = -3(-3) - 17$

$-27 + 19 = 9 - 17$

$-8 = -8$

23. $-2h + 17 - 8h = 15 - 7h - 10$
$$-10h + 17 = 5 - 7h$$
$$\underline{+7h \qquad +7h}$$
$$-3h + 17 = 5 + 0$$
$$-3h + 17 = \quad 5$$
$$\underline{-17 \quad -17}$$
$$-3h + 0 = -12$$
$$-3h = -12$$
$$\frac{-3h}{-3} = \frac{-12}{-3}$$
$$1h = 4$$
$$h = 4$$

Check: $-2 \cdot 4 + 17 - 8 \cdot 4 = 15 - 7 \cdot 4 - 10$
$$-8 + 17 - 32 = 15 - 28 - 10$$
$$9 - 32 = -13 - 10$$
$$-23 = -23$$

25. $13 - 9h - 15 = 7h + 22 - 8h$
$$-2 - 9h = -h + 22$$
$$\underline{+9h = +9h}$$
$$-2 + 0 = 8h + 22$$
$$-2 = 8h + 22$$
$$\underline{-22 = \qquad -22}$$
$$-24 = 8h + 0$$
$$\frac{-24}{8} = \frac{8h}{8}$$
$$-3 = 1h$$
$$-3 = h$$

Check: $13 - 9(-3) - 15 = 7(-3) + 22 - 8(-3)$
$$13 - (-27) - 15 = -21 + 22 - (-24)$$
$$13 + 27 - 15 = -21 + 22 + 24$$
$$40 - 15 = 1 + 24$$
$$25 = 25$$

27. $3(t - 4) = 5(t + 1) - 1$
$$3t - 12 = 5t + 5 - 1$$
$$3t - 12 = 5t + 4$$
$$\underline{-3t \qquad -3t}$$
$$0 - 12 = 2t + 4$$
$$-12 = 2t + 4$$
$$\underline{-4 \qquad -4}$$
$$-16 = 2t + 0$$
$$-16 = 2t$$
$$\frac{-16}{2} = \frac{2t}{2}$$
$$-8 = 1t$$
$$-8 = t$$

Check: $3(-8 - 4) = 5(-8 + 1) - 1$
$$3(-12) = 5(-7) - 1$$
$$-36 = -35 - 1$$
$$-36 = -36$$

29. $9 - 2(3x + 5) = 4x + 3(x - 9)$
$$9 - 6x - 10 = 4x + 3x - 27$$
$$-1 - 6x = \quad 7x - 27$$
$$\underline{+6x \quad +6x}$$
$$-1 + 0 = 13x - 27$$
$$-1 = 13x - 27$$
$$\underline{+27 \qquad +27}$$
$$26 = 13x - 0$$
$$26 = 13x$$
$$\frac{26}{13} = \frac{13x}{13}$$
$$2 = 1x$$
$$2 = x$$

Check: $9 - 2(3 \cdot 2 + 5) = 4 \cdot 2 + 3(2 - 9)$
$$9 - 2(6 + 5) = 4 \cdot 2 + 3(-7)$$
$$9 - 2 \cdot 11 = 8 + (-21)$$
$$9 - 22 = -13$$
$$-13 = -13$$

31. $15 + 3(3u - 1) = 16 - (2u - 7)$
$$15 + 9u - 3 = 16 - 2u + 7$$
$$12 + 9u = 23 - 2u$$
$$\underline{+2u \qquad +2u}$$
$$12 + 11u = 23 + 0$$
$$12 + 11u = \quad 23$$
$$\underline{-12 \qquad -12}$$
$$0 + 11u = \quad 11$$
$$11u = 11$$
$$\frac{11u}{11} = \frac{11}{11}$$
$$1u = 1$$
$$u = 1$$

Check: $15 + 3(3 \cdot 1 - 1) = 16 - (2 \cdot 1 - 7)$
$$15 + 3(3 - 1) = 16 - (2 - 7)$$
$$15 + 3 \cdot 2 = 16 - (-5)$$
$$15 + 6 = 16 + 5$$
$$21 = 21$$

33. Mistake: Did not write the minus sign to the left of 11.

Correct: $5x - 11 = \quad 7x - 9$
$$\underline{-5x \qquad -5x}$$
$$0 - 11 = 2x - 9$$
$$-11 = 2x - 9$$
$$\underline{+9 \qquad +9}$$
$$-2 = 2x$$
$$\frac{-2}{2} = \frac{2x}{2}$$
$$-1 = 1x$$
$$-1 = x$$

Check: $5(-1) - 11 = 7(-1) - 9$
$$-5 - 11 = -7 - 9$$
$$-16 = -16$$

35. Mistake: Did not change minus sign in $x - 8$.

Correct: $6 - 2(x + 5) = 3x - (x - 8)$
$$6 - 2x - 10 = 3x - x + 8$$
$$-4 - 2x = 2x + 8$$
$$\underline{+2x \quad\quad +2x}$$
$$-4 + 0 = 4x + 8$$
$$-4 = 4x + 8$$
$$\underline{-8 \quad\quad -8}$$
$$-12 = 4x + 0$$
$$-12 = 4x$$
$$\frac{-12}{4} = \frac{4x}{4}$$
$$-3 = 1x$$
$$-3 = x$$

Check: $6 - 2(-3 + 5) = 3(-3) - (-3 - 8)$
$$6 - 2(2) = 3(-3) - (-11)$$
$$6 - 4 = -9 + 11$$
$$2 = 2$$

37. **Understand:** We are given an area and length and must find width. Based on the given information, we can use $A = lw$.

Plan: Use $A = lw$. Replace A with 864 and l with 36 and solve for w.

Execute: $A = lw$
$$864 = 36w$$
$$\frac{864}{36} = \frac{36w}{36}$$
$$24 = 1w$$
$$24 = w$$
Answer: The width must be 24 in.

Check: Verify that the area of the blueprint sheet is 864 sq. in. when the length is 36 inches and the width is 24 inches.
$$864 = (36)(24)$$
$$864 = 864$$

39. **Understand:** We are given volume, width, and height and must find length. Based on the given information, we can use $V = lwh$.

Plan: Use $V = lwh$. Replace V with 1683, w with 11, h with 9 and solve for l.

Execute: $V = lwh$
$$1683 = l(11)(9)$$
$$1683 = 99l$$
$$\frac{1683}{99} = \frac{99l}{99}$$
$$17 = 1l$$
$$17 = l$$
Answer: The length of the chamber is 17 ft.

Check: Verify that the volume of the burial chamber is 1683 cubic feet when the length is 17 feet, width is 11 feet and height is 9 feet.
$$1683 = (17)(11)(9)$$
$$1683 = 1683$$

41. **Understand:** We are given distance and rate and must find time. Based on the given information, we can use $d = rt$.

Plan: Use $d = rt$. Replace d with 120, r with 30, and solve for t.

Execute: $d = rt$
$$120 = 30t$$
$$\frac{120}{30} = \frac{30t}{30}$$
$$4 = 1t$$
$$4 = t$$
Answer: She will reach the exit in 4 sec.

Check: Verify that an object traveling at 30 feet per second for 4 seconds will travel a distance of 120 feet. $\quad 120 = (30)(4)$
$$120 = 120$$

43. **Understand:** We are given voltage and current and must find resistance. Based on the given information, we can use $V = ir$.

Plan: Use $V = ir$. Replace V with -220, i with -5, and solve for r.

Execute: $V = ir$
$$-220 = -5r$$
$$\frac{-220}{-5} = \frac{-5r}{-5}$$
$$44 = 1r$$
$$44 = r$$
Answer: The resistance is 44 ohms.

Check: Verify that voltage is -220 volts when the current is -5 amps and the resistance is 44 ohms. $\quad -220 = (-5)(44)$
$$-220 = -220$$

45. Understand: We are given perimeter and width, and must find length. Based on the given information, we can use $P = 2l + 2w$.

Plan: Use $P = 2l + 2w$. Replace P with 1218, w with 88, and solve for l.

Execute:
$$P = 2l + 2w$$
$$1218 = 2l + 2(88)$$
$$1218 = 2l + 176$$
$$\underline{-176 \qquad -176}$$
$$1042 = 2l + \quad 0$$
$$1042 = 2l$$
$$\frac{1042}{2l} = \frac{2l}{2l}$$
$$521 = 1l$$
$$521 = l$$

Answer: The length of the memorial is 521 ft.

Check: Verify that the perimeter of the memorial is 1218 feet when the length is 521 feet and the width is 88 feet.
$$1218 = 2(521) + 2(88)$$
$$1218 = 1042 + 176$$
$$1218 = 1218$$

47. Understand: We are given the surface area, length and width of a box. We must find height. Based on the given information, we can use $SA = 2lw + 2lh + 2wh$.

Plan: Use $SA = 2lw + 2lh + 2wh$. Replace SA with 11,232, l with 36, w with 88, and solve for h.

Execute:
$$SA = 2lw + 2lh + 2wh$$
$$11,232 = 2 \cdot 36 \cdot 28 + 2 \cdot 36h + 2 \cdot 28h$$
$$11,232 = 2016 + 72h + 56h$$
$$11,232 = \quad 2016 + 128h$$
$$\underline{-2016 \qquad -2016}$$
$$9216 = \qquad 0 + 128h$$
$$9216 = 128h$$
$$\frac{9216}{128} = \frac{128h}{128}$$
$$72 = 1h$$
$$72 = h$$

Answer: The height of the box is 72 in.

Check: Verify that the surface area of the box is 11,232 cubic inches when the length is 36 inches, width is 28 inches and the height is 72 inches.

$$SA = 2lw + 2lh + 2wh$$
$$11,232 = (2 \cdot 36 \cdot 28) + (2 \cdot 36 \cdot 72) + (2 \cdot 28 \cdot 72)$$
$$11,232 = 2016 + 5184 + 4032$$
$$11,232 = \quad 11,232$$

49. Understand: We are given the amount that Li has to spend and must find how long the plumber can work.

Plan: Use $C = 28h + 40$. Replace C with 152, and solve for h.

Execute:
$$C = 28h + 40$$
$$152 = 28h + 40$$
$$\underline{-40 \qquad -40}$$
$$112 = 28h + 0$$
$$112 = 28h$$
$$\frac{112}{28} = \frac{28h}{28}$$
$$4 = 1h$$
$$4 = h$$

Answer: Li can afford the plumber for 4 hours.

Check: Verify that when the plumber is there for four hours the cost will be $152.
$$152 = 28(4) + 40$$
$$152 = 112 + 40$$
$$152 = 152$$

51. Understand: We are given the final velocity, initial velocity and acceleration. We must find the time.

Plan: Use $v = v_i + at$. Replace v with 40, v_i with 30, a with 2, and solve for t.

Execute:
$$v = v_i + at$$
$$40 = 30 + 2t$$
$$\underline{-30 \quad -30}$$
$$10 = 0 + 2t$$
$$10 = 2t$$
$$\frac{10}{2} = \frac{2t}{2}$$
$$5 = 1t$$
$$5 = t$$

Answer: It will take the car 5 sec.

Check: Verify that the final velocity will be 40 mph when the initial velocity is 30 mph, acceleration is 2 mph/sec. and time is 5 sec.
$$40 = 30 + 2(5)$$
$$40 = 30 + 10$$
$$40 = 40$$

53. Understand: We are given the total cost budget and must find the number of chips that can be produced.

Plan: Total cost = Material costs + Labor costs. Replace total cost with 54,000, material costs with $(54b + 1215)$, labor costs with $(25b + 4200)$ and solve for b.

Execute: $54{,}000 = (54b + 1215) + (25b + 4200)$

$$54{,}000 = 79b + 5415$$
$$\underline{-5415 \qquad\quad -5415}$$
$$48{,}585 = 79b + 0$$
$$48{,}585 = 79b$$
$$\frac{48{,}585}{79} = \frac{79b}{79}$$
$$615 = b$$

Answer: We can make 615 chips.

Check: Verify that when the number of chips is produced, the total costs are 54,000.

$$54{,}000 = (54 \cdot 615 + 1215) + (25 \cdot 615 + 4200)$$
$$54{,}000 = (33{,}210 + 1215)(15{,}375 + 4200)$$
$$54{,}000 = 34{,}425 + 19{,}575$$
$$54{,}000 = 54{,}000$$

Review Exercises

1. six million, seven hundred eighty-four thousand, two hundred nine

2. $7^0 = 1$

3. $x^2 - 9x + 7 = (-3)^2 - 9(-3) + 7$
$$= 9 - 9(-3) + 7$$
$$= 9 + 27 + 7$$
$$= 43$$

4. $240 = 2^4 \cdot 3 \cdot 5$

5. $24x^5 - 30x^4 + 18x^3$
$$= 6x^3 \left(\frac{24x^5 - 30x^4 + 18x^3}{6x^3} \right)$$
$$= 6x^3 \left(\frac{24x^5}{6x^3} - \frac{30x^4}{6x^3} + \frac{18x^3}{6x^3} \right)$$
$$= 6x^3 (4x^2 - 5x + 3)$$

Exercise Set 4.4

1. Answers may vary; three possibilities are add, sum, plus.

3. Answers may vary; three possibilities are multiply, times, product.

5. **Understand:** *More than* indicates addition, *is equal to* indicates an equals sign.

   Plan: Translate to an equation using the key words, then solve the equation.

   Execute: Translation: $n + 5 = -7$

Solve: $n + 5 = -7$
$$\underline{-5 \quad -5}$$
$$n + 0 = -12$$
$$n = -12$$

Answer: $n = -12$

Check: Make sure that -12 satisfies the original sentence. $n + 5 = -7$
$$-12 + 5 = -7$$
$$-7 = -7$$

7. **Understand:** *Less than* indicates subtraction, *is* indicates an equals sign.

   Plan: Translate to an equation using the key words, then solve the equation.

   Execute: Translation: $n - 6 = 15$

   Solve: $n - 6 = 15$
   $$\underline{+6 \quad +6}$$
   $$n + 0 = 21$$
   $$n = 21$$

   Answer: $n = 21$

   Check: Make sure that 21 satisfies the original sentence. $n - 6 = 15$
   $$21 - 6 = 15$$
   $$15 = 15$$

9. **Understand:** *Increased by* indicates addition, *is* indicates an equals sign.

   Plan: Translate to an equation using the key words, then solve the equation.

   Execute: Translation: $x + 17 = -8$

   Solve: $x + 17 = -8$
   $$\underline{-17 \quad -17}$$
   $$x + 0 = -25$$
   $$x = -25$$

   Answer: $x = -25$

   Check: Make sure that -25 satisfies the original sentence. $x + 17 = -8$
   $$-25 + 17 = -8$$
   $$-8 = -8$$

11. **Understand:** *Product* indicates multiplication, *is* indicates an equals sign.

    Plan: Translate to an equation using the key words, then solve the equation.

    Execute: Translation: $-3y = 21$

Solve: $-3y = 21$

$$\frac{-3y}{-3} = \frac{21}{-3}$$
$$1y = -7$$
$$y = -7$$

Answer: $y = -7$

Check: Make sure that -7 satisfies the original sentence. $-3y = 21$
$$-3(-7) = 21$$
$$21 = 21$$

13. **Understand:** *Multiplied by* indicates multiplication, *is* indicates an equals sign.

Plan: Translate to an equation using the key words, then solve the equation.

Execute: Translation: $9b = -36$

Solve: $9b = -36$
$$\frac{9b}{9} = \frac{-36}{9}$$
$$1b = -4$$
$$b = -4$$

Answer: $b = -4$

Check: Make sure that -4 satisfies the original sentence. $9b = -36$
$$9(-4) = -36$$
$$-36 = -36$$

15. **Understand:** *More than* indicates addition, *product* indicates multiplication, and *yields* indicates an equals sign.

Plan: Translate to an equation using the key words, then solve the equation.

Execute: Translation: $5x + 4 = 14$

Solve: $5x + 4 = 14$
$$\underline{-4 \quad -4}$$
$$5x + 0 = 10$$
$$5x = 10$$
$$\frac{5x}{5} = \frac{10}{5}$$
$$1x = 2$$
$$x = 2$$

Answer: $x = 2$

Check: Make sure that 2 satisfies the original sentence. $5x + 4 = 14$
$$5 \cdot 2 + 4 = 14$$
$$10 + 4 = 14$$
$$14 = 14$$

17. **Understand:** *Difference* indicates subtraction, *times* indicates multiplication, and *is* indicates an equals sign.

Plan: Translate to an equation using the key words, then solve the equation.

Execute: Translation: $-6m - 16 = 14$

Solve: $-6m - 16 = 14$
$$\underline{+16 \quad +16}$$
$$-6m + 0 = 30$$
$$-6m = 30$$
$$\frac{-6m}{-6} = \frac{30}{-6}$$
$$1m = -5$$
$$m = -5$$

Answer: $m = -5$

Check: Make sure that -5 satisfies the original sentence. $-6m - 16 = 14$
$$-6(-5) - 16 = 14$$
$$30 - 16 = 14$$
$$14 = 14$$

19. **Understand:** *Minus* indicates subtraction, *times* indicates multiplication, *is equal to* indicates an equals sign, and *product* indicates multiplication.

Plan: Translate to an equation using the key words, then solve the equation.

Execute: Translation: $39 - 5x = 8x$

Solve: $39 - 5x = 8x$
$$\underline{+5x \quad +5x}$$
$$39 + 0 = 13x$$
$$39 = 13x$$
$$\frac{39}{13} = \frac{13x}{13}$$
$$3 = 1x$$
$$3 = x$$

Answer: $3 = x$

Check: Make sure that 3 satisfies the original sentence. $39 - 5x = 8x$
$$39 - 5 \cdot 3 = 8 \cdot 3$$
$$39 - 15 = 24$$
$$24 = 24$$

21. **Understand:** *Sum* indicates addition, *times* indicates multiplication, *is the same as* indicates an equals sign, and *difference* indicates subtraction.

Plan: Translate to an equation using the key words, then solve the equation.

Execute: Translation: $17 + 4t = 6t - 9$

Solve: $17 + 4t = 6t - 9$

$$\underline{-4t \qquad -4t}$$
$$17 + 0 = 2t - 9$$
$$17 = 2t - 9$$
$$\underline{+9 \qquad +9}$$
$$26 = 2t + 0$$
$$26 = 2t$$
$$\frac{26}{2} = \frac{2t}{2}$$
$$13 = 1t$$
$$13 = t$$

Answer: $13 = t$

Check: Make sure that 13 satisfies the original sentence. $\quad 17 + 4t = 6t - 9$
$$17 + 4 \cdot 13 = 6 \cdot 13 - 9$$
$$17 + 52 = 78 - 9$$
$$69 = 69$$

23. **Understand:** *Times* indicates multiplication, *difference* indicates subtraction, *is equal to* indicates an equals sign, and *plus* indicates addition.

    Plan: Translate to an equation using the key words, then solve the equation.

    Execute: Translation: $2(b - 8) = 5 + 9b$

    Solve: $2(b - 8) = 5 + 9b$
    $$2b - 16 = 5 + 9b$$
    $$\underline{-2b \qquad\qquad -2b}$$
    $$0 - 16 = 5 + 7b$$
    $$-16 = 5 + 7b$$
    $$\underline{-5 \quad -5}$$
    $$-21 = 0 + 7b$$
    $$-21 = 7b$$
    $$\frac{-21}{7} = \frac{7b}{7}$$
    $$-3 = 1b$$
    $$-3 = b$$

    Answer: $-3 = b$

    Check: Make sure that -3 satisfies the original sentence. $\quad 2(b - 8) = 5 + 9b$
    $$2(-3 - 8) = 5 + 9(-3)$$
    $$2(-11) = 5 + (-27)$$
    $$-22 = -22$$

25. **Understand:** *Times* indicates multiplication, *difference* indicates subtraction, *is equal to* indicates an equals sign, *minus* indicates subtraction and *sum* indicates addition.

Plan: Translate to an equation using the key words, then solve the equation.

Execute: Translation:
$$6x + 5(x - 7) = 19 - (x + 6)$$

Solve: $6x + 5(x - 7) = 19 - (x + 6)$
$$6x + 5x - 35 = 19 - x - 6$$
$$11x - 35 = 13 - x$$
$$\underline{+x \qquad\qquad +x}$$
$$12x - 35 = 13 + 0$$
$$12x - 35 = 13$$
$$\underline{+35 \quad +35}$$
$$12x + 0 = 48$$
$$12x = 48$$
$$\frac{12x}{12} = \frac{48}{12}$$
$$1x = 4$$
$$x = 4$$

Answer: $x = 4$

Check: Make sure that 4 satisfies the original sentence. $\quad 6x + 5(x - 7) = 19 - (x + 6)$
$$6 \cdot 4 + 5(4 - 7) = 19 - (4 + 6)$$
$$6 \cdot 4 + 5(-3) = 19 - 10$$
$$24 + (-15) = 9$$
$$9 = 9$$

27. **Understand:** *Less than* indicates subtraction in the reverse order from what we read, *times* indicates multiplication, difference indicates subtraction, *is the same as* indicates an equals sign, *product* indicates multiplication and *subtracted from* indicates subtraction.

    Plan: Translate to an equation using the key words, then solve the equation.

    Execute: Translation:
    $$-8(y - 3) - 14 = -2y - (y - 5)$$

    Solve: $-8(y - 3) - 14 = -2y - (y - 5)$
    $$-8y + 24 - 14 = -2y - y + 5$$
    $$-8y + 10 = -3y + 5$$
    $$\underline{+8y \qquad\qquad +8y}$$
    $$0 + 10 = 5y + 5$$
    $$10 = 5y + 5$$
    $$\underline{-5 \qquad -5}$$
    $$5 = 5y + 0$$
    $$5 = 5y$$
    $$\frac{5}{5} = \frac{5y}{5}$$
    $$1 = 1y$$
    $$1 = y$$

Answer: $1 = y$

Check: Make sure that 1 satisfies the original sentence. $-8(y-3)-14 = -2y-(y-5)$

$$-8(1-3)-14 = -2\cdot 1-(1-5)$$
$$-8(-2)-14 = -2\cdot 1-(-4)$$
$$16-14 = -2+4$$
$$2 = 2$$

29. Mistake: incorrect subtraction order
    Correct: $n-7 = 15$

31. Mistake: multiplied 2 times x instead of the sum
    Correct: $2(x+13) = -10$

33. Mistake: incorrect order on the left side of the equation
    Correct: $16-6n = 2(n-4)$

Review Exercises:

1. $6xy^5$

2. $18b^5 - 27b^3 + 54b^2$
$$= 9b^2\left(\frac{18b^5 - 27b^3 + 54b^2}{9b^2}\right)$$
$$= 9b^2\left(\frac{18b^5}{9b^2} - \frac{27b^3}{9b^2} + \frac{54b^2}{9b^2}\right)$$
$$= 9b^2(2b^3 - 3b + 6)$$

3. $P = w+3w+w+3w$ 4. $P = 8\cdot 6$
   $P = 8w$ $P = 48$ ft

5. Let's let l be the large shrubs and s be the small shrubs. If she sold a total of 16 shrubs and 7 were large, she sold $16 - 7 = 9$ small shrubs.
   total profit $= 7l + 5s$
   $$= 7\cdot 7 + 5\cdot 9$$
   $$= 49 + 45$$
   $$= \$94$$

   Jasmine made a total profit of $94.

Exercise Set 4.5

1. An equilateral triangle is a triangle with all three sides of equal length.

3. Supplementary angles are two angles whose sum is 180°.

5. width; $w + 5$

7. **Understand:** We are to find two unknown numbers given two relationships.

   Plan: Select a variable for one of the unknowns, translate to an equation, then solve the equation.

Execute: Use the first relationship to determine which unknown will be represented by the variable.

Relationship 1: One number is four more than the other. Therefore, one number $= 4 + n$

Relationship 2: The sum of the numbers is 34.
$$(4 + n) + n = 34$$
$$4 + 2n = 34$$
$$\underline{-4 \qquad -4}$$
$$0 + 2n = 30$$
$$2n = 30$$
$$\frac{2n}{2} = \frac{30}{2}$$
$$1n = 15$$
$$n = 15$$

Answer: The second number is 15. To find the first number, we use relationship one. If $n = 15$ and one number $= 4 + n$, then one number $= 4 + 15 = 19$, so the two numbers are 15 and 19.

Check: Verify that 15 and 19 satisfy both relationships: 19 is 4 more than 15 and the sum of 15 and 19 is 34.

9. **Understand:** We are to find two unknown numbers given two relationships.

   Plan: Select a variable for one of the unknowns, translate to an equation, then solve the equation.

   Execute: Use the first relationship to determine which unknown will be represented by the variable.

   Relationship 1: One number is three times the other. Therefore, one number $= 3\cdot n$

   Relationship 2: The sum of the numbers is 32.
   $$(3\cdot n) + n = 32$$
   $$3n + n = 32$$
   $$4n = 32$$
   $$\frac{4n}{4} = \frac{32}{4}$$
   $$1n = 8$$
   $$n = 8$$

   Answer: The second number is 8. To find the first number, we use relationship one. If $n = 8$ and one number $= 3\cdot n$, then one number $= 3(8) = 24$, so the two numbers are 8 and 24.

Check: Verify that 8 and 24 satisfy both relationships: 24 is 3 times 8 and the sum of 8 and 24 is 32.

11. **Understand:** Draw a picture.

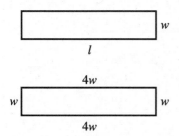

List information:
The total perimeter is 300 ft.
The length is four times the width.
There are two amounts missing, the length and the width. Because we only know how to solve linear equations in one variable, we must choose the length or width to be the variable and write our equation in terms of that variable.

Plan: Translate to an equation using the key words, then solve the equation.

Execute: Remember there are two amounts missing, length and width, and we'll have to select one of those amounts to be our variable.
Translation: Length is four times the width.
$$length = 4 \cdot width$$
The translation helps us decide which amount to choose as the variable.
Because width is multiplied by 4, let's choose the width to be our variable. Let w represent width. So we can now say:
$$width = w$$
$$length = 4w$$
To get our equation, we must use the other piece of information that we were given, which was the fact that there is a perimeter of 300 ft.
$$perimeter = 300 \text{ ft.}$$
Recall that to find a perimeter we must add the lengths of all the sides.

$$perimeter = 300$$
$$length + width + length + width = 300$$
$$4w + w + 4w + w = 300$$
$$10w = 300$$
$$\frac{10w}{10} = \frac{300}{10}$$
$$w = 30 \text{ ft.}$$

Answer: The width is 30 ft. Don't forget that we were asked to find both the length and the width. How can we get the length? Go back to our initial translation.

The length is four times the width.
$$length = 4w$$
$$length = 4 \cdot 30 = 120 \text{ ft.}$$
The building must be 30 ft. wide by 120 ft. long.

Check: Let's make sure everything in the original problem is satisfied. There were two restrictions in the problem. The length must be four times the width and the perimeter must be 300 ft. If the length is 120 ft. and the width is 30 ft., then the length is four times the width. Is the perimeter 300 ft.? Yes, because 30 + 120 + 30 + 120 = 300 ft.

13. **Understand:** Draw a picture

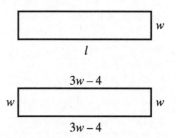

List information:
The total perimeter is 48 in.
The length is four less than three times the width.
There are two amounts missing, the length and the width. Because we only know how to solve linear equations in one variable, we must choose the length or width to be the variable and write our equation in terms of that variable.

Plan: Translate to an equation using the key words, then solve the equation.

Execute: Remember there are two amounts missing, length and width, and we'll have to select one of those amounts to be our variable.
Translation: Length is four less than three times the width.
$$length = 3 \cdot width - 4$$
The translation helps us decide which amount to choose as the variable.
Because width is multiplied by 3 and subtracted by 4, let's choose the width to be our variable. Let w represent width. So we can now say:
$$width = w$$
$$length = 3w - 4$$
To get our equation, we must use the other piece of information that we were given, which was the fact that there is a perimeter of 48 ft.
$$perimeter = 48 \text{ ft.}$$
Recall that to find a perimeter we must add the lengths of all the sides.

perimeter = 48

length + width + length + width = 48

$$\left(3w-4\right)+w+\left(3w-4\right)+w=48$$

$$8w-8=48$$

$$\underline{+8=+8}$$

$$8w+0=56$$

$$\frac{8w}{8}=\frac{56}{8}$$

$$w=7 \text{ in.}$$

Answer: The width is 7 in. Don't forget that we were asked to find both the length and the width. How can we get the length? Go back to our initial translation.
Length is four less than three times the width.

$$\text{length}=3w-4$$
$$\text{length}=3\cdot7-4=17 \text{ in.}$$
The frame must be 7 in. wide by 17 in. long.

Check: Let's make sure everything in the original problem is satisfied. There were two restrictions in the problem. The length is four less than three times the width and the perimeter must be 48 in. If the length is 17 in. and the width is 7 in., then the length is four less than three times the width. Is the perimeter 48 in.? Yes, because $17+7+17+7=48$ in.

15. **Understand:** Isosceles triangles have two equal length sides.
These two sides are equal in length. We are told they are 5 m less than twice the base (bottom). We are also told that the perimeter is 70 m.

Plan: Translate the relationship, write an equation, then solve.

Execute: Translation:
The equal length sides are 5 less than twice the base.
equal length sides = $2\cdot\text{base}-5$
Because the base is multiplied by 2, we will use the variable L to represent its length.

$$\text{equal length sides}=2L-5$$
$$\text{base}=L$$

The sentence concerning the perimeter of the triangle is our second relationship. We use it to create the equation that we will solve. Perimeter is the sum of all the side lengths.

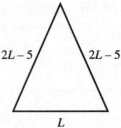

equal length sides + base = 70

$$\left(2L-5\right)+\left(2L-5\right)+L=70$$

$$5L-10=70$$

$$\underline{+10\quad+10}$$

$$5L=80$$

$$\frac{5L}{5}=\frac{80}{5}$$

$$L=16$$

Answer: The base is 16 m. To get the equal length sides, use the relationship:
equal length sides = $2L-5$

$$=2\cdot16-5$$
$$=32-5$$
$$=27 \text{ m}$$
The sides are 27 m and the base is 16m.

Check: Because 27 is 5 less than twice 16, the equal length sides are in fact twice the length of the other side. We also need to verify that the perimeter is 70. $27+27+16=70$ m.

17. **Understand:** Equilateral triangles have three equal length sides. We are told they are the same as the lengths of the rectangle on which the triangle sits.
We know to find the perimeter of a rectangle we add all the sides.
We are told the width of the rectangle is 10 ft. less than three times the length.
The perimeter for both figures is the same.

Plan: Translate the relationship, write an equation, then solve.

Execute: Translation:
The equal width sides of the triangle are equal to the length of the rectangle.
equal width sides = length of rectangle
Because the length of the rectangle is acted on, we will use the variable l to represent its length.

$$\text{length of rectangle}=l$$
$$\text{equal length sides of triangle}=l$$
$$\text{width of rectangle}=3l-10$$

The sentences concerning the perimeter of the triangle and the perimeter of the rectangle are our second relationships. We use them to create the equation that we will solve. Perimeter is the sum of all the side widths. Since the perimeters of both figures are equal, we will set the expressions equal to each other.

perimeter of triangle = perimeter of rectangle

$$l + l + l = l + (3l - 10) + l + (3l - 10)$$

$$3l = 8l - 20$$

$$\underline{-3l \quad -3l}$$

$$0 = 5l - 20$$

$$\underline{+20 \qquad +20}$$

$$20 = 5l$$

$$\frac{20}{5} = \frac{l}{5}$$

$$4 = l$$

Answer: The length of the rectangle is 4 ft. This is the same width of each side of the triangle. To find the width of the rectangle, we will use the relationship established.

$$\text{width of rectangle} = 3l - 10$$

$$= 3 \cdot 4 - 10$$

$$= 12 - 10$$

$$= 2$$

The width of the rectangle is 2 ft.

Check: The equal width sides of the triangle are the same as the length of the rectangle. The width of the rectangle is in fact ten less than three times the length. We also need to verify that the perimeters are equal. The triangle: $4 + 4 + 4 = 12$ ft. The perimeter of the rectangle is $4 + 2 + 4 + 2 = 12$ ft.

19. **Understand:** We must find the two angle measurements.
We are given a relationship about the angles: One angle is four times the other angle.
Because the problem involves two unknowns, we must have another relationship in order to solve it. However, this second relationship was not given in an obvious manner. We must resort to our own wits and knowledge for the second relationship. Because we are told that the angle formed by the walls is 90°, we know they are complementary and the sum of the two angles must equal 90°.

Plan: Translate the relationships to an equation, then solve.

Execute: Translation:
One angle is 4 times the other angle.

$$\text{angle } 1 = 4 \cdot \text{angle } 2$$

Because angle 2 is multiplied by 4, we should let the variable represent angle 2. Let's use the letter a.

$$\text{angle } 2 = a$$

$$\text{angle } 1 = 4a$$

Now we tie this translation into our other relationship. We must add the two angles to show a total of 90°. $a + 4a = 90$

$$5a = 90$$

$$\frac{5a}{5} = \frac{90}{5}$$

$$a = 18°$$

Answer: Because a represents angle 2, we can say that angle 2 measures 18°. To find the other angle, go back to our relationship.

Let angle $2 = 18°$

$$\text{angle } 1 = 4 \cdot \text{angle } 2$$

$$= 4 \cdot 18$$

$$= 72°$$

The angles are 18° and 72°.

Check: We must make sure that our answer satisfies all of the information in the problem. Angle 1 is 72° and is four times angle 2, which is 18°. Since the angles are complementary, their sum must be 90°. $72 + 18 = 90°$

21. **Understand:** We must find the two angle measurements.
We are given a relationship about the angles: One angle is 15° more than four times the other angle. Because the problem involves two unknowns, we must have another relationship in order to solve it. However, this second relationship was not given in an obvious manner. We must resort to our own wits and knowledge for the second relationship. Because we are told that the detector is flat, we know the angles are supplementary and the sum of the two angles must equal 180°.

Plan: Translate the relationships to an equation, then solve.

Execute: Translation:
One angle is 15° more than four times the other angle.

$$\text{angle } 1 = 4 \cdot \text{angle } 2 + 15$$

Because angle 2 is added to, we should let the variable represent angle 2. Let's use the letter a.

$$\text{angle } 2 = a$$

$$\text{angle } 1 = 4a + 15$$

Now we tie this translation into our other relationship. We must add the two angles to show a total of 180°. $a + (4a + 15) = 180$

$$5a + 15 = 180$$

$$\underline{-15 \quad -15}$$

$$5a = 165$$

$$\frac{5a}{5} = \frac{165}{5}$$

$$a = 33°$$

Answer: Because a represents angle 2, we can say that angle 2 measures 33°. To find the other angle, go back to our relationship.

angle 2 = 33°

$$\text{angle } 1 = 4 \cdot a + 15$$
$$= 4 \cdot 33 + 15$$
$$= 132 + 15$$
$$= 147°$$

The two angles are 33° and 147°.

Check: We must make sure that our answer satisfies all of the information in the problem. Angle 1 is 147° and is 15° more than four times angle 2 which is 33°. Since the angles are supplementary, their sum must be 180°.

$147 + 33 = 180°$

23. **Understand:** We must find the three angle measurements.

We are given a relationship about the angles: The angles formed by the entering and exiting beams are the same measurement. The third angle is 10° more than three times the other angle. Because we are told that the mirror is flat, we know that they are supplementary and the sum of these three angles is 180°.

Plan: Translate the relationships to an equation, then solve.

Execute: Translation:

Let a represent the first angle. Since the second angle measures the same as the first, it will also be represented by a.

angle 1 = a

angle 2 = a

The third angle is 10° more than three times the other angle.

$$\text{angle } 3 = 3 \cdot \text{angle } 1 + 10 = 3a + 10$$

Now we tie this translation into our other relationship. We must add the three angles to show a total of 180°.

$$a + a + (3a + 10) = 180$$
$$5a + 10 = 180$$
$$\underline{{-10} \quad {-10}}$$
$$5a = 170$$
$$\frac{5a}{5} = \frac{170}{5}$$
$$a = 34°$$

Answer: Because a represents angle 1 and angle 2, we can say that angle 1 and angle 2 each

measures 34°. To find the other angle, go back to our relationship. angle 1 = a

$$\text{angle } 3 = 3a + 10$$
$$= 3 \cdot 34 + 10$$
$$= 102 + 10$$
$$= 112°$$

The angles are 34°, 34°, and 112°.

Check: We must make sure that our answer satisfies all of the information in the problem. angle 1 and angle 2 are each 34°. Angle 3 is 112° and is 10° more than three times angle 1 which is 34°. The angles must sum to 180°.

$34 + 34 + 112 = 180°$

25. **Understand:** The key word *sum* means to add. We must find the relationships between the three angles in a triangle.

If we choose the third angle to be x, we can say:

first angle: 43

second angle: $2 + 2x$

third angle: x

Plan: Write an equation, then solve.

Execute: Translation:

The sum of the angles in any triangle is 180°.

first angle + second angle + third angle = 180°

$$43 + (2 + 2x) + x = 180$$
$$2x + x + 2 + 43 = 180$$
$$3x + 45 = 180$$
$$\underline{{-45} \quad {-45}}$$
$$3x = 135$$
$$\frac{3x}{3} = \frac{135}{3}$$
$$x = 45°$$

Answer: Because x represents the third angle, we can say:

first angle: 43°

second angle: $2 + 2x = 2 + 2(45) = 92°$

third angle: $x = 45°$

Check: We must verify that the three-angle sum is 180°. $43° + 92° + 45° = 180°$.

Angle 2 is 92° which is 2° more than twice the second angle.

27. **Understand:** The key word *sum* means to add. We must find the relationships between the three angles in a triangle.

If we choose the first angle to be x, we can say:

first angle: x

second angle: $10 + x$

third angle: $x - 7$

Plan: Write an equation, then solve.

Execute: Translation:

The sum of the angles in any triangle is 180°.

first angle + second angle + third angle = 180°

$$x + (10 + x) + (x - 7) = 180$$
$$x + x + x + 10 - 7 = 180$$
$$3x + 3 = 180$$
$$\underline{-3 \quad -3}$$
$$3x = 177$$
$$\frac{3x}{3} = \frac{177}{3}$$
$$x = 59°$$

Answer: Because x represents the first angle, we can say:

first angle: $x = 59°$

second angle: $10 + x = 10 + 59° = 69°$

third angle: $x - 7 = 59° - 7 = 52°$

Check: We must verify that the three-angle sum is 180°. $59° + 69° + 52° = 180°$.

29. Let L be the number of large ornaments sold.

Categories	Selling price	Number of ornaments sold	Income
Small ornaments	3	$L + 8$	$3(L + 8)$
Large ornaments	7	L	$7L$

31. Let n be the number of large charms sold.

Categories	Selling price	Number of charms sold	Income
Small charms	4	$36 - n$	$4(36 - n)$
Large charms	6	n	$6n$

33. **Understand:** The overall relationship here is that the total income is $260. From this we can say:

income small size + income large size = 260

How can we describe the income from each bottle? What do we know about the individual bottles? We know the selling price and that there were 5 more smaller bottles sold than there were larger bottles. If we know the number of bottles sold we could multiply that number times the

selling price to get the income from that size bottle.

Because of all the information we have, it is helpful to use a four-column table.

Categories	Value	Number of bottles	Amount
Small bottle	$8	$n + 5$	$8(n + 5)$
Large bottle	$12	n	$12n$

We were given these. We selected n to represent the number of the large bottles, then translate: "5 more smaller bottles than larger bottles."

We multiplied straight across because: selling price $\times$ number of bottles = amount from each.

The expressions in the last column in the table describe the income from the sale of each bottle.

Plan: Translate the information in the table into an equation, then solve.

Execute: Now we can use our initial relationship:

income small size + income large size = 260

$$8(n + 5) + 12n = 260$$
$$8n + 40 + 12n = 260$$
$$20n + 40 = 260$$
$$\underline{-40 \quad -40}$$
$$20n = 220$$
$$\frac{20n}{20} = \frac{220}{20}$$
$$n = 11$$

Answer: We go back to our table and use the relationships about the number of each bottle.

the number of small bottles: $n + 5 = 11 + 5 = 16$

the number of large bottles: $n = 11$

Check: Verify that 16 of the smaller $8 bottles and 11 of the larger $12 bottles make a total income of $260. $16 \cdot 8 + 11 \cdot 12 = 260$
$$128 + 132 = 260$$
$$260 = 260$$

35. **Understand:** We are given the value of two different bills. We are also given a total amount of money and a total amount of bills to split up. We can say: income of small bills + income of large bills = total money

Because we have two categories of bills along with a value and a number of bills, this problems lends itself to a four-column table. We are given the value of each bill: $5 and $10. The tricky part is filling in the number column.

We are told that Ian has a total of 19. We can say: Small bills + large bills = 19

Or, if we knew one of the numbers, we could find the other by subtracting from 19. We can write a related subtraction this way:

number of small bills = 19 − number of large bills

Let's choose L = number of large.

Categories	Value	Number	Amount
$5 bills	5	$19 - L$	$5(19 - L)$
$10 bills	10	L	$10L$

We were given these. Select one of the categories to be the variable. The other will be total number – variable.

We multiplied straight across because: value × number of bills = amount from each.

The expressions in the last column in the table describe the income from the value of each bill.

Plan: Translate the information in the table into an equation, then solve.

Execute: Now we can use our initial relationship:

small bills + large bills = total money
$$5(19 - L) + 10L = 125$$
$$95 - 5L + 10L = 125$$
$$95 + 5L = 125$$
$$\underline{-95 \qquad -95}$$
$$5L = 30$$
$$\frac{5L}{5} = \frac{30}{5}$$
$$L = 6$$

Answer: number of large bills: $L = 6$
number of small bills: $19 - L = 19 - 6 = 13$
Ian has 6 $10 bills and 13 $5 bills.

Check: Verify that 13 five dollar bills and 6 ten dollar bills will total $125. $13 \times \$5 + 6 \times \$10 = \$65 + \$60 = \$125$.

37. **Understand:** The overall relationship here is that the total value is $108. From this we can say:

income small bill + income more valuable bill = 108

How can we describe the income from each bill? What do we know about the individual bills? We know the number of each bill and that the more valuable bill is worth three times that of the other bill.

Because of all the information we have, it is helpful to use a four column table. The value column is the value of each bill. The amount column is the income from each bill.

Categories	Value	Number of bills	Amount
Other bill	x	18	$18x$
More valuable bill	$3x$	12	$12(3x)$

We were given these. We selected x to represent the number of the other bills, then translate: "three times that of the other bill."

We multiplied straight across because: value × number of bills = amount from each.

The expressions in the last column in the table describe the income from the exchange of each bill.

Plan: Translate the information in the table into an equation, then solve.

Execute: Now we can use our initial relationship:

income small bill + income more valuable bill = 108

$$18x + 12(3x) = 108$$
$$18x + 36x = 108$$
$$54x = 108$$
$$\frac{54x}{54} = \frac{108}{54}$$
$$x = 2$$

Answer: We go back to our table and use the relationships about the value of each bill.

the value of small bills: $x = \$2$

the value of more valuable bills: $x = 3 \times \$2 = \$6$

Check: Verify that 18 of the smaller $2 bills and 12 of the more valuable $6 bills make a total amount of $108. $18 \cdot 2 + 12 \cdot 6 = 108$

$$36 + 72 = 108$$
$$108 = 108$$

39. **Understand:** The overall relationship here is that the total amount is $1470. From this we can say:

income from hand-fed cockatiels + income from parent-raised cockatiels = 1470

How can we describe the income from each type of cockatiel? What do we know about the individual type of cockatiel? We know the number of each type of cockatiel and that the total value of the two different cockatiels together is $205.

Because of all the information we have, it is helpful to use a table. The value column is the value of each cockatiel. The income column is the income from each cockatiel.

Categories	Value	Number of cockatiels	Income
Hand-fed	x	6	$6x$
Parent-raised	$205 - x$	9	$9(205 - x)$

We were given these. Select one of the categories to be the variable. The other will be: total number – variable.

We multiplied straight across because: value × number of cockatiels = income from each.

The expressions in the last column in the table describe the income from the exchange of each cockatiel.

Plan: Translate the information in the table into an equation, then solve.

Execute: Now we can use our initial relationship:

income from hand-fed cockatiels + income from parent-raised cockatiels = 1470

$$6x + 9(205 - x) = 1470$$
$$6x + 1845 - 9x = 1470$$
$$-3x + 1845 = 1470$$
$$\underline{-1845 \quad -1845}$$
$$-3x = -375$$
$$\frac{-3x}{-3} = \frac{-375}{-3}$$
$$x = 125$$

Answer: We go back to our table and use the relationships about the value of each cockatiel.

the value of hand-fed: $x = \$125$

the value of parent-raised: $205 - x = 205 - 125$ =$80

Check: Verify that 6 of hand-fed cockatiels at $125 and 9 of parent-raised at $80 make a total income of $1470. $6 \cdot 125 + 9 \cdot 80 = 1470$

$$750 + 720 = 1470$$
$$1470 = 1470$$

Review Exercises

1. $4(-3) - 8 = 5(-3) - 6$
    $-12 - 8 \overset{?}{=} -15 - 6$
    $-20 \neq -21$

    -3 is not a solution.

2. No, because it has a degree higher than one.

3. $m + 19 = 12$ Check: $-7 + 19 = 12$
    $\underline{-19 \quad -19}$ $12 = 12$
    $m + 0 = -7$
    $m = -7$

4. $-15n = 75$ Check: $-15 \cdot 5 = -75$
    $\dfrac{-15n}{-15} = \dfrac{75}{-15}$ $-75 = -75$
    $1n = -5$
    $n = -5$

5. $9x - 7(x + 2) = -3x + 1$
    $9x - 7x - 14 = -3x + 1$
    $2x - 14 = -3x + 1$
    $\underline{+3x \qquad +3x}$
    $5x - 14 = 0 + 1$
    $5x - 14 = 1$
    $\underline{+14 \quad +14}$
    $5x = 15$
    $\dfrac{5x}{5} = \dfrac{15}{5}$
    $x = 3$

Check: $9 \cdot 3 - 7(3+2) = -3 \cdot 3 + 1$
$9 \cdot 3 - 7(5) = -9 + 1$
$27 - 35 = -8$
$-8 = -8$

Chapter 4 Review Exercises

1. false   2. true   3. true

4. true   5. false   6. false

7. Replace the variable(s) in the original equation with the number. If the resulting equation is true, the number is a solution.

8. same; added; subtracted

9. multiplying; dividing; same

10. 1. simplify; distribute; combine;
    2. addition/subtraction; 3. multiplication/division

11. $x - 7 = 12$ 5 is not a solution
$5 - 7 \stackrel{?}{=} 12$
$-2 \neq 12$

12. $3a = -12$ -4 is a solution
$3 \cdot -4 \stackrel{?}{=} -12$
$-12 = -12$

13. $-7n + 12 = 5$ 1 is a solution.
$-7 \cdot 1 + 12 \stackrel{?}{=} 5$
$-7 + 12 \stackrel{?}{=} 5$
$5 = 5$

14. $14 - 4y = 3y$ -2 is not a solution.
$14 - 4(-2) \stackrel{?}{=} 3(-2)$
$14 - (-8) \stackrel{?}{=} -6$
$14 + 8 \stackrel{?}{=} -6$
$22 \neq -6$

15. $b^2 - 15 = 2b$ -3 is a solution.
$(-3)^2 - 15 \stackrel{?}{=} 2(-3)$
$9 - 15 \stackrel{?}{=} -6$
$-6 = -6$

16. $9r - 17 = -r + 23$ 5 is not a solution.
$9 \cdot 5 - 17 \stackrel{?}{=} -5 + 23$
$45 - 17 \stackrel{?}{=} 18$
$28 \neq 18$

17. $5(x - 2) - 3 = 10 - 4(x - 1)$ 3 is a solution.
$5(3 - 2) - 3 \stackrel{?}{=} 10 - 4(3 - 1)$
$5 \cdot 1 - 3 \stackrel{?}{=} 10 - 4 \cdot 2$
$5 - 3 \stackrel{?}{=} 10 - 8$
$2 = 2$

18. $u^3 - 7 = u^2 + 6u$ -1 is not a solution.
$(-1)^3 - 7 \stackrel{?}{=} (-1)^2 + 6(-1)$
$-1 - 7 \stackrel{?}{=} 1 + 6(-1)$
$-8 \stackrel{?}{=} 1 - 6$
$-8 \neq -5$

19. $n + 19 = 27$ Check: $8 + 19 = 27$
$\underline{-19 \quad -19}$ $27 = 27$
$n + 0 = 8$
$n = 8$

20. $y - 6 = -8$ Check: $-2 - 6 = -8$
$\underline{+6 \quad +6}$ $-8 = -8$
$y + 0 = -2$
$y = -2$

21. $-4k = 36$ Check: $-4(-9) = 36$
$\dfrac{-4k}{-4} = \dfrac{36}{-4}$ $36 = 36$
$1k = -9$
$k = -9$

22. $-8m = -24$ Check: $-8 \cdot 3 = -24$
$\dfrac{-8m}{-8} = \dfrac{-24}{-8}$ $-24 = -24$
$1m = 3$
$m = 3$

23. $2x + 11 = -3$ Check: $2(-7) + 11 = -3$
$\underline{-11 \quad -11}$ $-14 + 11 = -3$
$2x + 0 = -14$ $-3 = -3$
$\dfrac{2x}{2} = \dfrac{-14}{2}$
$1x = -7$
$x = -7$

24. $-3h - 8 = 19$ Check: $-3(-9) - 8 = 19$
$\underline{+8 \quad +8}$ $27 - 8 = 19$
$-3h + 0 = 27$ $19 = 19$
$\dfrac{-3h}{-3} = \dfrac{27}{-3}$
$1h = -9$
$h = -9$

25. $9t - 14 = 3t + 4$

$\underline{-3t \qquad -3t}$

$6t - 14 = 0 + 4$

$6t - 14 = 4$

$\underline{+14 \quad +14}$

$6t + 0 = 18$

$\dfrac{6t}{6} = \dfrac{18}{6}$

$1t = 3$

$t = 3$

Check: $9 \cdot 3 - 14 = 3 \cdot 3 + 4$

$27 - 14 = 9 + 4$

$13 = 13$

26. $20 - 5y = 34 + 2y$

$\underline{+5y \qquad +5y}$

$20 + 0 = 34 + 7y$

$\underline{-34 \qquad -34}$

$-14 = 0 + 7y$

$\dfrac{-14}{7} = \dfrac{7y}{7}$

$-2 = 1y$

$-2 = y$

Check: $20 - 5(-2) = 34 + 2(-2)$

$20 - (-10) = 34 - 4$

$20 + 10 = 30$

$30 = 30$

27. $10m - 17 + m = 2 + 12m - 20$

$11m - 17 = 12m - 18$

$\underline{-11m \qquad -11m}$

$0 - 17 = m - 18$

$\underline{+18 \qquad +18}$

$1 = m + 0$

$1 = m$

Check: $10 \cdot 1 - 17 + 1 = 2 + 12 \cdot 1 - 20$

$10 - 17 + 1 = 2 + 12 - 20$

$-7 + 1 = 14 - 20$

$-6 = -6$

28. $-16v - 18 + 7v = 24 - v + 6$

$-9v - 18 = 30 - v$

$\underline{+9v \qquad +9v}$

$0 - 18 = 30 + 8v$

$\underline{-30 \quad -30}$

$-48 = 0 + 8v$

$\dfrac{-48}{8} = \dfrac{8v}{8}$

$-6 = 1v$

$-6 = v$

Check: $-16(-6) - 18 + 7(-6) = 24 - (-6) + 6$

$96 - 18 + (-42) = 24 + 6 + 6$

$78 + (-42) = 36$

$36 = 36$

29. $4x - 3(x + 5) = 16 - 7(x + 1)$

$4x - 3x - 15 = 16 - 7x - 7$

$x - 15 = 9 - 7x$

$\underline{+7x \qquad +7x}$

$8x - 15 = 9 + 0$

$\underline{+15 \quad +15}$

$8x + 0 = 24$

$\dfrac{8x}{8} = \dfrac{24}{8}$

$1x = 3$

$x = 3$

Check: $4 \cdot 3 - 3(3 + 5) = 16 - 7(3 + 1)$

$4 \cdot 3 - 3 \cdot 8 = 16 - 7 \cdot 4$

$12 - 24 = 16 - 28$

$-12 = -12$

30. $9y - (2y + 3) = 4(y - 5) + 2$

$9y - 2y - 3 = 4y - 20 + 2$

$7y - 3 = 4y - 18$

$\underline{-4y \qquad -4y}$

$3y - 3 = 0 - 18$

$\underline{+3 \qquad +3}$

$3y = -15$

$\dfrac{3y}{3} = \dfrac{-15}{3}$

$1y = -5$

$y = -5$

Check: $9(-5) - (2(-5) + 3) = 4(-5 - 5) + 2$

$9(-5) - (-10 + 3) = 4(-10) + 2$

$9(-5) - (-7) = -40 + 2$

$-45 + 7 = -38$

$-38 = -38$

31. $-15 + t = 28$

$\underline{+15 \qquad +15}$

$0 + t = 43$

$t = 43°$

The temperature rose by 43°F.

32. $-547 + p = -350$

$\underline{+547 \qquad +547}$

$0 + p = 197$

$p = \$197$

Kari must pay $197.

33. $V = lwh$

$$5000 = 50 \cdot 25 \cdot h$$
$$5000 = 1250h$$
$$\frac{5000}{1250} = \frac{1250h}{1250}$$
$$4 = 1h$$
$$4 = h$$

The height must be 4 ft.

34. $SA = 2lw + 2lh + 2wh$

$$136 = 2 \cdot 3w + 2 \cdot 3 \cdot 4 + 2 \cdot w \cdot 4$$
$$136 = 6w + 24 + 8w$$
$$136 = 14w + 24$$
$$\underline{-24 \qquad\quad -24}$$
$$112 = 14w + 0$$
$$\frac{112}{14} = \frac{14w}{14}$$
$$8 = 1w$$
$$8 = w$$

The width must be 8 ft.

35. $15 - 7n = 22$

$$\underline{-15 \qquad\quad -15}$$
$$-7n = 7$$
$$\frac{-7n}{-7} = \frac{7}{-7}$$
$$1n = -1$$
$$n = -1$$

36. $5x - 4 = 3x$

$$\underline{-3x \qquad\quad -3x}$$
$$2x - 4 = 0$$
$$\underline{+4 \quad +4}$$
$$2x + 0 = 4$$
$$\frac{2x}{2} = \frac{4}{2}$$
$$1x = 2$$
$$x = 2$$

37. $2(n + 12) = -6n - 8$

$$2n + 24 = -6n - 8$$
$$\underline{+6n \qquad\qquad +6n}$$
$$8n + 24 = 0 - 8$$
$$\underline{-24 \qquad\quad -24}$$
$$8n + 0 = -32$$
$$\frac{8n}{8} = \frac{-32}{8}$$
$$1n = -4$$
$$n = -4$$

38. $12 - 3(x - 7) = 6x - 3$

$$12 - 3x + 21 = 6x - 3$$
$$33 - 3x = 6x - 3$$
$$\underline{+3x \quad +3x}$$
$$33 + 0 = 9x - 3$$
$$\underline{+3 \qquad\qquad +3}$$
$$36 = 9x + 0$$
$$\frac{36}{9} = \frac{9x}{9}$$
$$4 = 1x$$
$$4 = x$$

39.   one number $= n$
second number $= n + 12$

$$n + (n + 12) = 42$$
$$2n + 12 = 42$$
$$\underline{-12 \quad -12}$$
$$2n + 0 = 30$$
$$\frac{2n}{2} = \frac{30}{2}$$
$$1n = 15$$
$$n = 15$$

One number is 15 and the other is 15 + 12 = 27.

40.   one number $= n$
second number $= 5 \cdot n$

$$n + (5 \cdot n) = 36$$
$$n + 5n = 36$$
$$6n = 36$$
$$\frac{6n}{6} = \frac{36}{6}$$
$$1n = 6$$
$$n = 6$$

One number is 6 and the other is $5 \cdot 6 = 30$.

41.   To find the perimeter of a rectangle, we must use the formula $P = 2l + 2w$. We will let the width be w. The length is "2m less than three times the width" so we say $3w - 2$. We also know that the perimeter is 188.

$$P = 2l + 2w$$
$$188 = 2(3w - 2) + 2w$$
$$188 = 6w - 4 + 2w$$
$$188 = 8w - 4$$
$$\underline{+4 \qquad\quad +4}$$
$$192 = 8w + 0$$
$$\frac{192}{8} = \frac{8w}{8}$$
$$24 = 1w$$
$$24 = w$$

The width is 24m and the length, $3w - 2$, is $3 \times 24 - 2 = 70$ m.

42. An isosceles triangle has two equal sides. Let's call the base b. Since the two equal sides are "38 in. longer than the base," we can say the equal sides are $b + 38$. To find the perimeter of any figure, we add the sides. We also know the total perimeter is 256 in.

$$P = b + (b + 38) + (b + 38)$$
$$256 = 3b + 76$$
$$\underline{-76 \qquad -76}$$
$$180 = 3b + 0$$
$$\frac{180}{3} = \frac{3b}{3}$$
$$60 = 1b$$
$$60 = b$$

The base of the triangle is 60 in. and the two equal sides are $b + 38 = 60 + 38 = 98$ in.

43. We will assume that the two angles form a 180° angle. Let's call Angle 1 x and "the angle made with the wall on one side of the beam is 16° more than the other angle," we'll say is $x + 16$.

$$x + x + 16 = 180$$
$$2x + 16 = 180$$
$$\underline{-16 \quad -16}$$
$$2x + 0 = 164$$
$$\frac{2x}{2} = \frac{164}{2}$$
$$1x = 82$$
$$x = 82°$$

Angle 1 is 82° and Angle 2 is $82 + 16 = 98°$.

44. Since a basketball goal is flat, we conclude that we are dealing with supplementary angles that will total 180°. Let's call the angle made on the goal side x and the angle facing away from the goal "is to be 15° less than twice the angle made on the goal side," so we will call it $2x - 15$.

$$x + (2x - 15) = 180 \text{ The angle on the goal side is}$$
$$3x - 15 = 180$$
$$\underline{+15 \quad +15}$$
$$3x + 0 = 195$$
$$\frac{3x}{3} = \frac{195}{3}$$
$$1x = 65$$
$$x = 65°$$

65° and the angle facing away from the goal side is $2 \times 65 - 15 = 115°$.

45. Let n be the number of $20 bills.

Categories	Value	Number	Amount
$10 bills	10	$n + 9$	$10(n + 9)$
$20 bills	20	n	$20n$

$$10(n + 9) + 20n = 330$$
$$10n + 90 + 20n = 330$$
$$30n + 90 = 330$$
$$\underline{-90 \quad -90}$$
$$30n + 0 = 240$$
$$\frac{30n}{30} = \frac{240}{30}$$
$$1n = 8$$
$$n = 8$$

She has 8 twenty-dollar bills and 17 ten-dollar bills.

46. Let n be the number of small bags.

Categories	Value	Number of bags	Income
Large bags	6	n	$6n$
Small bags	4	$27 - n$	$4(27 - n)$

$$6n + 4(27 - n) = 146$$
$$6n + 108 - 4n = 146$$
$$2n + 108 = 146$$
$$\underline{-108 \quad -108}$$
$$2n + 0 = 38$$
$$\frac{2n}{2} = \frac{38}{2}$$
$$1n = 19$$
$$n = 19$$

19 large bags were sold. $27 - 19 = 8$ small bags were sold.

Chapter 4 Practice Test

1. It is an equation because it contains an equals sign.

2. $$-9 = 3x + 5 \qquad -4 \text{ is not a solution.}$$
$$-9 \overset{?}{=} 3(-4) + 5$$
$$-9 \overset{?}{=} -12 + 5$$
$$-9 \neq -7$$

3. $3x - 11 = 2(x - 2) - 5$ 2 is a solution.
    $3 \cdot 2 - 11 = 2(2 - 2) - 5$
    $6 - 11 \overset{?}{=} 2 \cdot 0 - 5$
    $-5 \overset{?}{=} 0 - 5$
    $-5 = -5$

4. It is not linear; there is a variable term with a degree other than 1.

5. $n - 15 = -7$ Check: $8 - 15 = -7$
    $\underline{+15 \quad +15}$ $-7 = -7$
    $n + 0 = 8$
    $n = 8$

6. $-9m = 54$ Check: $-9 \cdot -6 = 54$
    $\dfrac{-9m}{-9} = \dfrac{54}{-9}$ $54 = 54$
    $1m = -6$
    $m = -6$

7. $-6y + 3 = -21$ Check: $-6 \cdot 4 + 3 = -21$
    $\underline{\quad -3 \quad\quad -3}$ $-24 + 3 = -21$
    $-6y + 0 = -24$ $-21 = -21$
    $-6y = -24$
    $\dfrac{-6y}{-6} = \dfrac{-24}{-6}$
    $1y = 4$
    $y = 4$

8. $9k + 5 = 17 - 3k$
    $\underline{+3k \quad\quad\quad +3k}$
    $12k + 5 = 17 + 0$
    $\underline{\quad -5 \quad -5}$
    $12k + 0 = 12$
    $12k = 12$
    $\dfrac{12k}{12} = \dfrac{12}{12}$
    $1k = 1$
    $k = 1$

    Check: $9 \cdot 1 + 5 = 17 - 3 \cdot 1$
    $9 + 5 = 17 - 3$
    $14 = 14$

9. $-13x + 26 = 11 - 8x$
    $\underline{+13x \quad\quad\quad +13x}$
    $0 + 26 = 11 + 5x$
    $26 = 11 + 5x$
    $\underline{\quad -11 \quad -11}$
    $15 = 0 + 5x$
    $15 = 5x$
    $\dfrac{15}{5} = \dfrac{5x}{5}$
    $3 = 1x$
    $3 = x$

Check: $-13 \cdot 3 + 26 = 11 - 8 \cdot 3$
$-39 + 26 = 11 - 24$
$-13 = -13$

10. $4t - 13 + t = 11 + 6t - 3$
    $5t - 13 = 8 + 6t$
    $\underline{-5t \quad\quad\quad -5t}$
    $0 - 13 = 8 + t$
    $\underline{\quad -8 \quad -8}$
    $-21 = 0 + t$
    $-21 = t$

Check: $4(-21) - 13 + (-21) = 11 + 6(-21) - 3$
$-84 - 13 + (-21) = 11 + (-126) - 3$
$-97 + (-21) = -115 - 3$
$-118 = -118$

11. $7(u - 2) + 12 = 4(u - 2)$
    $7u - 14 + 12 = 4u - 8$
    $7u - 2 = 4u - 8$
    $\underline{-4u \quad\quad\quad -4u}$
    $3u - 2 = 0 - 8$
    $3u - 2 = -8$
    $\underline{\quad +2 \quad\quad +2}$
    $3u + 0 = -6$
    $3u = -6$
    $\dfrac{3u}{3} = \dfrac{-6}{3}$
    $1u = -2$
    $u = -2$

    Check: $7(-2 - 2) + 12 = 4(-2 - 2)$
    $7(-4) + 12 = 4(-4)$
    $-28 + 12 = -16$
    $-16 = -16$

12. $6 + 3(k + 4) = 5k + (k - 9)$
    $6 + 3k + 12 = 5k + k - 9$
    $18 + 3k = 6k - 9$
    $\underline{\quad -3k \quad -3k}$
    $18 + 0 = 3k - 9$
    $\underline{+9 \quad\quad\quad +9}$
    $27 = 3k + 0$
    $27 = 3k$
    $\dfrac{27}{3} = \dfrac{3k}{3}$
    $9 = 1k$
    $9 = k$

    Check: $6 + 3(9 + 4) = 5(9) + (9 - 9)$
    $6 + 3 \cdot 13 = 5(9) + 0$
    $6 + 39 = 45 + 0$
    $45 = 45$

13. $375 + x = 458$
$\underline{-375 \qquad -375}$
$0 + x = 83$
$ x = \$83$ \qquad Daryl needs \$83.

14. $ A = lw$
$96,800 = l \cdot 242$
$96,800 = 242l$
$\dfrac{96,800}{242} = \dfrac{242l}{242}$
$400 = 1l$
$400 = l$ \qquad The length must be 400 yd.

15. $4x - 9 = 23$
$\underline{+9 \quad +9}$
$4x + 0 = 32$
$\dfrac{4x}{4} = \dfrac{32}{4}$
$1x = 8$
$x = 8$

16. $3(x - 5) = 4x - 9$
$3x - 15 = 4x - 9$
$\underline{-3x -3x}$
$0 - 15 = x - 9$
$-15 = x - 9$
$\underline{+9 \qquad +9}$
$-6 = x + 0$
$-6 = x$

17. One number $= 3x$
Second number $= x$

$3x + x = 44$
$4x = 44$
$\dfrac{4x}{4} = \dfrac{44}{4}$
$1x = 11$
$x = 11$

One number is 33 and the other is 11.

18. An isosceles triangle has two equal sides. Let's call the base b. Since the two equal sides are "9 in. more than twice the base," we can say the equal sides are $2b + 9$. To find the perimeter of any figure, we add the sides. We also know the total perimeter is 258 in. $\qquad P = b + (2b + 9) + (2b + 9)$
$ 258 = 5b + 18$
$\underline{-18 \qquad -18}$
$ 240 = 5b + 0$
$\dfrac{240}{5} = \dfrac{5b}{5}$
$48 = b$

The base of the triangle is 48 in. and the two equal sides are each $2b + 9 = 2 \times (48) + 9 = 105$ in.

19. Since the floor of the stage is flat, we conclude that we are dealing with supplementary angles that will total 180°. Let's call the angle made on the other side x and the angle made by the microphone and the floor "is 30° less than the angle made on the other side," so we will call it; $x - 30$.

$x + (x - 30) = 180$
$2x - 30 = 180$
$\underline{+30 \qquad +30}$
$2x + 0 = 210$
$\dfrac{2x}{2} = \dfrac{210}{2}$
$x = 105°$

The larger angle is 105° and the angle made by the microphone and the floor is $x - 30 = 105 - 30 = 75°$.

20.

Categories	Value	Number of guitars	Income
Model A	450	$12 - x$	$450(12 - x)$
Model B	675	x	$675x$

$450(12 - x) + 675x = 6300$
$5400 - 450x + 675x = 6300$
$225x + 5400 = 6300$
$\underline{-5400 \quad -5400}$
$225x + 0 = 900$
$\dfrac{225x}{225} = \dfrac{900}{225}$
$x = 4$

The music store sold 8 Model A guitars and 4 Model B guitars.

Chapter 4 Cumulative Review Exercises

1. true \qquad 2. false \qquad 3. false

4. false \qquad 5. true \qquad 6. false

7. positive \qquad 8. add

9. subtract; divisor's; dividend's

10. Replace the variables with the given values, then calculate.

11. $36,097 = 3 \times 10,000 + 6 \times 1000 + 9 \times 10 + 7 \times 1$

12.

13. $30,000$: The digit 9 is in the thousands place. Look to the 5 on the right. Since 5 is greater than or equal to 5, we round the 9 up.

14. $6000 \div 200 = 30$: Think $6 \div 2$ and cancel 2 zeros.

15. -8 16. 0 17. 5

18. $|2 - 3(7)| = |2 - 21|$
$\qquad = |-19|$
$\qquad = 19$

19. $45 - (-52) = 45 + 52$
$\qquad\qquad = 97$

20. $19 - 5(9) = 19 - 45$ 21. $3 + (-4)^3 = 3 + (-64)$
$\qquad\qquad = -26$ $\qquad\qquad = -61$

22. $-2^4 = -(2 \cdot 2 \cdot 2 \cdot 2) = -16$

23. $2(-9) - 20 \div (5) = -18 - 4$
$\qquad\qquad\qquad = -22$

24. $7^2 + [8 - 5(6)][6 + 2(7)] = 7^2 + [8 - 30][6 + 14]$
$\qquad\qquad\qquad\qquad = 7^2 + (-22)(20)$
$\qquad\qquad\qquad\qquad = 49 + (-22)(20)$
$\qquad\qquad\qquad\qquad = 49 + (-440)$
$\qquad\qquad\qquad\qquad = -391$

25. $2 \cdot 9 - 5\sqrt{9 + 16} = 2 \cdot 9 - 5\sqrt{25}$
$\qquad\qquad\qquad = 2 \cdot 9 - 5 \cdot 5$
$\qquad\qquad\qquad = 18 - 25$
$\qquad\qquad\qquad = -7$

26. $7t^3 + t^3 + 14t^2 - 10t - 15t - 7$
$= 8t^3 + 14t^2 - 25t - 7$

27. $(10x^3 - 7x^2 - 15) - (4x^3 + x - 8)$
$= (10x^3 - 7x^2 - 15) + (-4x^3 - x + 8)$
$= 10x^3 - 4x^3 - 7x^2 - x - 15 + 8$
$= 6x^3 - 7x^2 - x - 7$

28. $(-6a^4)(9a^2) = -54a^{4+2}$
$\qquad\qquad = -54a^6$

29. $(3x - 5)(3x + 5) = 9x^2 - 25$

30. $2^2 \cdot 3 \cdot 5 \cdot 7$ 31. $36m^2$

32. $\dfrac{28k^6}{-4k^5} = \dfrac{28}{-4}k^{6-5} = -7k$

33. $20n^5 + 15mn^3 - 10n^2$
$= 5n^2\left(\dfrac{20n^5 + 15mn^3 - 10n^2}{5n^2}\right)$
$= 5n^2\left(\dfrac{20n^5}{5n^2} + \dfrac{15mn^3}{5n^2} - \dfrac{10n^2}{5n^2}\right)$
$= 5n^2(4n^3 + 3mn - 2)$

34. $x - 11 = 28$ Check: $39 - 11 = 28$
$\underline{+11 \quad +11}$ $\qquad\qquad 28 = 28$
$x + 0 = 39$
$\quad x = 39$

35. $-7t = 42$ $\qquad\qquad$ Check: $-7(-6) = 42$
$\dfrac{-7t}{-7} = \dfrac{42}{-7}$ $\qquad\qquad\qquad 42 = 42$
$1t = -6$
$\quad t = -6$

36. $30 = 2b - 14$ $\qquad$ Check: $30 = 2 \cdot 22 - 14$
$\underline{+14 \qquad +14}$ $\qquad\qquad 30 = 44 - 14$
$44 = 2b + 0$ $\qquad\qquad\quad 30 = 30$
$\dfrac{44}{2} = \dfrac{2b}{2}$
$22 = b$

37. $9y - 13 = 4y + 7$
$\underline{-4y \qquad\quad -4y}$
$5y - 13 = 0 + 7$
$\underline{\quad +13 \quad +13}$
$5y + 0 = 20$
$\dfrac{5y}{5} = \dfrac{20}{5}$
$\quad y = 4$

Check: $9 \cdot 4 - 13 = 4 \cdot 4 + 7$
$\qquad 36 - 13 = 16 + 7$
$\qquad\quad 23 = 23$

38. $6(n + 2) = 3n - 9$
$6n + 12 = 3n - 9$
$\underline{-3n \qquad\quad -3n}$
$3n + 12 = 0 - 9$
$3n + 12 = -9$
$\underline{\quad -12 \quad -12}$
$3n + 0 = -21$
$3n = -21$
$\dfrac{3n}{3} = \dfrac{-21}{3}$
$\quad n = -7$

Check: $6(-7+2) = 3 \cdot -7 - 9$
$6(-5) = -21 - 9$
$-30 = -30$

39. $278 + x = 500$
$\underline{-278 \qquad -278}$
$0 + x = 222$
$x = 222$

Carlos must complete 222 more hours.

40. We must find the total of the assets and debts to find the net worth.
$985 + 1862 + 12,006 + (-2345) + (-75,189) + (-4500)$
$= 14,853 + (-82,034)$
$= -\$67,181$
The Goodman family's net worth is –\$67,181.

41. $P = R - C$
$45,698 = 96,408 - C$
$\underline{-96,408 \qquad -96,408}$
$-50,710 = 0 - C$
$\dfrac{-50,710}{-1} = \dfrac{-C}{-1}$
$50,710 = C$

The total cost was \$50,710.

42. Adrian drove for a total of 2 hours.
$d = rt$
$124 = r \cdot 2$
$124 = 2r$
$\dfrac{124}{2} = \dfrac{2r}{2}$
$62 = r$ His average speed was 62 mph.

43. This figure is a composite of two parallelograms of the same size and shape put together. We will find the area of one and then multiply by 2 to get the total area. $A = bh$
$A = 18 \cdot 6$
$A = 108$ in.^{2}

Now we will multiply by 2 to get the total area.
$2 \times 108 = 216$ in^2.

44. $A = bh$
$A = (x+2)(x-6)$
$A = x \cdot x + x \cdot (-6) + 2 \cdot x + 2 \cdot (-6)$
$A = x^2 - 6x + 2x - 12$
$A = x^2 - 4x - 12$

45. $V = lwh$
$630 = 10 \cdot 9 \cdot h$
$630 = 90h$
$\dfrac{630}{90} = \dfrac{90h}{90}$
$7 = h$ The height is 7 ft.

46. $SA = 2lw + 2lh + 2wh$
$314 = 2 \cdot 9 \cdot 8 + 2 \cdot 9h + 2 \cdot 8h$
$314 = 144 + 18h + 16h$
$314 = 144 + 34h$
$\underline{-144 \qquad -144}$
$170 = 34h$
$\dfrac{170}{34} = \dfrac{34h}{34}$
$5 = h$

The height of the box is 5 in.

47. $15 + 6n = 3$
$\underline{-15 \qquad -15}$
$6n = -12$
$\dfrac{6n}{6} = \dfrac{-12}{6}$
$n = -2$

48. one number $= 2n$
second number $= n$

$2n + n = 96$
$3n = 96$
$\dfrac{3n}{3} = \dfrac{96}{3}$
$n = 32$

One number is 64 and the other is 32.

49. To find the perimeter of a rectangle, we add all of the sides. We will let the width be w. The length is "4 ft. more than the width" so we say $w + 4$. We also know that the perimeter is 44.
$P = (w+4) + w + (w+4) + w$
$44 = 4w + 8$
$\underline{-8 \qquad -8}$
$36 = 4w + 0$
$\dfrac{36}{4} = \dfrac{4w}{4}$
$9 = w$

The width is 9 ft. and the length $9 + 4 = 13$ ft.

50.

Categories	Value	Number of blankets	Income
Small size	45	$16 - x$	$45(16 - x)$
Large size	75	x	$75x$

$$45(16 - x) + 75x = 930$$
$$720 - 45x + 75x = 930$$
$$30x + 720 = 930$$
$$\underline{-720 \qquad -720}$$
$$30x + 0 = 210$$
$$\frac{30x}{30} = \frac{210}{30}$$
$$x = 7$$

Kedra sold 7 large blankets and $16 - 7 = 9$ small blankets.

Chapter 5 Fractions and Rational Expressions

Exercise Set 5.1

1. A rational number is a number that can be expressed in the form $\dfrac{a}{b}$, where a and b are integers and $b \neq 0$.

3. undefined

5. 1. Divide the denominator into the numerator.
   2. Write the results in the form:
   $$\text{quotient}\ \dfrac{\text{remainder}}{\text{original denominator}}$$

7. $\dfrac{1}{3}$ 9. $\dfrac{5}{8}$ 11. $\dfrac{2}{3}$

13. a) $\dfrac{5}{16}$ b) $\dfrac{15}{16}$

   c) The sense of smell seems to contribute more information about foods than taste.

   d) If she is trying the same foods in the same order she may be guessing better the second time.

15. The fraction of eggs that survive to adulthood is $\dfrac{17}{800}$. Because the egg can either survive or not, there are only two possible categories. This means that if 17 out of 800 eggs survive, then the rest are not surviving. We can subtract 17 from 800 to get the number of eggs that do not survive. There are $800 - 17 = 783$ eggs that do not survive out of the 800 total eggs, therefore the fraction is $\dfrac{783}{800}$.

17. The fraction of students that pass is $\dfrac{249}{258}$.

   Because the student can either pass or not, there are only two possible categories. This means that if 249 out of 258 students pass, then the rest are not passing. We can subtract 249 from 258 to get the number of students that do not pass. There are $258 - 249 = 9$ students that do not pass out of the 258 total students, therefore the fraction is $\dfrac{9}{258}$.

   Very few of Mrs. Jones's students do not pass the course.

19. $\dfrac{451}{600}$ 21. $\dfrac{149}{600}$

23.

25.

27.

29.

31. 23   33. 0   35. 1   37. undefined

39. $\dfrac{5}{9} = \dfrac{?}{27}$
    $\dfrac{5 \cdot 3}{9 \cdot 3} = \dfrac{15}{27}$

41. $\dfrac{21}{36} = \dfrac{?}{12}$
    $\dfrac{21 \div 3}{36 \div 3} = \dfrac{7}{12}$

43. $-\dfrac{9}{15} = -\dfrac{18}{?}$
    $-\dfrac{9 \cdot 2}{15 \cdot 2} = -\dfrac{18}{30}$

45. $\dfrac{-6}{16} = \dfrac{?}{80}$
    $\dfrac{-6 \cdot 5}{16 \cdot 5} = \dfrac{-30}{80}$

47. Write equivalent fractions that have common denominators.

   $\dfrac{4}{9} ? \dfrac{2}{5}$

   $\dfrac{4}{9} = \dfrac{4 \cdot 5}{9 \cdot 5} = \dfrac{20}{45}$

   $\dfrac{2}{5} = \dfrac{2 \cdot 9}{5 \cdot 9} = \dfrac{18}{45}$

   Comparing numerators, we see that $20 > 18$. Therefore $\dfrac{4}{9} > \dfrac{2}{5}$.

49. Write equivalent fractions that have common denominators.

   $\dfrac{12}{18} ? \dfrac{9}{16}$

   $\dfrac{12}{18} = \dfrac{12 \cdot 8}{18 \cdot 8} = \dfrac{96}{144}$

   $\dfrac{9}{16} = \dfrac{9 \cdot 9}{16 \cdot 9} = \dfrac{81}{144}$

   Comparing numerators, we see that $96 > 81$. Therefore $\dfrac{12}{18} > \dfrac{9}{16}$.

51. Write equivalent fractions that have common denominators.

$$-\frac{4}{15} \; ? -\frac{6}{17}$$

$$-\frac{4}{15} = \frac{-4 \cdot 17}{15 \cdot 17} = \frac{-68}{255}$$

$$-\frac{6}{17} = \frac{-6 \cdot 15}{17 \cdot 15} = \frac{-90}{255}$$

Comparing numerators, we see that $-68 > -91$.

Therefore $-\frac{4}{15} > -\frac{6}{17}$.

53. Write equivalent fractions that have common denominators.

$$-\frac{9}{12} \; ? -\frac{15}{20}$$

$$-\frac{9}{12} = \frac{-9 \cdot 5}{12 \cdot 5} = \frac{-45}{60}$$

$$-\frac{15}{20} = \frac{-15 \cdot 3}{20 \cdot 3} = \frac{-45}{60}$$

Comparing numerators, we see that $-45 = -45$.

Therefore $-\frac{9}{12} = -\frac{15}{20}$.

55. $\dfrac{30}{7} = 7{\overline{\smash{\big)}\,30}} = 4\dfrac{2}{7}$
$\underline{28}$
2

57. $\dfrac{85}{4} = 4{\overline{\smash{\big)}\,85}} = 21\dfrac{1}{4}$
$\underline{8}$
05
$\underline{4}$
1

59. $\dfrac{-64}{5} = 5{\overline{\smash{\big)}\,64}} = -12\dfrac{4}{5}$
$\underline{5}$
14
$\underline{10}$
4

61. $\dfrac{103}{-8} = 8{\overline{\smash{\big)}\,103}} = -12\dfrac{7}{8}$
$\underline{8}$
23
$\underline{16}$
7

63. $59 \div 6 = 6{\overline{\smash{\big)}\,59}} = 9\dfrac{5}{6}$
$\underline{54}$
5

65. $140 \div 11 = 11{\overline{\smash{\big)}\,140}} = 12\dfrac{8}{11}$
$\underline{11}$
30
$\underline{22}$
8

67. $-839 \div 8 = 8{\overline{\smash{\big)}\,839}} = -104\dfrac{7}{8}$
$\underline{8}$
3
$\underline{0}$
39
$\underline{32}$
7

69. $-5629 \div (-14) = 14{\overline{\smash{\big)}\,5629}} = 402\dfrac{1}{14}$
$\underline{56}$
2
$\underline{0}$
29
$\underline{28}$
1

71. $5\dfrac{1}{6} = \dfrac{6 \cdot 5 + 1}{6} = \dfrac{30 + 1}{6} = \dfrac{31}{6}$

73. $11 = \dfrac{11}{1}$

75. $-9\dfrac{7}{8} = -\dfrac{8 \cdot 9 + 7}{8} = -\dfrac{72 + 7}{8} = -\dfrac{79}{8}$

77. $-1\dfrac{9}{20} = -\dfrac{20 \cdot 1 + 9}{20} = -\dfrac{20 + 9}{20} = -\dfrac{29}{20}$

Review Exercises

1. $-48 \div (-12) = 4$ 　　 2. $840 = 2^3 \cdot 3 \cdot 5 \cdot 7$

3. $24 : 1, 2, 3, 4, 6, 8, 12, 24$
   $60 : 1, 2, 3, 4, 5, 6, 10, 12, 15, 20, 30, 60$
   GCF $= 12$

4. $40x^5 = 2^3 \cdot 5 \cdot x^5$
   $56x^2 = 2^3 \cdot 7 \cdot x^2$
   GCF $= 2^3 \cdot x^2 = 8x^2$

5. $40x^5 - 56x^2$

$= 8x^2 \left(\dfrac{40x^5 - 56x^2}{8x^2} \right)$

$= 8x^2 \left(\dfrac{40x^5}{8x^2} - \dfrac{56x^2}{8x^2} \right)$

$= 8x^2(5x^3 - 7)$

Exercise Set 5.2

1. greatest common factor; 1

3. $\dfrac{25}{30} = \dfrac{25 \div 5}{30 \div 5} = \dfrac{5}{6}$

5. $-\dfrac{14}{35} = -\dfrac{14 \div 7}{35 \div 7} = -\dfrac{2}{5}$

7. $\dfrac{26}{52} = \dfrac{26 \div 26}{52 \div 26} = \dfrac{1}{2}$

9. $-\dfrac{24}{40} = -\dfrac{24 \div 8}{40 \div 8} = -\dfrac{3}{5}$

11. $\dfrac{66}{88} = \dfrac{66 \div 22}{88 \div 22} = \dfrac{3}{4}$

13. $-\dfrac{57}{76} = -\dfrac{57 \div 19}{76 \div 19} = -\dfrac{3}{4}$

15. $\dfrac{120}{140} = \dfrac{120 \div 20}{140 \div 20} = \dfrac{6}{7}$

17. $\dfrac{182}{234} = \dfrac{\cancel{2} \cdot 7 \cdot \cancel{13}}{\cancel{2} \cdot 3 \cdot 3 \cdot \cancel{13}} = \dfrac{7}{3 \cdot 3} = \dfrac{7}{9}$

19. $-\dfrac{121}{187} = -\dfrac{\cancel{11} \cdot 11}{\cancel{11} \cdot 17} = -\dfrac{11}{17}$

21. $-\dfrac{360}{480} = -\dfrac{360 \div 120}{480 \div 120} = -\dfrac{3}{4}$

23. Since there are 60 seconds in a minute, we say that 35 seconds is $\dfrac{35}{60} = \dfrac{35 \div 5}{60 \div 5} = \dfrac{7}{12}$.

25. Since there are 24 hours in a day, we say that 9 hours is $\dfrac{9}{24} = \dfrac{9 \div 3}{24 \div 3} = \dfrac{3}{8}$ of a day.

27. a) Since 96 out of 248 use a cosmetic, we say $\dfrac{96}{248} = \dfrac{96 \div 8}{248 \div 8} = \dfrac{12}{31}$.

b) 12 out of every 31 women can be expected to use this product.

c) 248 may not be a large enough sample.

29. Since 48 out of 500 transactions are mistakes, we say $\dfrac{48}{500} = \dfrac{48 \div 4}{500 \div 4} = \dfrac{12}{125}$. Because 48 mistakes were made, we subtract 48 from 500 to find the number of transactions that were not mistakes,

then write as a fraction and simplify. $500 - 48 = 452$ which is $\dfrac{452}{500} = \dfrac{452 \div 4}{500 \div 4} = \dfrac{113}{125}$.

31. a) total travel expenses = food + airfare + hotel + ground transportation

total travel expenses = $1028 + 3515 + 2849 + 2450$

$= \$9842$

b) $\dfrac{1028}{9842} = \dfrac{1028 \div 2}{9842 \div 2} = \dfrac{514}{4921}$

c) $\dfrac{2450}{9842} = \dfrac{2450 \div 14}{9842 \div 14} = \dfrac{175}{703}$

d) $\dfrac{2849}{9842} = \dfrac{2849 \div 259}{9842 \div 259} = \dfrac{11}{38}$

e) $\dfrac{3515}{9842} = \dfrac{3515 \div 703}{9842 \div 703} = \dfrac{5}{14}$

33. $\dfrac{30}{8} = \dfrac{\cancel{2} \cdot 3 \cdot 5}{\cancel{2} \cdot 2 \cdot 2} = \dfrac{3 \cdot 5}{2 \cdot 2} = \dfrac{15}{4} = 3\dfrac{3}{4}$

35. $-\dfrac{50}{15} = -\dfrac{2 \cdot \cancel{5} \cdot 5}{3 \cdot \cancel{5}} = -\dfrac{2 \cdot 5}{3} = -\dfrac{10}{3} = -3\dfrac{1}{3}$

37. $\dfrac{116}{28} = \dfrac{\cancel{2} \cdot \cancel{2} \cdot 29}{\cancel{2} \cdot \cancel{2} \cdot 7} = \dfrac{29}{7} = 4\dfrac{1}{7}$

39. $\dfrac{-186}{36} = \dfrac{-\cancel{2} \cdot \cancel{3} \cdot 31}{\cancel{2} \cdot 2 \cdot \cancel{3} \cdot 3} = \dfrac{-31}{2 \cdot 3} = \dfrac{-31}{6} = -5\dfrac{1}{6}$

41. $\dfrac{10x}{32} = \dfrac{\cancel{2} \cdot 5 \cdot x}{\cancel{2} \cdot 2 \cdot 2 \cdot 2 \cdot 2} = \dfrac{5 \cdot x}{2 \cdot 2 \cdot 2 \cdot 2} = \dfrac{5x}{16}$

43. $-\dfrac{x^3}{xy} = -\dfrac{\cancel{x} \cdot x \cdot x}{\cancel{x} \cdot y} = -\dfrac{x \cdot x}{y} = -\dfrac{x^2}{y}$

45. $-\dfrac{6m^4 n}{15m^7} = -\dfrac{2 \cdot \cancel{3} \cdot \cancel{m} \cdot \cancel{m} \cdot \cancel{m} \cdot \cancel{m} \cdot n}{\cancel{3} \cdot 5 \cdot \cancel{m} \cdot \cancel{m} \cdot \cancel{m} \cdot \cancel{m} \cdot m \cdot m \cdot m}$

$= -\dfrac{2 \cdot n}{5 \cdot m \cdot m \cdot m} = -\dfrac{2n}{5m^3}$

47. $\dfrac{9t^2 u}{36t^5 u^2} = \dfrac{\cancel{3} \cdot \cancel{3} \cdot \cancel{t} \cdot \cancel{t} \cdot \cancel{u}}{2 \cdot 2 \cdot \cancel{3} \cdot \cancel{3} \cdot \cancel{t} \cdot \cancel{t} \cdot t \cdot t \cdot t \cdot \cancel{u} \cdot u}$

$= \dfrac{1}{2 \cdot 2 \cdot t \cdot t \cdot t \cdot u} = \dfrac{1}{4t^3 u}$

49. $\dfrac{21a^6 bc}{35a^5 b^2} = \dfrac{3 \cdot \cancel{7} \cdot \cancel{a} \cdot \cancel{a} \cdot \cancel{a} \cdot \cancel{a} \cdot \cancel{a} \cdot a \cdot \cancel{b} \cdot c}{5 \cdot \cancel{7} \cdot \cancel{a} \cdot \cancel{a} \cdot \cancel{a} \cdot \cancel{a} \cdot \cancel{a} \cdot \cancel{b} \cdot b}$

$= \dfrac{3 \cdot a \cdot c}{5 \cdot b} = \dfrac{3ac}{5b}$

51.

$$\frac{-14a^4b^5c^2}{28a^{10}bc^7}$$

$$= \frac{-\cancel{2}\cdot\cancel{7}\cdot \cancel{a}\cdot\cancel{a}\cdot\cancel{a}\cdot\cancel{a}\cdot\cancel{b}\cdot b\cdot b\cdot b\cdot b\cdot\cancel{c}\cdot\cancel{c}}{\cancel{2}\cdot 2\cdot\cancel{7}\cdot \cancel{a}\cdot\cancel{a}\cdot\cancel{a}\cdot\cancel{a}\cdot a\cdot a\cdot a\cdot a\cdot a\cdot a\cdot\cancel{b}\cdot\cancel{c}\cdot\cancel{c}\cdot c\cdot c\cdot c\cdot c\cdot c}$$

$$= \frac{-1\cdot b\cdot b\cdot b\cdot b}{2\cdot a\cdot a\cdot a\cdot a\cdot a\cdot a\cdot c\cdot c\cdot c\cdot c\cdot c}$$

$$= \frac{-b^4}{2a^6c^5}$$

Review Exercises

1. $165\cdot(-91) = -15{,}015$ 2. $(-12)(-6) = 72$

3. $(5x^2)(7x^3) = 5\cdot 7\cdot x^{2+3}$ 4. $(-2x^3)^5 = (-2)^5 x^{3\cdot 5}$
$\qquad\qquad\quad = 35x^5$ $\qquad\qquad\qquad = -32x^{15}$

5. $A = bh$
$A = 11\cdot 9$
$A = 99 \text{ m}^2$

Exercise Set 5.3

1. 1. Divide out any numerator factor with any like denominator factor.
2. Multiply numerator by numerator and denominator by denominator.
3. Simplify.

3. The formula is used to calculate the area of a triangle. A represents the area, b represents the base, and h represents the height.

5. The circumference of a circle is the distance around the circle.

7. π represents an irrational number representing the ratio of the circumference of a circle to its diameter. It can be approximated using the fraction $\frac{22}{7}$.

9. $\frac{2}{5}\cdot\frac{4}{7} = \frac{2\cdot 4}{5\cdot 7} = \frac{8}{35}$

11. $-\frac{1}{6}\cdot\frac{1}{9} = -\frac{1\cdot 1}{6\cdot 9} = -\frac{1}{54}$

13. $-\frac{3}{4}\cdot\left(-\frac{5}{7}\right) = \frac{-3\cdot(-5)}{4\cdot 7} = \frac{15}{28}$

15. $\frac{7}{100}\cdot\frac{7}{10} = \frac{7\cdot 7}{100\cdot 10} = \frac{49}{1000}$

17. $\frac{6}{7}\cdot\frac{14}{15} = \frac{84}{105} = \frac{84\div 21}{105\div 21} = \frac{4}{5}$

19. $-\frac{8}{12}\cdot\frac{10}{20} = -\frac{80}{240} = -\frac{80\div 80}{240\div 80} = -\frac{1}{3}$

21. $\frac{24}{32}\cdot\frac{26}{30} = \frac{\cancel{2}\cdot\cancel{2}\cdot\cancel{2}\cdot\cancel{3}}{\cancel{2}\cdot\cancel{2}\cdot\cancel{2}\cdot 2\cdot 2}\cdot\frac{\cancel{2}\cdot 13}{\cancel{2}\cdot\cancel{3}\cdot 5}$

$\qquad = \frac{13}{2\cdot 2\cdot 5} = \frac{13}{20}$

23. $\frac{16}{38}\cdot\left(-\frac{57}{80}\right) = \frac{\cancel{2}\cdot\cancel{2}\cdot\cancel{2}\cdot 2}{\cancel{2}\cdot 19}\cdot\frac{-3\cdot\cancel{19}}{\cancel{2}\cdot\cancel{2}\cdot\cancel{2}\cdot 2\cdot 5}$

$\qquad = \frac{-3}{2\cdot 5} = -\frac{3}{10}$

25. $\frac{19}{20}\cdot\frac{16}{38} = \frac{\cancel{19}}{\cancel{2}\cdot\cancel{2}\cdot 5}\cdot\frac{\cancel{2}\cdot\cancel{2}\cdot\cancel{2}\cdot 2}{\cancel{2}\cdot\cancel{19}}$

$\qquad = \frac{2}{5}$

27. $-\frac{18}{40}\cdot\left(-\frac{28}{30}\right) = -\frac{\cancel{2}\cdot\cancel{3}\cdot 3}{\cancel{2}\cdot\cancel{2}\cdot\cancel{2}\cdot 5}\cdot\left(-\frac{\cancel{2}\cdot\cancel{2}\cdot 7}{2\cdot\cancel{3}\cdot 5}\right)$

$\qquad = \frac{3\cdot 7}{5\cdot 2\cdot 5}$

$\qquad = \frac{21}{50}$

29. $\frac{36}{-40}\cdot\left(\frac{-30}{54}\right) = \frac{\cancel{2}\cdot\cancel{2}\cdot\cancel{3}\cdot\cancel{3}}{-\cancel{2}\cdot\cancel{2}\cdot\cancel{2}\cdot\cancel{5}}\cdot\frac{-\cancel{2}\cdot\cancel{3}\cdot\cancel{5}}{2\cdot\cancel{3}\cdot\cancel{3}\cdot\cancel{3}}$

$\qquad = \frac{1}{2}$

31. Estimate: $3\cdot 2 = 6$

Actual: $3\frac{1}{5}\cdot 1\frac{3}{4} = \frac{16}{5}\cdot\frac{7}{4} = \frac{\cancel{2}\cdot\cancel{2}\cdot 2\cdot 2}{5}\cdot\frac{7}{\cancel{2}\cdot\cancel{2}}$

$\qquad = \frac{2\cdot 2\cdot 7}{5} = \frac{28}{5} = 5\frac{3}{5}$

33. Estimate: $1\cdot 28 = 28$

Actual: $\frac{5}{6}(28) = \frac{5}{6}\cdot\frac{28}{1} = \frac{5}{\cancel{2}\cdot 3}\cdot\frac{\cancel{2}\cdot 2\cdot 7}{1}$

$\qquad = \frac{5\cdot 2\cdot 7}{3} = \frac{70}{3} = 23\frac{1}{3}$

35. Estimate: $5\cdot 16 = 80$

Actual: $4\frac{5}{8}\cdot 16 = \frac{37}{8}\cdot\frac{16}{1} = \frac{37}{\cancel{2}\cdot\cancel{2}\cdot\cancel{2}}\cdot\frac{\cancel{2}\cdot\cancel{2}\cdot\cancel{2}\cdot 2}{1}$

$\qquad = \frac{37\cdot 2}{1} = \frac{74}{1} = 74$

37. Estimate: $-6\cdot\frac{1}{2} = \frac{-\cancel{2}\cdot 3}{1}\cdot\frac{1}{\cancel{2}} = \frac{-3}{1} = -3$

Actual: $-6\dfrac{1}{8}\cdot\dfrac{4}{7}=-\dfrac{49}{8}\cdot\dfrac{4}{7}=-\dfrac{7\cdot\cancel{7}}{\cancel{2}\cdot\cancel{2}\cdot 2}\cdot\dfrac{\cancel{2}\cdot\cancel{2}}{\cancel{7}}$

$\qquad = -\dfrac{7}{2}=-3\dfrac{1}{2}$

39. Estimate: $-6\cdot\left(-7\right)=42$

Actual: $-5\dfrac{2}{3}\cdot\left(-7\dfrac{1}{5}\right)=-\dfrac{17}{3}\cdot-\dfrac{36}{5}$

$\qquad = -\dfrac{17}{\cancel{3}}\cdot\left(\dfrac{-2\cdot 2\cdot\cancel{3}\cdot 3}{5}\right)=\dfrac{17\cdot 2\cdot 2\cdot 3}{5}$

$\qquad = \dfrac{204}{5}=40\dfrac{4}{5}$

41. Estimate: $-\dfrac{1}{\cancel{3}}\cdot\dfrac{\overset{15}{\cancel{45}}}{1}=-15$

Actual: $\dfrac{-3}{10}(45)=\dfrac{-3}{10}\cdot\dfrac{45}{1}=\dfrac{-3}{2\cdot\cancel{5}}\cdot\dfrac{3\cdot 3\cdot\cancel{5}}{1}$

$\qquad = \dfrac{-3\cdot 3\cdot 3}{2}=\dfrac{-27}{2}=-13\dfrac{1}{2}$

43. $\dfrac{x^2}{5}\cdot\dfrac{2}{3}=\dfrac{2x^2}{15}$

45. $\dfrac{4x}{9}\cdot\dfrac{3x}{8}=\dfrac{\cancel{2}\cdot\cancel{2}\cdot x}{\cancel{3}\cdot 3}\cdot\dfrac{\cancel{3}\cdot x}{\cancel{2}\cdot\cancel{2}\cdot 2}=\dfrac{x^2}{6}$

47. $\dfrac{xy}{10}\cdot\dfrac{4y^3}{9}=\dfrac{x\cdot y}{\cancel{2}\cdot 5}\cdot\dfrac{\cancel{2}\cdot 2\cdot y\cdot y\cdot y}{3\cdot 3}=\dfrac{2xy^4}{45}$

49. $-\dfrac{5hk^3}{9}\cdot\dfrac{3}{4h}=-\dfrac{5\cdot\cancel{h}\cdot k\cdot k\cdot k}{3\cdot\cancel{3}}\cdot\dfrac{\cancel{3}}{2\cdot 2\cdot\cancel{h}}=-\dfrac{5k^3}{12}$

51. $-\dfrac{10x^4y}{11z}\cdot\left(-\dfrac{22z}{14x^2}\right)$

$\qquad = -\dfrac{\cancel{2}\cdot 5\cdot\cancel{x}\cdot\cancel{x}\cdot x\cdot x\cdot y}{\cancel{11}\cdot\cancel{z}}\cdot\left(-\dfrac{2\cdot\cancel{11}\cdot\cancel{z}}{\cancel{2}\cdot 7\cdot\cancel{x}\cdot\cancel{x}}\right)$

$\qquad = \dfrac{10x^2y}{7}$

53. $\dfrac{9m^3}{25n^4}\cdot\dfrac{-15n}{18m^2}$

$\qquad = \dfrac{\cancel{3}\cdot\cancel{3}\cdot\cancel{m}\cdot\cancel{m}\cdot m}{\cancel{5}\cdot 5\cdot\cancel{n}\cdot n\cdot n\cdot n}\cdot\dfrac{-3\cdot\cancel{5}\cdot\cancel{n}}{2\cdot\cancel{3}\cdot\cancel{3}\cdot\cancel{m}\cdot\cancel{m}}$

$\qquad = \dfrac{-3m}{10n^3}$

55. $\left(\dfrac{5}{6}\right)^2=\dfrac{5^2}{6^2}=\dfrac{25}{36}$

57. $\left(-\dfrac{3}{4}\right)^3=\dfrac{(-3)^3}{4^3}=-\dfrac{27}{64}$

59. $\left(\dfrac{-1}{2}\right)^6=\dfrac{(-1)^6}{2^6}=\dfrac{1}{64}$

61. $\left(\dfrac{x}{2}\right)^3=\dfrac{x^3}{2^3}=\dfrac{x^3}{8}$

63. $\left(\dfrac{2x^2}{3}\right)^3=\dfrac{2^3 x^{2\cdot 3}}{3^3}=\dfrac{8x^6}{27}$

65. $\left(\dfrac{-m^3 n}{3p^2}\right)^4=\dfrac{m^{3\cdot 4}n^4}{3^4 p^{2\cdot 4}}=\dfrac{m^{12}n^4}{81p^8}$

67. a) Heating and cooling represented $\dfrac{2}{5}$ of the total home energy spending. We can translate this to

$\dfrac{2}{5}\cdot 1640=\dfrac{2}{5}\cdot\dfrac{1640}{1}=\dfrac{2}{1\,\cancel{5}}\cdot\dfrac{\overset{328}{\cancel{1640}}}{1}=\dfrac{656}{1}=\$656$

b) Lighting and appliances represented $\dfrac{7}{20}$ of the total home energy spending. We can translate this to

$\dfrac{7}{20}\cdot 1640=\dfrac{7}{20}\cdot\dfrac{1640}{1}=\dfrac{7}{1\,20}\cdot\dfrac{\overset{82}{\cancel{1640}}}{1}=\dfrac{574}{1}=\$574$

c) Heating water represented $\dfrac{3}{20}$ of the total home energy spending. We can translate this to

$\dfrac{3}{20}\cdot 1640=\dfrac{3}{20}\cdot\dfrac{1640}{1}=\dfrac{3}{1\,20}\cdot\dfrac{\overset{82}{\cancel{1640}}}{1}=\dfrac{246}{1}=\$246$

d) Refrigeration represented $\dfrac{1}{10}$ of the total home energy spending. We can translate this to

$\dfrac{1}{10}\cdot 1640=\dfrac{1}{10}\cdot\dfrac{1640}{1}=\dfrac{1}{1\,10}\cdot\dfrac{\overset{164}{\cancel{1640}}}{1}=\dfrac{164}{1}=\$164$

69. We will translate *of* to mean multiplication.
$\dfrac{1}{8}\cdot 12,480=\dfrac{12,480}{8}=\dfrac{12,480\div 8}{8\div 8}=\dfrac{1560}{1}=1560$

1560 pieces of fruit can be expected to go bad during shipment.

71. We will translate *of* to mean multiplication. $\dfrac{5}{6}\cdot\dfrac{3}{4}=\dfrac{15}{24}=\dfrac{15\div 3}{24\div 3}=\dfrac{5}{8}$

$\dfrac{5}{8}$ of the employees live within a 10-mile radius.

73. The fraction representing the long-distance calls to her parents is $\frac{5}{6} \cdot \frac{24}{30} = \frac{120 \div 60}{180 \div 60} = \frac{2}{3}$. We will translate *of* to mean multiplication. We must find $\frac{5}{6}$ of the 24 in state calls.

$$\frac{5}{6} \cdot 24 = \frac{5}{6} \cdot \frac{24}{1} = \frac{120}{6} = \frac{120 \div 6}{6 \div 6} = \frac{20}{1} = 20$$

Tanya made 20 calls to her parents.

75. $\frac{1}{2} \cdot 25\frac{1}{2} = \frac{1}{2} \cdot \frac{51}{2} = \frac{51}{4} = 12\frac{3}{4}$ in.

77. $d = rt$

$d = 16 \cdot 2\frac{2}{3}$

$d = \frac{16}{1} \cdot \frac{8}{3}$

$d = \frac{128}{3}$

$d = 42\frac{2}{3}$ miles

79. $A = \frac{1}{2}bh$

$A = \frac{1}{2} \cdot 12 \cdot 8\frac{1}{2}$

$A = \frac{1}{2} \cdot \frac{12}{1} \cdot \frac{17}{2}$

$A = \frac{1}{\cancel{2}} \cdot \frac{\cancel{2} \cdot \cancel{2} \cdot 3}{1} \cdot \frac{17}{\cancel{2}}$

$A = \frac{51}{1}$

$A = 51$ ft.^{2}

81. a) The radius is $\frac{1}{2}$ of the diameter.

$\frac{1}{2} \cdot 9 = \frac{1}{2} \cdot \frac{9}{1} = \frac{9}{2} = 4\frac{1}{2}$ m

b) $C = 2\pi r$

$C \approx 2 \cdot \frac{22}{7} \cdot \frac{9}{2}$

$C \approx \frac{\cancel{2}}{1} \cdot \frac{2 \cdot 11}{7} \cdot \frac{3 \cdot 3}{\cancel{2}}$

$C \approx \frac{198}{7}$

$C \approx 28\frac{2}{7}$ m

83. a) The diameter is two times the radius.

$2 \cdot \frac{273}{440} = \frac{\cancel{2}}{1} \cdot \frac{273}{\cancel{440}_{220}} = \frac{273}{220} = 1\frac{53}{220}$ mi.

b) $C = \pi d$

$C \approx \frac{22}{7} \cdot 1\frac{53}{220}$

$C \approx \frac{\cancel{22}}{\cancel{7}} \cdot \frac{\cancel{273}^{39}}{\cancel{220}_{10}}$

$C \approx \frac{39}{10}$

$C \approx 3\frac{9}{10}$ mi.

Review Exercises

1. $1836 \div (-9) = -204$ 2. $\sqrt{256} = 16$

3. $9x^2 - 12x = 3x\left(\frac{9x^2 - 12x}{3x}\right)$

$= 3x\left(\frac{9x^2}{3x} - \frac{12x}{3x}\right)$

$= 3x(3x - 4)$

4. $-6x = 30$ Check: $-6 \cdot (-5) = 30$

$\frac{-6x}{-6} = \frac{30}{-6}$ $30 = 30$

$1x = -5$

$x = -5$

5. $12 \div 6 = 2$ Each piece will be 2 ft.

Exercise Set 5.4

1. Reciprocals are two numbers whose product is 1. Examples may vary, but one pair is $\frac{2}{3}$ and $\frac{3}{2}$.

3. Write each mixed number as an improper fraction, then follow the process for dividing fractions.

5. $\frac{3}{2}$ 7. 6   9. $-\frac{1}{15}$ 11. $-\frac{4}{5}$

13. $\frac{1}{2} \div \frac{1}{6} = \frac{1}{2} \cdot \frac{6}{1}$ 15. $\frac{3}{10} \div \frac{5}{6} = \frac{3}{10} \cdot \frac{6}{5}$

$= \frac{1}{\cancel{2}} \cdot \frac{\cancel{2} \cdot 3}{1}$ $= \frac{3}{\cancel{2} \cdot 5} \cdot \frac{\cancel{2} \cdot 3}{5}$

$= \frac{3}{1}$ $= \frac{9}{25}$

$= 3$

17. $\dfrac{14}{15} \div \left(-\dfrac{7}{12}\right) = \dfrac{14}{15} \cdot -\dfrac{12}{7}$

$= \dfrac{2 \cdot \cancel{7}}{\cancel{3} \cdot 5} \cdot \left(-\dfrac{\cancel{3} \cdot 2 \cdot 2}{\cancel{7}}\right)$

$= -\dfrac{8}{5}$

$= -1\dfrac{3}{5}$

19. $-\dfrac{7}{12} \div (-14) = -\dfrac{7}{12} \div \left(\dfrac{-14}{1}\right)$

$= -\dfrac{7}{12} \cdot \left(\dfrac{1}{-14}\right)$

$= \dfrac{-\cancel{7}}{2 \cdot 2 \cdot 3} \cdot \dfrac{1}{-2 \cdot \cancel{7}}$

$= \dfrac{1}{24}$

21. $\dfrac{\frac{2}{5}}{\frac{4}{15}} = \dfrac{2}{5} \div \dfrac{4}{15}$

$= \dfrac{2}{5} \cdot \dfrac{15}{4}$

$= \dfrac{\cancel{2}}{\cancel{5}} \cdot \dfrac{3 \cdot \cancel{5}}{\cancel{2} \cdot 2}$

$= \dfrac{3}{2}$

$= 1\dfrac{1}{2}$

23. $\dfrac{12}{\frac{-2}{3}} = 12 \div \left(\dfrac{-2}{3}\right)$

$= \dfrac{12}{1} \cdot \left(\dfrac{3}{-2}\right)$

$= \dfrac{\cancel{2} \cdot 2 \cdot 3}{1} \cdot \dfrac{3}{-\cancel{2}}$

$= \dfrac{18}{-1}$

$= -18$

25. Estimate: $5 \div \dfrac{1}{2} = \dfrac{5}{1} \cdot \dfrac{2}{1}$

$= 10$

Actual: $5\dfrac{1}{6} \div \dfrac{4}{9} = \dfrac{31}{6} \div \dfrac{4}{9}$

$= \dfrac{31}{6} \cdot \dfrac{9}{4}$

$= \dfrac{31}{2 \cdot \cancel{3}} \cdot \dfrac{\cancel{3} \cdot 3}{2 \cdot 2}$

$= \dfrac{93}{8}$

$= 11\dfrac{5}{8}$

27. Estimate: $10 \div 4 = \dfrac{10}{1} \cdot \dfrac{1}{4} = \dfrac{\overset{5}{\cancel{10}}}{1} \cdot \dfrac{1}{\cancel{4}_2} = \dfrac{5}{2} = 2\dfrac{1}{2}$

Actual: $9\dfrac{1}{2} \div 3\dfrac{3}{4} = \dfrac{19}{2} \div \dfrac{15}{4}$

$= \dfrac{19}{2} \cdot \dfrac{4}{15}$

$= \dfrac{19}{\cancel{2}} \cdot \dfrac{\cancel{2} \cdot 2}{3 \cdot 5}$

$= \dfrac{38}{15}$

$= 2\dfrac{8}{15}$

29. Estimate: $-3 \div 10 = -\dfrac{3}{10}$

Actual: $-2\dfrac{9}{16} \div 10\dfrac{1}{4} = -\dfrac{41}{16} \div \dfrac{41}{4}$

$= -\dfrac{41}{16} \cdot \dfrac{4}{41}$

$= -\dfrac{\cancel{41}}{4 \cdot \cancel{4}} \cdot \dfrac{\cancel{4}}{\cancel{41}}$

$= -\dfrac{1}{4}$

31. Estimate: $-13 \div -7 = \dfrac{-13}{1} \cdot \dfrac{1}{-7}$

$= \dfrac{13}{7}$

$= 1\dfrac{6}{7}$

Actual: $\dfrac{-12\frac{1}{2}}{-6\frac{3}{4}} = -12\dfrac{1}{2} \div -6\dfrac{3}{4}$

$= -\dfrac{25}{2} \div -\dfrac{27}{4}$

$= -\dfrac{25}{2} \cdot -\dfrac{4}{27}$

$= -\dfrac{5 \cdot 5}{\cancel{2}} \cdot -\dfrac{\cancel{2} \cdot 2}{3 \cdot 3 \cdot 3}$

$= \dfrac{50}{27} = 1\dfrac{23}{27}$

33. $\dfrac{7}{15x} \div \dfrac{1}{6x^3} = \dfrac{7}{15x} \cdot \dfrac{6x^3}{1}$

$= \dfrac{7}{\cancel{3} \cdot 5 \cdot \cancel{x}} \cdot \dfrac{2 \cdot \cancel{3} \cdot \cancel{x} \cdot x \cdot x}{1}$

$= \dfrac{14x^2}{5}$

35. $\dfrac{9m^5}{20n} \div \dfrac{12m^2 n}{25}$

$= \dfrac{9m^5}{20n} \cdot \dfrac{25}{12m^2 n}$

$= \dfrac{\cancel{3} \cdot 3 \cdot \cancel{m} \cdot \cancel{m} \cdot m \cdot m \cdot m}{2 \cdot 2 \cdot \cancel{5} \cdot n} \cdot \dfrac{\cancel{5} \cdot 5}{\cancel{3} \cdot 2 \cdot 2 \cdot \cancel{m} \cdot \cancel{m} \cdot n}$

$= \dfrac{15m^3}{16n^2}$

37. $\dfrac{14x^6}{28y^4 z} \div \left(\dfrac{-8x^2 y}{18z} \right)$

$= \dfrac{14x^6}{28y^4 z} \cdot \left(\dfrac{18z}{-8x^2 y} \right)$

$= \dfrac{\cancel{2} \cdot \cancel{7} \cdot \cancel{x} \cdot \cancel{x} \cdot x \cdot x \cdot x \cdot x}{\cancel{2} \cdot \cancel{2} \cdot \cancel{7} \cdot y \cdot y \cdot y \cdot y \cdot z} \cdot \left(\dfrac{\cancel{2} \cdot 3 \cdot 3 \cdot \cancel{z}}{-2 \cdot 2 \cdot 2 \cdot \cancel{x} \cdot \cancel{x} \cdot y} \right)$

$= -\dfrac{9x^4}{8y^5}$

39. $\sqrt{\dfrac{64}{81}} = \dfrac{\sqrt{64}}{\sqrt{81}} = \dfrac{8}{9}$ 41. $\sqrt{\dfrac{121}{36}} = \dfrac{\sqrt{121}}{\sqrt{36}} = \dfrac{11}{6}$

43. $\sqrt{\dfrac{180}{5}} = \sqrt{36} = 6$ 45. $\sqrt{\dfrac{1053}{13}} = \sqrt{81} = 9$

47. $\dfrac{3}{4}x = 12$ Check: $\dfrac{3}{4} \cdot 16 = 12$

$\dfrac{\cancel{4}}{\cancel{3}} \cdot \dfrac{\cancel{3}}{\cancel{4}} x = 12 \cdot \dfrac{4}{3}$ $\dfrac{3}{4} \cdot \dfrac{16}{1} = 12$

$1x = \dfrac{2 \cdot 2 \cdot \cancel{3}}{1} \cdot \dfrac{2 \cdot 2}{\cancel{3}}$ $\dfrac{3}{\cancel{4}} \cdot \dfrac{\cancel{4} \cdot 4}{1} = 12$

$x = \dfrac{16}{1}$ $\dfrac{12}{1} = 12$

$x = 16$ $12 = 12$

49. $\dfrac{5}{8} = \dfrac{-3}{16}y$ Check: $\dfrac{5}{8} = -\dfrac{3}{16} \cdot -3\dfrac{1}{3}$

$\dfrac{-16}{3} \cdot \dfrac{5}{8} = \dfrac{-\cancel{16}}{\cancel{3}} \cdot \dfrac{-\cancel{3}}{\cancel{16}}y$ $\dfrac{5}{8} = \dfrac{3}{16} \cdot \dfrac{10}{3}$

$\dfrac{-2 \cdot \cancel{8}}{3} \cdot \dfrac{5}{\cancel{8}} = y$ $\dfrac{5}{8} = \dfrac{\cancel{3}}{\cancel{2} \cdot 8} \cdot \dfrac{\cancel{2} \cdot 5}{\cancel{3}}$

$-\dfrac{10}{3} = y$ $\dfrac{5}{8} = \dfrac{5}{8}$

$-3\dfrac{1}{3} = y$

51. $\dfrac{n}{12} = -\dfrac{3}{20}$ Check: $\dfrac{-1\dfrac{4}{5}}{12} = -\dfrac{3}{20}$

$\dfrac{\cancel{12}}{1} \cdot \dfrac{n}{\cancel{12}} = -\dfrac{3}{20} \cdot \dfrac{12}{1}$ $\dfrac{-\dfrac{9}{5}}{12} = -\dfrac{3}{20}$

$1n = -\dfrac{3}{\cancel{4} \cdot 5} \cdot \dfrac{3 \cdot \cancel{4}}{1}$ $-\dfrac{9}{5} \div \dfrac{12}{1} = -\dfrac{3}{20}$

$n = -\dfrac{9}{5}$ $-\dfrac{9}{5} \cdot \dfrac{1}{12} = -\dfrac{3}{20}$

$n = -1\dfrac{4}{5}$ $-\dfrac{3}{20} = -\dfrac{3}{20}$

53. $\dfrac{-5m}{6} = \dfrac{-5}{21}$

$\dfrac{-6}{5} \cdot \dfrac{-5m}{6} = \dfrac{-5}{21} \cdot \left(\dfrac{-6}{5} \right)$

$1m = \dfrac{-\cancel{5}}{\cancel{5} \cdot 7} \cdot \left(\dfrac{-2 \cdot \cancel{3}}{\cancel{3}} \right)$

$m = \dfrac{2}{7}$

Check: $\dfrac{-5 \cdot \dfrac{2}{7}}{6} = \dfrac{-5}{21}$

$-\dfrac{10}{7} \div 6 = \dfrac{-5}{21}$

$-\dfrac{10}{7} \div \dfrac{6}{1} = \dfrac{-5}{21}$

$-\dfrac{\cancel{2} \cdot 5}{7} \cdot \dfrac{1}{\cancel{2} \cdot 3} = \dfrac{-5}{21}$

$\dfrac{-5}{21} = \dfrac{-5}{21}$

55. $1\dfrac{1}{3} \cdot n = 40$

$\dfrac{4}{3}n = 40$

$\dfrac{\cancel{3}}{\cancel{4}} \cdot \dfrac{\cancel{4}}{\cancel{3}} m = 40 \cdot \dfrac{3}{4}$

$1m = \dfrac{\cancel{4} \cdot 10}{1} \cdot \dfrac{3}{\cancel{4}}$

$m = \dfrac{30}{1}$

$m = 30$

No, there are 30 doses. Twice a day for 14 days would be $2 \times 14 = 28$ doses, so there are 2 extra doses.

57.
$$12n = 10\frac{1}{2}$$
$$\frac{1}{\cancel{12}} \cdot \cancel{12}n = 10\frac{1}{2} \cdot \frac{1}{12}$$
$$1n = \frac{21}{2} \cdot \frac{1}{12}$$
$$n = \frac{\cancel{3} \cdot 7}{2} \cdot \frac{1}{\cancel{3} \cdot 4}$$
$$n = \frac{7}{8}$$

The baker must put $\frac{7}{8}$ cup in each of the baking cups.

59. We must use $A = lw$.
$$2805\frac{1}{2} = l \cdot 30\frac{1}{2}$$
$$\frac{5611}{2} = \frac{61}{2}l$$
$$\frac{\cancel{2}}{61} \cdot \frac{5611}{\cancel{2}} = \frac{\cancel{61}}{\cancel{2}}l \cdot \frac{\cancel{2}}{\cancel{61}}$$
$$\frac{5611}{61} = 1l$$
$$91\frac{60}{61} = l$$

The length is $91\frac{60}{61}$ ft.

61.
$$A = \frac{1}{2}bh$$
$$14\frac{1}{2} = \frac{1}{2} \cdot 6\frac{1}{2} \cdot h$$
$$\frac{29}{2} = \frac{1}{2} \cdot \frac{13}{2} \cdot h$$
$$\frac{29}{2} = \frac{13}{4}h$$
$$\frac{\cancel{2} \cdot 2}{13} \cdot \frac{29}{\cancel{2}} = \frac{\cancel{13}}{\cancel{4}}h \cdot \frac{\cancel{4}}{\cancel{13}}$$
$$\frac{58}{13} = 1h$$
$$4\frac{6}{13}\text{ ft.} = h$$

The height must be $4\frac{6}{13}$ ft.

63. We must use $C = \pi d$.
$$3111\frac{2}{5} = \frac{22}{7}d$$
$$\frac{7}{22} \cdot 3111\frac{2}{5} = \frac{\cancel{7}}{\cancel{22}} \cdot \frac{\cancel{22}}{\cancel{7}}d$$
$$\frac{7}{22} \cdot \frac{15{,}557}{5} = 1d$$
$$\frac{108{,}899}{110} = d$$
$$989\frac{109}{110} = d$$

The diameter is $989\frac{109}{110}$ ft.

65. We must use $d = rt$.
$$18\frac{3}{10} = r \cdot \frac{2}{5}$$
$$\frac{5}{2} \cdot \frac{183}{10} = r \cdot \frac{\cancel{2}}{\cancel{5}} \cdot \frac{\cancel{5}}{\cancel{2}}$$
$$\frac{\cancel{5}}{2} \cdot \frac{183}{2 \cdot \cancel{5}} = r$$
$$\frac{183}{4} = r$$
$$45\frac{3}{4} = r$$

His average rate is $45\frac{3}{4}$ mph.

Review Exercises

1.
$$5t^3 - 8t + 9t^2 + 4t^3 - 16 - 5t$$
$$= 5t^3 + 4t^3 + 9t^2 - 8t - 5t - 16$$
$$= 9t^3 + 9t^2 - 13t - 16$$

2.
$$2^2 \cdot 3 \cdot 5 \cdot x^3 \cdot y = 4 \cdot 3 \cdot 5 \cdot x^3 \cdot y$$
$$= 60x^3y$$

3. $2 \cdot 3^3 \cdot 7$

4.
$$32x^2 + 16x = 16x\left(\frac{32x^2 + 16x}{16x}\right)$$
$$= 16x\left(\frac{32x^2}{16x} + \frac{16x}{16x}\right)$$
$$= 16x(2x + 1)$$

5. We can get common denominators for the fractions.
$$\frac{5}{6} = \frac{5 \cdot 2}{6 \cdot 2} = \frac{10}{12}$$

Since both numerators and denominators are the same the fractions must be equal;

so: $\dfrac{5}{6} = \dfrac{10}{12}$

Exercise Set 5.5

1. The LCM of a set of numbers is the smallest natural number that is evenly divisible by all the given numbers.

3. 1. Find the prime factorization of each given number.
   2. Write a factorization containing each prime factor raised to its greatest exponent in the factorizations.
   3. Multiply to get the LCM.

5. 4

7. $10:10,20,30$
   $6:6,12,18,24,30$
   $LCM = 30$

9. $12:12,24,36$
   $36:36$
   $LCM = 36$

11. $20:20,40,60$
    $30:30,60$
    $LCM = 60$

13. $6:6,12,18,24,30,36,42,48,54,60,66,72,$
    $78,84,90$
    $9:9,18,27,36,45,54,63,72,81,90$
    $15:15,30,45,60,75,90$
    $LCM = 90$

15. $\quad 18 = 2 \cdot 3^2$
    $\quad 24 = 2^3 \cdot 3$
    $LCM(18,24) = 2^3 \cdot 3^2 = 8 \cdot 9 = 72$

17. $\quad 63 = 3^2 \cdot 7$
    $\quad 28 = 2^2 \cdot 7$
    $LCM(63,28) = 2^2 \cdot 3^2 \cdot 7 = 4 \cdot 9 \cdot 7 = 252$

19. $\quad 52 = 2^2 \cdot 13$
    $\quad 28 = 2^2 \cdot 7$
    $LCM(28,52) = 2^2 \cdot 7 \cdot 13 = 4 \cdot 7 \cdot 13 = 364$

21. $\quad 180 = 2^2 \cdot 3^2 \cdot 5$
    $\quad 200 = 2^3 \cdot 5^2$
    $LCM(180,200) = 2^3 \cdot 3^2 \cdot 5^2 = 8 \cdot 9 \cdot 25 = 1800$

23. $\quad 26 = 2 \cdot 13$
    $\quad 30 = 2 \cdot 3 \cdot 5$
    $\quad 39 = 3 \cdot 13$
    $LCM(26,30,36) = 2 \cdot 3 \cdot 5 \cdot 13 = 390$

25. $\quad 28 = 2^2 \cdot 7$
    $\quad 32 = 2^5$
    $\quad 60 = 2^2 \cdot 3 \cdot 5$
    $LCM(28,32,60) = 2^5 \cdot 3 \cdot 5 \cdot 7 = 32 \cdot 3 \cdot 5 \cdot 7 = 3360$

27. $\quad 12x = 2^2 \cdot 3 \cdot x$
    $\quad 8y = 2^3 \cdot y$
    $LCM(12x,8y) = 2^3 \cdot 3 \cdot x \cdot y = 8 \cdot 3xy = 24xy$

29. $\quad 10y^3 = 2 \cdot 5 \cdot y^3$
    $\quad 6y = 2 \cdot 3 \cdot y$
    $LCM(10y^3,6y) = 2 \cdot 3 \cdot 5 \cdot y^3 = 30y^3$

31. $\quad 16mn = 2^4 \cdot m \cdot n$
    $\quad 8m = 2^3 \cdot m$
    $LCM(16mn,8m) = 2^4 \cdot m \cdot n = 16mn$

33. $\quad 18x^2y = 2 \cdot 3^2 \cdot x^2 \cdot y$
    $\quad 12xy^3 = 2^2 \cdot 3 \cdot x \cdot y^3$
    $LCM(18x^2y,12xy^3) = 2^2 \cdot 3^2 \cdot x^2 \cdot y^3$
    $\qquad\qquad\qquad = 4 \cdot 9x^2y^3$
    $\qquad\qquad\qquad = 36x^2y^3$

35. $LCM(10,6) = 30$
    $\dfrac{3 \cdot 3}{10 \cdot 3} = \dfrac{9}{30}$
    $\dfrac{5 \cdot 5}{6 \cdot 5} = \dfrac{25}{30}$

37. $LCM(12,36) = 36$
    $\dfrac{7 \cdot 3}{12 \cdot 3} = \dfrac{21}{36}$
    $\dfrac{11 \cdot 1}{36 \cdot 1} = \dfrac{11}{36}$

39. $LCM(20,30) = 60$
    $\dfrac{1 \cdot 3}{20 \cdot 3} = \dfrac{3}{60}$
    $\dfrac{17 \cdot 2}{30 \cdot 2} = \dfrac{34}{60}$

41. $LCM(4,6,9) = 36$
    $\dfrac{3 \cdot 9}{4 \cdot 9} = \dfrac{27}{36}$
    $\dfrac{1 \cdot 6}{6 \cdot 6} = \dfrac{6}{36}$
    $\dfrac{7 \cdot 4}{9 \cdot 4} = \dfrac{28}{36}$

43. $LCM(12x,4) = 12x$
    $\dfrac{7 \cdot 1}{12x \cdot 1} = \dfrac{7}{12x}$
    $\dfrac{3 \cdot 3x}{4 \cdot 3x} = \dfrac{9x}{12x}$

45. $LCM(16mn,8m) = 16mn$
    $\dfrac{9 \cdot 1}{16mn \cdot 1} = \dfrac{9}{16mn}$
    $\dfrac{3n \cdot 2n}{8m \cdot 2n} = \dfrac{6n^2}{16mn}$

47. $LCM(10y^3z,6y) = 30y^3z$
    $\dfrac{7 \cdot 3}{10y^3z \cdot 3} = \dfrac{21}{30y^3z}$
    $\dfrac{5z \cdot 5y^2z}{6y \cdot 5y^2z} = \dfrac{25y^2z^2}{30y^3z}$

49. $\text{LCM}(18x^2y, 12xy^3) = 36x^2y^3$

$$\frac{z \cdot 2y^2}{18x^2y \cdot 2y^2} = \frac{2y^2z}{36x^2y^3}$$

$$\frac{-5 \cdot 3x}{12xy^3 \cdot 3x} = \frac{-15x}{36x^2y^3}$$

Review Exercises

1. $16 + (-28) = -12$ 2. $-8 - (-19) = -8 + 19$
                                                    $= 11$

3. $14x^2 - 16x + 25x^2 - 12 + 4x^3 + 9$
    $= 4x^3 + 14x^2 + 25x^2 - 16x - 12 + 9$
    $= 4x^3 + 39x^2 - 16x - 3$

4. $m + 14 = 6$ Check: $-8 + 14 = 6$
    $\underline{-14 \quad -14}$ $6 = 6$
    $m + 0 = -8$
    $m = -8$

5. $x - 37 = 16$ Check: $53 - 37 = 16$
    $\underline{+37 \quad +37}$ $16 = 16$
    $x + 0 = 53$
    $x = 53$

6. $-3a = \dfrac{12}{5}$ Check: $-3\left(-\dfrac{4}{5}\right) = \dfrac{12}{5}$

$$\frac{-3a}{-3} = \frac{\frac{12}{5}}{-3} \qquad\qquad -\frac{3}{1} \cdot \left(-\frac{4}{5}\right) = \frac{12}{5}$$

$$1a = \frac{12}{5} \div (-3) \qquad\qquad \frac{3 \cdot 4}{1 \cdot 5} = \frac{12}{5}$$

$$a = \frac{12}{5} \cdot \left(-\frac{1}{3}\right) \qquad\qquad \frac{12}{5} = \frac{12}{5}$$

$$a = \frac{2 \cdot 2 \cdot \cancel{3}}{5} \cdot \left(-\frac{1}{\cancel{3}}\right)$$

$$a = -\frac{4}{5}$$

Exercise Set 5.6

1. 1. Add or subtract numerators and keep the same denominators.
    2. Simplify.

3. Multiply the numerator and denominator of $\dfrac{3}{4}$ by 3 and multiply the numerator and denominator of $\dfrac{5}{6}$ by 2.

5. $\dfrac{2}{7} + \dfrac{3}{7} = \dfrac{2+3}{7} = \dfrac{5}{7}$

7. $-\dfrac{4}{9} + \left(-\dfrac{2}{9}\right) = \dfrac{-4 + (-2)}{9} = \dfrac{-6}{9} = \dfrac{-2}{3}$

9. $\dfrac{10}{17} - \dfrac{4}{17} = \dfrac{10-4}{17} = \dfrac{6}{17}$

11. $\dfrac{6}{35} - \dfrac{13}{35} = \dfrac{6-13}{35}$

$$= \frac{-7}{35}$$

$$= -\frac{1}{5}$$

13. $\dfrac{9}{x} + \dfrac{3}{x} = \dfrac{9+3}{x} = \dfrac{12}{x}$

15. $\dfrac{8x^2}{9} - \dfrac{2x^2}{9} = \dfrac{8x^2 - 2x^2}{9}$

$$= \frac{6x^2}{9}$$

$$= \frac{2x^2}{3}$$

17. $\dfrac{4m}{n} - \dfrac{6m}{n} = \dfrac{4m - 6m}{n} = -\dfrac{2m}{n}$

19. $\dfrac{3x^2 + 4x}{7y} + \dfrac{x^2 - 7x + 1}{7y}$

$$= \frac{\left(3x^2 + 4x\right) + \left(x^2 - 7x + 1\right)}{7y}$$

$$= \frac{3x^2 + x^2 + 4x - 7x + 1}{7y}$$

$$= \frac{4x^2 - 3x + 1}{7y}$$

21. $\dfrac{7n^2 + 12}{5m} - \dfrac{2n^2 + 3}{5m} = \dfrac{\left(7n^2 + 12\right) + \left(-2n^2 - 3\right)}{5m}$

$$= \frac{7n^2 - 2n^2 + 12 - 3}{5m}$$

$$= \frac{5n^2 + 9}{5m}$$

23. $\dfrac{3}{10} + \dfrac{5}{6} = \dfrac{3 \cdot 3}{10 \cdot 3} + \dfrac{5 \cdot 5}{6 \cdot 5}$

$$= \frac{9}{30} + \frac{25}{30}$$

$$= \frac{34}{30}$$

$$= 1\frac{4}{30}$$

$$= 1\frac{2}{15}$$

25. $\dfrac{7}{12} - \dfrac{11}{36} = \dfrac{7 \cdot 3}{12 \cdot 3} - \dfrac{11}{36}$

$= \dfrac{21}{36} - \dfrac{11}{36}$

$= \dfrac{10}{36}$

$= \dfrac{5}{18}$

27. $\dfrac{1}{20} + \dfrac{17}{30} = \dfrac{1 \cdot 3}{20 \cdot 3} + \dfrac{17 \cdot 2}{30 \cdot 2}$

$= \dfrac{3}{60} + \dfrac{34}{60}$

$= \dfrac{37}{60}$

29. $\dfrac{3}{4} + \dfrac{1}{6} + \dfrac{7}{9} = \dfrac{3 \cdot 9}{4 \cdot 9} + \dfrac{1 \cdot 6}{6 \cdot 6} + \dfrac{7 \cdot 4}{9 \cdot 4}$

$= \dfrac{27}{36} + \dfrac{6}{36} + \dfrac{28}{36}$

$= \dfrac{61}{36}$

$= 1\dfrac{25}{36}$

31. $\dfrac{7x}{12} + \dfrac{3x}{8} = \dfrac{7x \cdot 2}{12 \cdot 2} + \dfrac{3x \cdot 3}{8 \cdot 3}$

$= \dfrac{14x}{24} + \dfrac{9x}{24}$

$= \dfrac{23x}{24}$

33. $\dfrac{9}{16m} - \dfrac{3}{8m} = \dfrac{9}{16m} - \dfrac{3 \cdot 2}{8m \cdot 2}$

$= \dfrac{9}{16m} - \dfrac{6}{16m}$

$= \dfrac{3}{16m}$

35. $\dfrac{13}{20} + \dfrac{4}{5h} = \dfrac{13 \cdot h}{20 \cdot h} + \dfrac{4 \cdot 4}{5h \cdot 4}$

$= \dfrac{13h}{20h} + \dfrac{16}{20h}$

$= \dfrac{13h + 16}{20h}$

37. $\dfrac{2}{3n^2} - \dfrac{7}{9n} = \dfrac{2 \cdot 3}{3n^2 \cdot 3} - \dfrac{7 \cdot n}{9n \cdot n}$

$= \dfrac{6}{9n^2} - \dfrac{7n}{9n^2}$

$= \dfrac{6 - 7n}{9n^2}$

39. $3\dfrac{4}{9} + 7 = 3 + \dfrac{4}{9} + 7$

$= 10 + \dfrac{4}{9}$

$= 10\dfrac{4}{9}$

41. $2\dfrac{1}{4} + 5\dfrac{1}{4} = 2 + \dfrac{1}{4} + 5 + \dfrac{1}{4}$

$= 2 + 5 + \dfrac{1}{4} + \dfrac{1}{4}$

$= 7 + \dfrac{2}{4}$

$= 7\dfrac{2}{4} = 7\dfrac{1}{2}$

43. $5\dfrac{5}{6} + 1\dfrac{2}{3} = 5\dfrac{5}{6} + 1\dfrac{2 \cdot 2}{3 \cdot 2}$

$= 5\dfrac{5}{6} + 1\dfrac{4}{6}$

$= 6\dfrac{9}{6}$

$= 6 + 1\dfrac{3}{6}$

$= 7\dfrac{3}{6} = 7\dfrac{1}{2}$

45. $6\dfrac{5}{8} + 3\dfrac{7}{12} = 6\dfrac{5 \cdot 3}{8 \cdot 3} + 3\dfrac{7 \cdot 2}{12 \cdot 2}$

$= 6\dfrac{15}{24} + 3\dfrac{14}{24}$

$= 9\dfrac{29}{24}$

$= 9 + 1\dfrac{5}{24}$

$= 10\dfrac{5}{24}$

47. $9\dfrac{7}{8} - \dfrac{1}{8} = 9\dfrac{6}{8}$ 49. $11\dfrac{7}{12} - 2\dfrac{5}{12} = 9\dfrac{2}{12}$

$= 9\dfrac{3}{4}$ $= 9\dfrac{1}{6}$

51. $5\dfrac{1}{8} - 1\dfrac{5}{6} = 5\dfrac{1 \cdot 3}{8 \cdot 3} - 1\dfrac{5 \cdot 4}{6 \cdot 4}$

$= 5\dfrac{3}{24} - 1\dfrac{20}{24}$

$= 4\dfrac{27}{24} - 1\dfrac{20}{24}$

$= 3\dfrac{7}{24}$

53.

$$8\frac{1}{3} - 7\frac{5}{6} = 8\frac{1\cdot 2}{3\cdot 2} - 7\frac{5}{6}$$

$$= 8\frac{2}{6} - 7\frac{5}{6}$$

$$= 7\frac{8}{6} - 7\frac{5}{6}$$

$$= \frac{3}{6} = \frac{1}{2}$$

55.

$$6\frac{3}{4} + \left(-2\frac{1}{3}\right) = \frac{27}{4} - \frac{7}{3}$$

$$= \frac{27\cdot 3}{4\cdot 3} - \frac{7\cdot 4}{3\cdot 4}$$

$$= \frac{81}{12} - \frac{28}{12}$$

$$= \frac{53}{12}$$

$$= 4\frac{5}{12}$$

57.

$$-7\frac{5}{6} - 2\frac{2}{3} = -\frac{47}{6} - \frac{8}{3}$$

$$= -\frac{47}{6} - \frac{8\cdot 2}{3\cdot 2}$$

$$= -\frac{47}{6} - \frac{16}{6}$$

$$= -\frac{63}{6}$$

$$= -10\frac{3}{6} = -10\frac{1}{2}$$

59.

$$\frac{5}{8} - 4\frac{1}{4} = \frac{5}{8} - \frac{17}{4}$$

$$= \frac{5}{8} - \frac{17\cdot 2}{4\cdot 2}$$

$$= \frac{5}{8} - \frac{34}{8}$$

$$= -\frac{29}{8} = -3\frac{5}{8}$$

61.

$$-7\frac{3}{4} - \left(-1\frac{1}{8}\right) = -\frac{31}{4} + \frac{9}{8}$$

$$= -\frac{31\cdot 2}{4\cdot 2} + \frac{9}{8}$$

$$= -\frac{62}{8} + \frac{9}{8}$$

$$= -\frac{53}{8}$$

$$= -6\frac{5}{8}$$

63.

$$x + \frac{3}{5} = \frac{7}{10}$$

$$x + \frac{3}{5} - \frac{3}{5} = \frac{7}{10} - \frac{3}{5}$$

$$x + 0 = \frac{7}{10} - \frac{3\cdot 2}{5\cdot 2}$$

$$x = \frac{7}{10} - \frac{6}{10}$$

$$x = \frac{1}{10}$$

Check:

$$\frac{1}{10} + \frac{3}{5} = \frac{7}{10}$$

$$\frac{1}{10} + \frac{3\cdot 2}{5\cdot 2} = \frac{7}{10}$$

$$\frac{1}{10} + \frac{6}{10} = \frac{7}{10}$$

$$\frac{7}{10} = \frac{7}{10}$$

65.

$$n - \frac{4}{5} = \frac{1}{4}$$

$$n - \frac{4}{5} + \frac{4}{5} = \frac{1}{4} + \frac{4}{5}$$

$$n + 0 = \frac{1\cdot 5}{4\cdot 5} + \frac{4\cdot 4}{5\cdot 4}$$

$$n = \frac{5}{20} + \frac{16}{20}$$

$$n = \frac{21}{20}$$

$$n = 1\frac{1}{20}$$

Check:

$$\frac{21}{20} - \frac{4}{5} = \frac{1}{4}$$

$$\frac{21}{20} - \frac{4\cdot 4}{5\cdot 4} = \frac{1}{4}$$

$$\frac{21}{20} - \frac{16}{20} = \frac{1}{4}$$

$$\frac{5}{20} = \frac{1}{4}$$

$$\frac{1}{4} = \frac{1}{4}$$

67.

$$-\frac{1}{6} = b + \frac{3}{4}$$

$$-\frac{1}{6} - \frac{3}{4} = b + \frac{3}{4} - \frac{3}{4}$$

$$-\frac{1\cdot 2}{6\cdot 2} - \frac{3\cdot 3}{4\cdot 3} = b + 0$$

$$-\frac{2}{12} - \frac{9}{12} = b$$

$$-\frac{11}{12} = b$$

Check:

$$-\frac{1}{6} = -\frac{11}{12} + \frac{3}{4}$$

$$-\frac{1}{6} = -\frac{11}{12} + \frac{3\cdot 3}{4\cdot 3}$$

$$-\frac{1}{6} = -\frac{11}{12} + \frac{9}{12}$$

$$-\frac{1}{6} = -\frac{2}{12}$$

$$-\frac{1}{6} = -\frac{1}{6}$$

69.

$$-3\frac{1}{2} + t = -4\frac{1}{5}$$

$$-3\frac{1}{2} + 3\frac{1}{2} + t = -4\frac{1}{5} + 3\frac{1}{2}$$

$$0 + t = -\frac{21}{5} + \frac{7}{2}$$

$$t = -\frac{21 \cdot 2}{5 \cdot 2} + \frac{7 \cdot 5}{2 \cdot 5}$$

$$t = -\frac{42}{10} + \frac{35}{10}$$

$$t = -\frac{7}{10}$$

Check:

$$-\frac{7}{2} - \frac{7}{10} = -\frac{21}{5}$$

$$-\frac{7 \cdot 5}{2 \cdot 5} - \frac{7}{10} = -\frac{21}{5}$$

$$-\frac{35}{10} - \frac{7}{10} = -\frac{21}{5}$$

$$\frac{-42}{10} = -\frac{21}{5}$$

$$-\frac{21}{5} = -\frac{21}{5}$$

71. $\dfrac{1}{4} + \dfrac{5}{8} = \dfrac{1 \cdot 2}{4 \cdot 2} + \dfrac{5}{8} = \dfrac{2}{8} + \dfrac{5}{8} = \dfrac{7}{8}$

The total thickness will be $\dfrac{7}{8}$ in.

73. $\dfrac{3}{4} - \dfrac{1}{16} = \dfrac{3 \cdot 4}{4 \cdot 4} - \dfrac{1}{16} = \dfrac{12}{16} - \dfrac{1}{16} = \dfrac{11}{16}$

The tabletop is $\dfrac{11}{16}$ in.

75. a) $\dfrac{1}{3} + \dfrac{3}{8} = \dfrac{1 \cdot 8}{3 \cdot 8} + \dfrac{3 \cdot 3}{8 \cdot 3} = \dfrac{8}{24} + \dfrac{9}{24} = \dfrac{17}{24}$

$\dfrac{17}{24}$ of the students will earn an A or B.

b) Set up an equation where n is the fraction of students who were unsuccessful in the course.

$$\frac{1}{3} + \frac{3}{8} + \frac{1}{6} + n = 1$$

$$\frac{1 \cdot 8}{3 \cdot 8} + \frac{3 \cdot 3}{8 \cdot 3} + \frac{1 \cdot 4}{6 \cdot 4} + n = \frac{1 \cdot 24}{1 \cdot 24}$$

$$\frac{8}{24} + \frac{9}{24} + \frac{4}{24} + n = \frac{24}{24}$$

$$\frac{21}{24} + n = \frac{24}{24}$$

$$\frac{21}{24} - \frac{21}{24} + n = \frac{24}{24} - \frac{21}{24}$$

$$n = \frac{3}{24} = \frac{1}{8}$$

$\dfrac{1}{8}$ of the students were unsuccessful.

77.

$$P = 20\frac{3}{4} + 13\frac{1}{2} + 20\frac{3}{4} + 13\frac{1}{2}$$

$$P = 20\frac{3}{4} + 13\frac{1 \cdot 2}{2 \cdot 2} + 20\frac{3}{4} + 13\frac{1 \cdot 2}{2 \cdot 2}$$

$$P = 20\frac{3}{4} + 13\frac{2}{4} + 20\frac{3}{4} + 13\frac{2}{4}$$

$$P = 66\frac{10}{4}$$

$$P = 66 + 2\frac{2}{4}$$

$$P = 68\frac{2}{4}$$

$$P = 68\frac{1}{2} \text{ in.}$$

The total length of wood needed is $68\dfrac{1}{2}$ in.

79. We must first add the sides we know together.

$$12\frac{1}{2} + 15\frac{1}{4} = 12\frac{1 \cdot 2}{2 \cdot 2} + 15\frac{1}{4}$$

$$= 12\frac{2}{4} + 15\frac{1}{4}$$

$$= 27\frac{3}{4}$$

Now we need to subtract this from the opposite side length given. $31\dfrac{3}{4} - 27\dfrac{3}{4} = 4$

The hallway is 4 ft.

Review Exercises

1. $3^2 - 5[2 + (18 \div (-2))] + \sqrt{49}$

$= 3^2 - 5[2 + (-9)] + \sqrt{49}$

$= 3^2 - 5(-7) + 7$

$= 9 - 5(-7) + 7$

$= 9 - (-35) + 7$

$= 9 + 35 + 7$

$= 51$

2. $x^2 - 2x + \sqrt{x} = 4^2 - 2 \cdot 4 + \sqrt{4}$

$= 16 - 2 \cdot 4 + 2$

$= 16 - 8 + 2$

$= 8 + 2$

$= 10$

3. $y^2 + 2y + 3 = (-2)^2 + 2(-2) + 3$
$$= 4 + 2(-2) + 3$$
$$= 4 - 4 + 3$$
$$= 0 + 3$$
$$= 3$$

4. $(4x^2 + 5x - 6) + (6x^2 - 2x + 1)$
$$= 10x^2 + 3x - 5$$

5. $(8y^3 - 4y + 2) - (9y^3 + y^2 + 7)$
$$= (8y^3 - 4y + 2) + (-9y^3 - y^2 - 7)$$
$$= -y^3 - y^2 - 4y - 5$$

6. $(-6b^3)(9b) = -6 \cdot 9 \cdot b^{3+1} = -54b^4$

7. $(x+3)(2x-5) = x \cdot 2x - 5 \cdot x + 3 \cdot 2x + 3(-5)$
$$= 2x^2 - 5x + 6x - 15$$
$$= 2x^2 + x - 15$$

8. $\dfrac{-16x^7}{4x^5} = \dfrac{-16}{4} x^{7-5} = -4x^2$

9. $V = lwh$
$$V = y \cdot 3y \cdot (y+1)$$
$$= 3y^{1+1}(y+1)$$
$$= 3y^2(y+1)$$
$$= 3y^2 \cdot y + 3y^2 \cdot 1$$
$$= 3y^{2+1} + 3y^2$$
$$= 3y^3 + 3y^2$$

Exercise Set 5.7

1. A trapezoid is a four-sided figure with one pair of parallel sides.

3. In the formula for the area of a trapezoid, a and b represent the lengths of the parallel sides.

5. $\dfrac{1}{4} + 2 \cdot \dfrac{5}{8} = \dfrac{1}{4} + \dfrac{2}{1} \cdot \dfrac{5}{8}$
$$= \dfrac{1}{4} + \dfrac{10}{8}$$
$$= \dfrac{1}{4} + \dfrac{5}{4}$$
$$= \dfrac{6}{4}$$
$$= \dfrac{3}{2}$$
$$= 1\dfrac{1}{2}$$

7. $3\dfrac{1}{5} - \dfrac{1}{2} \div \dfrac{5}{6} = 3\dfrac{1}{5} - \dfrac{1}{\cancel{2}_1} \cdot \dfrac{\cancel{6}^3}{5}$
$$= 3\dfrac{1}{5} - \dfrac{3}{5}$$
$$= 2\dfrac{6}{5} - \dfrac{3}{5}$$
$$= 2\dfrac{3}{5}$$

9. $2\dfrac{1}{2} - \left(\dfrac{3}{4}\right)^2 = \dfrac{5}{2} - \dfrac{9}{16}$
$$= \dfrac{5 \cdot 8}{2 \cdot 8} - \dfrac{9}{16}$$
$$= \dfrac{40}{16} - \dfrac{9}{16}$$
$$= \dfrac{31}{16}$$
$$= 1\dfrac{15}{16}$$

11. $\dfrac{1}{4} - \dfrac{2}{3}\left(6 - 1\dfrac{1}{2}\right) = \dfrac{1}{4} - \dfrac{2}{3}\left(5\dfrac{2}{2} - 1\dfrac{1}{2}\right)$
$$= \dfrac{1}{4} - \dfrac{2}{3}\left(4\dfrac{1}{2}\right)$$
$$= \dfrac{1}{4} - \dfrac{\cancel{2}}{\cancel{3}} \cdot \dfrac{\cancel{9}^3}{\cancel{2}}$$
$$= \dfrac{1}{4} - \dfrac{3}{1}$$
$$= \dfrac{1}{4} - \dfrac{3 \cdot 4}{1 \cdot 4}$$
$$= \dfrac{1}{4} - \dfrac{12}{4}$$
$$= -\dfrac{11}{4}$$
$$= -2\dfrac{3}{4}$$

13. $5\dfrac{3}{4} - 2\sqrt{\dfrac{80}{5}} = 5\dfrac{3}{4} - 2\sqrt{16}$
$$= 5\dfrac{3}{4} - 2 \cdot 4$$
$$= \dfrac{23}{4} - \dfrac{8}{1}$$
$$= \dfrac{23}{4} - \dfrac{8 \cdot 4}{1 \cdot 4}$$
$$= \dfrac{23}{4} - \dfrac{32}{4}$$
$$= -\dfrac{9}{4}$$
$$= -2\dfrac{1}{4}$$

15. $\left(\dfrac{1}{2}\right)^3 + 2\dfrac{1}{4} - 5\left(\dfrac{3}{10} + \dfrac{1}{5}\right)$

$= \left(\dfrac{1}{2}\right)^3 + 2\dfrac{1}{4} - 5\left(\dfrac{3}{10} + \dfrac{2}{10}\right)$

$= \left(\dfrac{1}{2}\right)^3 + 2\dfrac{1}{4} - {}^1\!\!\not{5}\left(\dfrac{5}{\not{10}_2}\right)$

$= \dfrac{1}{8} + \dfrac{9}{4} - \dfrac{5}{2}$

$= \dfrac{1}{8} + \dfrac{9 \cdot 2}{4 \cdot 2} - \dfrac{5 \cdot 4}{2 \cdot 4}$

$= \dfrac{1}{8} + \dfrac{18}{8} - \dfrac{20}{8}$

$= -\dfrac{1}{8}$

17. $5\left(\dfrac{1}{2} - 3\dfrac{4}{5}\right) - 2\left(\dfrac{1}{6} + 4\right)$

$= 5\left(\dfrac{1}{2} - \dfrac{19}{5}\right) - 2\left(\dfrac{1}{6} + \dfrac{24}{6}\right)$

$= 5\left(\dfrac{1 \cdot 5}{2 \cdot 5} - \dfrac{19 \cdot 2}{5 \cdot 2}\right) - 2\left(\dfrac{1}{6} + \dfrac{24}{6}\right)$

$= 5\left(\dfrac{5}{10} - \dfrac{38}{10}\right) - {}^1\!\!\not{2} \cdot \dfrac{25}{\not{6}_3}$

$= \left({}^1\!\!\not{5} \cdot \dfrac{-33}{\not{10}_2}\right) - \dfrac{25}{3}$

$= \dfrac{-33}{2} - \dfrac{25}{3}$

$= \dfrac{-33 \cdot 3}{2 \cdot 3} - \dfrac{25 \cdot 2}{3 \cdot 2}$

$= \dfrac{-99}{6} - \dfrac{50}{6}$

$= -\dfrac{149}{6}$

$= -24\dfrac{5}{6}$

19. a) $\left(5\dfrac{1}{2} + 10\dfrac{1}{4} + 8\dfrac{3}{4} + 9\dfrac{3}{4} + 12 + 9\dfrac{1}{4}\right) \div 6$

$= \left(5\dfrac{2}{4} + 10\dfrac{1}{4} + 8\dfrac{3}{4} + 9\dfrac{3}{4} + 12 + 9\dfrac{1}{4}\right) \div 6$

$= 53\dfrac{10}{4} \div 6$

$= \dfrac{222}{4} \cdot \dfrac{1}{6}$

$= \dfrac{{}^{37}\!\not{111}}{2} \cdot \dfrac{1}{\not{6}_2}$

$= \dfrac{37}{4}$ or $9\dfrac{1}{4}$

The mean length is $9\dfrac{1}{4}$ in.

$\left(\dfrac{3}{4} + 2\dfrac{1}{4} + 1\dfrac{1}{2} + 2 + 3\dfrac{1}{4} + 1\dfrac{3}{4}\right) \div 6$

$= \left(\dfrac{3}{4} + 2\dfrac{1}{4} + 1\dfrac{2}{4} + 2 + 3\dfrac{1}{4} + 1\dfrac{3}{4}\right) \div 6$

$= 9\dfrac{10}{4} \cdot \dfrac{1}{6}$

$= \dfrac{46}{4} \cdot \dfrac{1}{6}$

$= \dfrac{23}{2} \cdot \dfrac{1}{6}$

$= \dfrac{23}{12}$ or $1\dfrac{11}{12}$

The mean weight is $1\dfrac{11}{12}$ lbs.

b) $5\dfrac{1}{2}, \ 8\dfrac{3}{4}, \ 9\dfrac{1}{4}, 9\dfrac{3}{4}, 10\dfrac{1}{4}, 12$

The median is the mean of the middle two numbers.

$\left(9\dfrac{1}{4} + 9\dfrac{3}{4}\right) \div 2$

$= 19 \div 2$

$= \dfrac{19}{2}$ or $9\dfrac{1}{2}$

The median length is $9\dfrac{1}{2}$ in.

$\dfrac{3}{4}, 1\dfrac{1}{2}, 1\dfrac{3}{4}, 2, 2\dfrac{1}{4}, 3\dfrac{1}{4}$

The median is the mean of the middle two numbers.

$\left(1\dfrac{3}{4} + 2\right) \div 2$

$= 3\dfrac{3}{4} \div 2$

$= \dfrac{15}{4} \cdot \dfrac{1}{2}$

$= \dfrac{15}{8}$ or $1\dfrac{7}{8}$

The median weight is $1\dfrac{7}{8}$ lbs.

21. a) Since there are 5 quarter miles, we multiply 5 times $\dfrac{1}{4}$ which is $5 \cdot \dfrac{1}{4} = \dfrac{5}{4} = 1\dfrac{1}{4}$ mi.

b) We need to sum the times for each quarter

mile. $25\frac{1}{5} + 24 + 23\frac{4}{5} + 23\frac{2}{5} + 23$

$$= 118\frac{7}{5}$$

$$= 119\frac{2}{5} \text{ sec.}$$

c) The average time is the same as the mean. Since we found the sum of the times in part b) we can take that answer and divide it by 5.

$$119\frac{2}{5} \div 5$$

$$= \frac{597}{5} \cdot \frac{1}{5}$$

$$= \frac{597}{25} \text{ or } 23\frac{22}{25} \text{ sec.}$$

d) $23, 23\frac{2}{5}, 23\frac{4}{5}, 24, 25\frac{1}{5}$

The median time is $23\frac{4}{5}$ sec.

23. $2\frac{1}{2} + 30\left(\frac{1}{4}\right)$

$$= \frac{5}{2} + \frac{30}{4}$$

$$= \frac{5}{2} + \frac{15}{2}$$

$$= \frac{20}{2}$$

$$= 10$$

25. $1\frac{1}{4}\left(\frac{3}{4}\right)^2 = \frac{5}{4} \cdot \frac{9}{16}$

$$= \frac{45}{64}$$

27. $\frac{1}{2} \cdot \left(-9\frac{4}{5}\right)\left(\frac{1}{2}\right)^2 = \frac{1}{2} \cdot \left(\frac{-49}{5}\right) \cdot \frac{1}{4}$

$$= \frac{-49}{40}$$

$$= -1\frac{9}{40}$$

29. $2 \cdot \frac{1}{6} \cdot \left(-1\frac{3}{4}\right) \cdot \left(-\frac{8}{7}\right)$

$$= \frac{\overset{1}{\cancel{2}}}{1} \cdot \frac{1}{\cancel{6}_3} \cdot \left(-\frac{\cancel{7}}{_1\cancel{4}}\right) \cdot \left(-\frac{\cancel{8}^2}{\cancel{7}}\right)$$

$$= \frac{2}{3}$$

31. $A = \frac{1}{2}h(a+b)$

$$A = \frac{1}{2} \cdot 5\frac{1}{2}(8+11)$$

$$A = \frac{1}{2} \cdot \frac{11}{2} \cdot 19$$

$$A = \frac{209}{4}$$

$$A = 52\frac{1}{4} \text{ m}^2$$

33. $A = \frac{1}{2} \cdot 3\frac{2}{3}\left(6 + 10\frac{1}{2}\right)$

$$A = \frac{1}{2} \cdot \frac{11}{3}\left(\frac{12}{2} + \frac{21}{2}\right)$$

$$A = \frac{1}{2} \cdot \frac{11}{\cancel{3}} \cdot \frac{\overset{11}{\cancel{33}}}{2}$$

$$A = \frac{121}{4}$$

$$A = 30\frac{1}{4} \text{ ft.}^2$$

35. $A = \pi r^2$

$$A = \frac{22}{7} \cdot 14^2$$

$$A = \frac{22}{7} \cdot 196$$

$$A = \frac{4312}{7}$$

$$A = 616 \text{ in.}^2$$

37. $r = \frac{1}{2}d$ so: $A = \pi r^2$

$r = \frac{1}{2} \cdot 2\frac{1}{2}$ $A = \frac{22}{7} \cdot \left(\frac{5}{4}\right)^2$

$r = \frac{1}{2} \cdot \frac{5}{2}$ $A = \frac{22}{7} \cdot \frac{25}{16}$

$r = \frac{5}{4}$ $A = \frac{550}{112}$

$\phantom{r = \frac{5}{4}}$ $A = 4\frac{51}{56} \text{ m}^2$

39. To find the area of the entire figure, we must find the area of the rectangle and trapezoid, then add for the total area.

$A_{\text{rectangle}} = bh$ $A_{\text{trapezoid}} = \frac{1}{2}h(a+b)$

$\phantom{A_{\text{rectangle}}} = 38 \cdot 3$ $\phantom{A_{\text{trapezoid}}} = \frac{1}{2} \cdot 4\frac{7}{10}(38+16)$

$\phantom{A_{\text{rectangle}}} = 114 \text{ m}^2$ $\phantom{A_{\text{trapezoid}}} = \frac{1}{2} \cdot \frac{47}{10} \cdot \frac{54}{1}$

$$= \frac{2538}{20}$$

$$= 126\frac{9}{10} \text{ m}^2$$

total area: $126\frac{9}{10} + 114 = 240\frac{9}{10} \text{ m}^2$

41. $\dfrac{1}{2}x^2 - \dfrac{3}{4}x - \dfrac{1}{3}x^2 - 6x$

$= \dfrac{1}{2}x^2 - \dfrac{1}{3}x^2 - \dfrac{3}{4}x - 6x$

$= \dfrac{1\cdot 3}{2\cdot 3}x^2 - \dfrac{1\cdot 2}{3\cdot 2}x^2 - \dfrac{3}{4}x - \dfrac{6\cdot 4}{1\cdot 4}x$

$= \dfrac{3}{6}x^2 - \dfrac{2}{6}x^2 - \dfrac{3}{4}x - \dfrac{24}{4}x$

$= \dfrac{1}{6}x^2 - \dfrac{27}{4}x$

43. $\left(5y^3 - \dfrac{4}{5}y^2 + y - \dfrac{1}{6}\right) + \left(y^3 + 3y^2 - \dfrac{2}{3}\right)$

$= 5y^3 + y^3 - \dfrac{4}{5}y^2 + 3y^2 + y - \dfrac{1}{6} - \dfrac{2}{3}$

$= 5y^3 + y^3 - \dfrac{4}{5}y^2 + \dfrac{15}{5}y^2 + y - \dfrac{1}{6} - \dfrac{2\cdot 2}{3\cdot 2}$

$= 6y^3 - \dfrac{4}{5}y^2 + \dfrac{15}{5}y^2 + y - \dfrac{1}{6} - \dfrac{4}{6}$

$= 6y^3 + \dfrac{11}{5}y^2 + y - \dfrac{5}{6}$

45. $\left(\dfrac{4}{5}t^3 + \dfrac{2}{3}t^2 - \dfrac{1}{6}\right) - \left(\dfrac{7}{10}t^3 + \dfrac{1}{4}t^2 + \dfrac{1}{2}\right)$

$= \left(\dfrac{4}{5}t^3 + \dfrac{2}{3}t^2 - \dfrac{1}{6}\right) + \left(-\dfrac{7}{10}t^3 - \dfrac{1}{4}t^2 - \dfrac{1}{2}\right)$

$= \dfrac{4}{5}t^3 - \dfrac{7}{10}t^3 + \dfrac{2}{3}t^2 - \dfrac{1}{4}t^2 - \dfrac{1}{6} - \dfrac{1}{2}$

$= \dfrac{4\cdot 2}{5\cdot 2}t^3 - \dfrac{7}{10}t^3 + \dfrac{2\cdot 4}{3\cdot 4}t^2 - \dfrac{1\cdot 3}{4\cdot 3}t^2 - \dfrac{1}{6} - \dfrac{1\cdot 3}{2\cdot 3}$

$= \dfrac{8}{10}t^3 - \dfrac{7}{10}t^3 + \dfrac{8}{12}t^2 - \dfrac{3}{12}t^2 - \dfrac{1}{6} - \dfrac{3}{6}$

$= \dfrac{1}{10}t^3 + \dfrac{5}{12}t^2 - \dfrac{4}{6}$

$= \dfrac{1}{10}t^3 + \dfrac{5}{12}t^2 - \dfrac{2}{3}$

47. $\left(-\dfrac{1}{6}m\right)\left(\dfrac{3}{5}m^3\right) = -\dfrac{1}{{}_2\cancel{6}}\cdot\dfrac{\cancel{3}^1}{5}m^{1+3}$

$\qquad = -\dfrac{1}{10}m^4$

49. $\dfrac{5}{8}\left(\dfrac{4}{5}t^2 - \dfrac{2}{3}t\right)$

$= \dfrac{{}^1\cancel{5}}{{}_2\cancel{8}}\cdot\dfrac{\cancel{4}^1}{\cancel{5}_1}t^2 + \dfrac{5}{{}_4\cancel{8}}\cdot\left(-\dfrac{\cancel{2}^1}{3}t\right)$

$= \dfrac{1}{2}t^2 - \dfrac{5}{12}t$

51. $\left(x - \dfrac{1}{2}\right)\left(x + \dfrac{1}{4}\right)$

$= x\cdot x + x\cdot\dfrac{1}{4} - \dfrac{1}{2}\cdot x - \dfrac{1}{2}\cdot\dfrac{1}{4}$

$= x^2 + \dfrac{1}{4}x - \dfrac{1}{2}x - \dfrac{1}{8}$

$= x^2 + \dfrac{1}{4}x - \dfrac{1\cdot 2}{2\cdot 2}x - \dfrac{1}{8}$

$= x^2 + \dfrac{1}{4}x - \dfrac{2}{4}x - \dfrac{1}{8}$

$= x^2 - \dfrac{1}{4}x - \dfrac{1}{8}$

Review Exercises

1. $4x - 9 = 11$ $\qquad$ Check: $4\cdot 5 - 9 = 11$
    $\underline{+9 \quad +9}$ $\qquad\qquad\qquad 20 - 9 = 11$
    $4x + 0 = 20$ $\qquad\qquad\qquad\quad 11 = 11$
    $\dfrac{4x}{4} = \dfrac{20}{4}$
    $1x = 5$
    $x = 5$

2. $5(n - 3) = 3n + 7$
    $5n - 15 = 3n + 7$
    $\underline{-3n \qquad\quad -3n}$
    $2n - 15 = 0 + 7$
    $\underline{+15 \qquad +15}$
    $2n + 0 = 22$
    $\dfrac{2n}{2} = \dfrac{22}{2}$
    $n = 11$

    Check: $5(11 - 3) = 3\cdot 11 + 7$
    $\qquad\qquad 5\cdot 8 = 33 + 7$
    $\qquad\qquad\quad 40 = 40$

3. $d = rt$
    $240 = 60\cdot t$
    $\dfrac{240}{60} = \dfrac{60t}{60}$
    $\quad 4 = t$
    Florence will drive 4 hr.

4. $6(n + 2) = n - 3$
    $6n + 12 = n - 3$
    $\underline{-n \qquad\quad -n}$
    $5n + 12 = 0 - 3$
    $\underline{-12 \qquad -12}$
    $5n + 0 = -15$
    $\dfrac{5n}{5} = \dfrac{-15}{5}$
    $n = -3$

5.

Categories	Value	Number	Amount
$10 bills	10	x	$10x$
$5 bills	5	$15 - x$	$5(15 - x)$

$$10x + 5(15 - x) = 105$$
$$10x + 75 - 5x = 105$$
$$5x + 75 = 105$$
$$\underline{-75 \quad -75}$$
$$5x + 0 = 30$$
$$\frac{5x}{5} = \frac{30}{5}$$
$$x = 6$$

There are 6 ten-dollar bills and 9 five dollar bills.

Exercise Set 5.8

1. Multiply both sides of the equation by the LCD of the fractions.

3. $10t$ represents a greater distance than $5t$, so object 2 travels the greater distance.

5.
$$x + \frac{3}{4} = \frac{1}{2}$$
$$4 \cdot \left(x + \frac{3}{4} \right) = 4 \cdot \left(\frac{1}{2} \right)$$
$$4 \cdot x + \frac{4}{1} \cdot \frac{3}{4} = \frac{4}{1} \cdot \frac{1}{2}$$
$$4 \cdot x + \frac{\overset{1}{\cancel{4}}}{1} \cdot \frac{3}{\cancel{4}_1} = \frac{\overset{2}{\cancel{4}}}{1} \cdot \frac{1}{\cancel{2}_1}$$
$$4x + 3 = 2$$
$$\underline{-3 \quad -3}$$
$$4x + 0 = -1$$
$$\frac{4x}{4} = \frac{-1}{4}$$
$$x = -\frac{1}{4}$$

Check: $-\dfrac{1}{4} + \dfrac{3}{4} = \dfrac{1}{2}$
$$\frac{-1 + 3}{4} = \frac{1}{2}$$
$$\frac{2}{4} = \frac{1}{2}$$
$$\frac{2 \div 2}{4 \div 2} = \frac{1}{2}$$
$$\frac{1}{2} = \frac{1}{2}$$

7.
$$p - \frac{1}{6} = \frac{1}{3}$$
$$6 \cdot \left(p - \frac{1}{6} \right) = 6 \cdot \left(\frac{1}{3} \right)$$
$$6 \cdot p - \frac{6}{1} \cdot \frac{1}{6} = \frac{6}{1} \cdot \frac{1}{3}$$
$$6 \cdot p - \frac{\overset{1}{\cancel{6}}}{1} \cdot \frac{1}{\cancel{6}_1} = \frac{\overset{2}{\cancel{6}}}{1} \cdot \frac{1}{\cancel{3}_1}$$
$$6p - 1 = 2$$
$$\underline{+1 \qquad +1}$$
$$6p + 0 = 3$$
$$\frac{6p}{6} = \frac{3}{6}$$
$$p = \frac{3}{6}$$
$$p = \frac{1}{2}$$

Check:
$$\frac{1}{2} - \frac{1}{6} = \frac{1}{3}$$
$$6 \left(\frac{1}{2} - \frac{1}{6} \right) = \left(\frac{1}{3} \right) 6$$
$$\frac{\overset{3}{\cancel{6}}}{1} \cdot \frac{1}{\cancel{2}_1} - \frac{\overset{1}{\cancel{6}}}{1} \cdot \frac{1}{\cancel{6}_1} = \frac{\overset{2}{\cancel{6}}}{1} \cdot \frac{1}{\cancel{3}_1}$$
$$3 - 1 = 2$$
$$2 = 2$$

9.
$$\frac{5}{6} + c = \frac{3}{5}$$
$$30 \cdot \left(\frac{5}{6} + c \right) = 30 \cdot \left(\frac{3}{5} \right)$$
$$\frac{30}{1} \cdot \frac{5}{6} + 30 \cdot c = \frac{30}{1} \cdot \frac{3}{5}$$
$$\frac{\overset{5}{\cancel{30}}}{1} \cdot \frac{5}{\cancel{6}_1} + 30 \cdot c = \frac{\overset{6}{\cancel{30}}}{1} \cdot \frac{3}{\cancel{5}_1}$$
$$25 + 30c = 18$$
$$\underline{-25 \qquad\quad -25}$$
$$0 + 30c = -7$$
$$\frac{30c}{30} = \frac{-7}{30}$$
$$c = -\frac{7}{30}$$

Check:

$$\frac{5}{6} + \left(-\frac{7}{30}\right) = \frac{3}{5}$$

$$30\left(\frac{5}{6} - \frac{7}{30}\right) = 30 \cdot \frac{3}{5}$$

$$\frac{30}{1} \cdot \frac{5}{6} - \frac{30}{1} \cdot \frac{7}{30} = \frac{30}{1} \cdot \frac{3}{5}$$

$$\frac{\overset{5}{\cancel{30}}}{1} \cdot \frac{5}{\underset{1}{\cancel{6}}} - \frac{\overset{1}{\cancel{30}}}{1} \cdot \frac{7}{\underset{1}{\cancel{30}}} = \frac{\overset{6}{\cancel{30}}}{1} \cdot \frac{3}{\underset{1}{\cancel{5}}}$$

$$25 - 7 = 18$$

$$18 = 18$$

Check:

$$\frac{6}{35} \cdot 3\frac{1}{9} = \frac{8}{15}$$

$$\frac{6}{35} \cdot \frac{28}{9} = \frac{8}{15}$$

$$\frac{\overset{2}{\cancel{6}}}{\underset{5}{\cancel{35}}} \cdot \frac{\overset{4}{\cancel{28}}}{\underset{3}{\cancel{9}}} = \frac{8}{15}$$

$$\frac{2 \cdot 4}{5 \cdot 3} = \frac{8}{15}$$

$$\frac{8}{15} = \frac{8}{15}$$

11.

$$\frac{3}{4}a = \frac{1}{2}$$

$$4 \cdot \left(\frac{3}{4}a\right) = 4 \cdot \left(\frac{1}{2}\right)$$

$$\frac{4}{1} \cdot \frac{3}{4}a = \frac{4}{1} \cdot \frac{1}{2}$$

$$\frac{\overset{1}{\cancel{4}}}{1} \cdot \frac{3}{\underset{1}{\cancel{4}}}a = \frac{\overset{2}{\cancel{4}}}{1} \cdot \frac{1}{\underset{1}{\cancel{2}}}$$

$$3a = 2$$

$$\frac{3a}{3} = \frac{2}{3}$$

$$a = \frac{2}{3}$$

Check:

$$\frac{3}{4} \cdot \frac{2}{3} = \frac{1}{2}$$

$$\frac{\overset{1}{\cancel{3}}}{\underset{2}{\cancel{4}}} \cdot \frac{\overset{1}{\cancel{2}}}{\underset{1}{\cancel{3}}} = \frac{1}{2}$$

$$\frac{1}{2} = \frac{1}{2}$$

13.

$$\frac{6}{35}f = \frac{8}{15}$$

$$\frac{35}{6} \cdot \frac{6}{35}f = \frac{35}{6} \cdot \frac{8}{15}$$

$$\frac{\overset{1}{\cancel{35}}}{\underset{1}{\cancel{6}}} \cdot \frac{\overset{1}{\cancel{6}}}{\underset{1}{\cancel{35}}}f = \frac{\overset{7}{\cancel{35}}}{\underset{3}{\cancel{6}}} \cdot \frac{\overset{4}{\cancel{8}}}{\underset{3}{\cancel{15}}}$$

$$1f = \frac{7 \cdot 4}{3 \cdot 3}$$

$$f = \frac{28}{9}$$

$$f = 3\frac{1}{9}$$

15.

$$3b - \frac{1}{4} = \frac{1}{2}$$

$$4 \cdot \left(3b - \frac{1}{4}\right) = 4 \cdot \left(\frac{1}{2}\right)$$

$$4 \cdot 3b - \frac{\overset{1}{\cancel{4}}}{1} \cdot \frac{1}{\underset{1}{\cancel{4}}} = \frac{\overset{2}{\cancel{4}}}{1} \cdot \frac{1}{\underset{1}{\cancel{2}}}$$

$$12b - 1 = 2$$

$$\underline{ +1 \quad +1}$$

$$12b + 0 = 3$$

$$\frac{12b}{12} = \frac{3}{12}$$

$$b = \frac{1}{4}$$

Check: $3 \cdot \dfrac{1}{4} - \dfrac{1}{4} = \dfrac{1}{2}$

$$\frac{3}{4} - \frac{1}{4} = \frac{1}{2}$$

$$\frac{2}{4} = \frac{1}{2}$$

$$\frac{1}{2} = \frac{1}{2}$$

17.

$$\frac{3}{4}x + \frac{1}{6} = \frac{1}{2}$$

$$12 \cdot \left(\frac{3}{4}x + \frac{1}{6}\right) = 12 \cdot \left(\frac{1}{2}\right)$$

$$\frac{\overset{3}{\cancel{12}}}{1} \cdot \frac{3}{\underset{1}{\cancel{4}}}x + \frac{\overset{2}{\cancel{12}}}{1} \cdot \frac{1}{\underset{1}{\cancel{6}}} = \frac{\overset{6}{\cancel{12}}}{1} \cdot \frac{1}{\underset{1}{\cancel{2}}}$$

$$9x + 2 = 6$$

$$\underline{ -2 \quad -2}$$

$$9x + 0 = 4$$

$$\frac{9x}{9} = \frac{4}{9}$$

$$x = \frac{4}{9}$$

Check: $\dfrac{\cancel{2}}{\cancel{4}} \cdot \dfrac{\cancel{4}}{\cancel{9}_3} + \dfrac{1}{6} = \dfrac{1}{2}$

$$\dfrac{1}{3} + \dfrac{1}{6} = \dfrac{1}{2}$$

$$\dfrac{2}{6} + \dfrac{1}{6} = \dfrac{1}{2}$$

$$\dfrac{3}{6} = \dfrac{1}{2}$$

$$\dfrac{1}{2} = \dfrac{1}{2}$$

19. $\dfrac{4}{5} - \dfrac{a}{2} = \dfrac{3}{4} - 1$

$$20 \cdot \left(\dfrac{4}{5} - \dfrac{a}{2}\right) = 20 \cdot \left(\dfrac{3}{4} - 1\right)$$

$$\dfrac{\overset{4}{\cancel{20}}}{1} \cdot \dfrac{4}{\cancel{5}} - \dfrac{\overset{10}{\cancel{20}}}{1} \cdot \dfrac{a}{\cancel{2}} = \dfrac{\overset{5}{\cancel{20}}}{1} \cdot \dfrac{3}{\cancel{4}} - 20 \cdot 1$$

$$16 - 10a = 15 - 20$$

$$16 - 10a = -5$$

$$\underline{-16 -16}$$

$$0 - 10a = -21$$

$$\dfrac{-10a}{-10} = \dfrac{-21}{-10}$$

$$a = \dfrac{21}{10}$$

$$a = 2\dfrac{1}{10}$$

Check: $\dfrac{4}{5} - \dfrac{2\frac{1}{10}}{2} = \dfrac{3}{4} - 1$

$$\dfrac{4}{5} - 2\dfrac{1}{10} \cdot \dfrac{1}{2} = \dfrac{3}{4} - 1$$

$$\dfrac{4}{5} - \dfrac{21}{10} \cdot \dfrac{1}{2} = \dfrac{3}{4} - \dfrac{4}{4}$$

$$\dfrac{4}{5} - \dfrac{21}{20} = -\dfrac{1}{4}$$

$$\dfrac{16}{20} - \dfrac{21}{20} = -\dfrac{1}{4}$$

$$-\dfrac{5}{20} = -\dfrac{1}{4}$$

$$-\dfrac{1}{4} = -\dfrac{1}{4}$$

21. $\dfrac{1}{8} + n = \dfrac{5}{6}n - \dfrac{2}{3}$

$$24 \cdot \left(\dfrac{1}{8} + n\right) = 24 \cdot \left(\dfrac{5}{6}n - \dfrac{2}{3}\right)$$

$$\dfrac{\overset{3}{\cancel{24}}}{1} \cdot \dfrac{1}{\cancel{8}} + 24 \cdot n = \dfrac{\overset{4}{\cancel{24}}}{1} \cdot \dfrac{5}{\cancel{6}}n - \dfrac{\overset{8}{\cancel{24}}}{1} \cdot \dfrac{2}{\cancel{3}}$$

$$3 + 24n = 20n - 16$$

$$\underline{-20n -20n}$$

$$3 + 4n = 0 - 16$$

$$\underline{-3 -3}$$

$$4n = -19$$

$$\dfrac{4n}{4} = \dfrac{-19}{4}$$

$$n = -4\dfrac{3}{4}$$

Check: $\dfrac{1}{8} + \left(-4\dfrac{3}{4}\right) = \dfrac{5}{6} \cdot \left(-4\dfrac{3}{4}\right) - \dfrac{2}{3}$

$$\dfrac{1}{8} - \dfrac{19}{4} = \dfrac{5}{6} \cdot -\dfrac{19}{4} - \dfrac{2}{3}$$

$$\dfrac{1}{8} - \dfrac{38}{8} = -\dfrac{95}{24} - \dfrac{2}{3}$$

$$-\dfrac{37}{8} = -\dfrac{95}{24} - \dfrac{16}{24}$$

$$-\dfrac{37}{8} = -\dfrac{111}{24}$$

$$-\dfrac{37}{8} = -\dfrac{37}{8}$$

23. $\dfrac{1}{2}(x - 6) = \dfrac{1}{4}x - \dfrac{2}{5}$

$$\dfrac{1}{2}x - \dfrac{1}{2} \cdot 6 = \dfrac{1}{4}x - \dfrac{2}{5}$$

$$\dfrac{1}{2}x - 3 = \dfrac{1}{4}x - \dfrac{2}{5}$$

$$20\left(\dfrac{1}{2}x - 3\right) = 20\left(\dfrac{1}{4}x - \dfrac{2}{5}\right)$$

$$\dfrac{\overset{10}{\cancel{20}}}{1} \cdot \dfrac{1}{\cancel{2}}x + \dfrac{20}{1} \cdot \dfrac{-3}{1} = \dfrac{\overset{5}{\cancel{20}}}{1} \cdot \dfrac{1}{\cancel{4}}x - \dfrac{\overset{4}{\cancel{20}}}{1} \cdot \dfrac{2}{\cancel{5}}$$

$$10x - 60 = 5x - 8$$

$$\underline{-5x -5x}$$

$$5x - 60 = 0 - 8$$

$$\underline{+60 +60}$$

$$5x + 0 = 52$$

$$\dfrac{5x}{5} = \dfrac{52}{5}$$

$$x = 10\dfrac{2}{5}$$

Check: $\dfrac{1}{2}\left(10\dfrac{2}{5}-6\right)=\dfrac{1}{4}\cdot 10\dfrac{2}{5}-\dfrac{2}{5}$

$$\dfrac{1}{2}\cdot\left(4\dfrac{2}{5}\right)=\dfrac{1}{4}\cdot\dfrac{52}{5}-\dfrac{2}{5}$$

$$\dfrac{1}{2}\cdot\dfrac{22}{5}=\dfrac{52}{20}-\dfrac{2}{5}$$

$$\dfrac{22}{10}=\dfrac{13}{5}-\dfrac{2}{5}$$

$$\dfrac{11}{5}=\dfrac{11}{5}$$

25. $\dfrac{3}{8}\cdot n=4\dfrac{5}{6}$

$$\dfrac{3}{8}n=\dfrac{29}{6}$$

$$\dfrac{\cancel{8}}{\cancel{3}}\cdot\dfrac{\cancel{3}}{\cancel{8}}n=\dfrac{29}{\cancel{6}_3}\cdot\dfrac{\cancel{8}^4}{3}$$

$$1n=\dfrac{116}{9}$$

$$n=12\dfrac{8}{9}$$

27. $y-8\dfrac{7}{10}=-2\dfrac{1}{2}$

$$y-\dfrac{87}{10}=-\dfrac{5}{2}$$

$$y-\dfrac{87}{10}+\dfrac{87}{10}=\dfrac{-25}{10}+\dfrac{87}{10}$$

$$y+0=\dfrac{62}{10}$$

$$y=6\dfrac{1}{5}$$

29. $3\dfrac{1}{4}+2n=-\dfrac{1}{6}$

$$\dfrac{13}{4}+2n=-\dfrac{1}{6}$$

$$12\cdot\left(\dfrac{13}{4}+2n\right)=12\cdot\left(-\dfrac{1}{6}\right)$$

$$\dfrac{\cancel{12}^3}{1}\cdot\dfrac{13}{\cancel{4}}+12\cdot 2n=\dfrac{\cancel{12}^2}{1}\cdot\left(-\dfrac{1}{\cancel{6}}\right)$$

$$39+24n=-2$$

$$\underline{-39\qquad\qquad -39}$$

$$0+24n=-41$$

$$\dfrac{24n}{24}=\dfrac{-41}{24}$$

$$n=-\dfrac{41}{24}$$

$$n=-1\dfrac{17}{24}$$

31. $\dfrac{3}{4}(b+10)=1\dfrac{1}{6}+b$

$$\dfrac{3}{4}(b+10)=\dfrac{7}{6}+b$$

$$12\cdot\dfrac{3}{4}(b+10)=12\cdot\left(\dfrac{7}{6}+b\right)$$

$$9(b+10)=\dfrac{12}{1}\cdot\dfrac{7}{6}+12\cdot b$$

$$9b+90=\dfrac{\cancel{12}^2}{1}\cdot\dfrac{7}{\cancel{6}_1}+12b$$

$$9b+90=14+12b$$

$$\underline{-9b\qquad\qquad -9b}$$

$$90=14+3b$$

$$\underline{-14\quad -14}$$

$$76=3b$$

$$\dfrac{76}{3}=\dfrac{3b}{3}$$

$$25\dfrac{1}{3}=b$$

33. $A=\dfrac{1}{2}bh$

$$7\dfrac{41}{50}=\dfrac{1}{2}\cdot b\cdot 3\dfrac{2}{5}$$

$$\dfrac{391}{50}=\dfrac{1}{2}\cdot\dfrac{17}{5}b$$

$$\dfrac{391}{50}=\dfrac{17}{10}b$$

$$\dfrac{10}{17}\cdot\dfrac{391}{50}=\dfrac{\cancel{17}}{\cancel{10}}b\cdot\dfrac{\cancel{10}}{\cancel{17}}$$

$$\dfrac{3910}{850}=1b$$

$$4\dfrac{3}{5}\text{ cm }=b$$

35. Let x be the fifth test score. The average (mean) is the sum of the test scores divided by the number of test scores.

$$\dfrac{86+96+90+88+x}{5}=90$$

$$5\cdot\left(\dfrac{86+96+90+88+x}{5}\right)=5\cdot 90$$

$$86+96+90+88+x=450$$

$$360+x=450$$

$$\underline{-360\qquad\quad =-360}$$

$$0+x=90$$

$$x=90$$

Daniel must have at least a 90 on the fifth test to receive an A.

37. Let x be the fourth test score. The average (mean) is the sum of the test scores divided by the number of test scores.

$$\frac{88+92+90+x}{4}=94$$

$$4\cdot\left(\frac{88+92+90+x}{4}\right)=4\cdot94$$

$$88+92+90+x=376$$

$$270+x=376$$

$$\underline{-270\qquad\quad=-270}$$

$$0+x=106$$

$$x=106$$

Lonnie must have at least a 106 on the fourth test to be exempt from the final exam.

39. Let x be the amount raised by the team, so Mary is $2\frac{1}{2}x$. The sum of the amounts is \$31,542.

$$x+2\frac{1}{2}x=31,542$$

$$3\frac{1}{2}x=31,542$$

$$\frac{7}{2}x=31,542$$

$$\frac{\cancel{2}}{\cancel{7}}\cdot\frac{\cancel{7}}{\cancel{2}}x=31,542\cdot\frac{2}{7}$$

$$1x=\frac{63,084}{7}$$

$$x=9012$$

The team raised \$9012, so Mary raised

$$2\frac{1}{2}x=\frac{5}{2}\cdot9012=\$22,530.$$

41.
$$\text{tank A}=\frac{3}{4}x$$

$$\text{tank B}=\frac{2}{3}x$$

$$\text{tank A} + \text{tank B} = 1700$$

$$\frac{3}{4}x+\frac{2}{3}x=1700$$

$$\frac{9}{12}x+\frac{8}{12}x=1700$$

$$\frac{17}{12}x=1700$$

$$\frac{\cancel{12}}{\cancel{17}}\cdot\frac{\cancel{17}}{\cancel{12}}x=1700\cdot\frac{12}{17}$$

$$1x=\frac{20,400}{17}$$

$$x=1200\text{ gal.}$$

43. If $\frac{1}{5}$ answered yes, then $\frac{4}{5}$ answered no. So,

$$\frac{4}{5}\cdot\frac{3}{4}=\frac{3}{5}$$ of the people contacted said no.

45. The angles shown are supplementary. If we call the larger angle x, then the smaller angle is $\frac{1}{3}x$.

They must total $180°$. $x+\frac{1}{3}x=180$

$$\frac{3}{3}x+\frac{1}{3}x=180$$

$$\frac{4}{3}x=180$$

$$\frac{\cancel{3}}{\cancel{4}}\cdot\frac{\cancel{4}}{\cancel{3}}x=\cancel{180}^{\,45}\cdot\frac{3}{\cancel{4}}$$

$$x=\frac{45\cdot3}{1}$$

$$x=135°$$

The larger angle is $135°$ and the smaller angle is $\frac{1}{3}x=\frac{1}{3}\cdot135=45°.$

47. a)

Categories	Rate	Time	Distance
First cyclist	$10\frac{1}{2}$	t	$10\frac{1}{2}t$
Second cyclist	$12\frac{3}{4}$	t	$12\frac{3}{4}t$

b) $10\frac{1}{2}t+12\frac{3}{4}t$

c) $10\frac{1}{2}t+12\frac{3}{4}t$

$$=10\frac{2}{4}t+12\frac{3}{4}t$$

$$=22\frac{5}{4}t\text{ or }23\frac{1}{4}t$$

The coefficient is the combined rate of the two cyclists.

49.

Categories	Rate	Time	Distance
F-16	$1483\frac{3}{5}$	t	$1483\frac{3}{5}t$
Mig-29	$741\frac{4}{5}$	t	$741\frac{4}{5}t$

$$1483\frac{3}{5}t + 741\frac{4}{5}t = 20$$

$$2224\frac{7}{5}t = 20$$

$$\frac{11,127}{5}t = 20$$

$$\frac{\cancel{5}}{11,127} \cdot \frac{\cancel{11,127}}{\cancel{5}}t = 20 \cdot \frac{5}{11,127}$$

$$t = \frac{100}{11,127} \text{ hr.}$$

51.

Categories	Rate	Time	Distance
Greg	2	t	$2t$
Debbie	$2\frac{1}{2}$	t	$2\frac{1}{2}t$

$$2t + 2\frac{1}{2}t = \frac{1}{2}$$

$$4\frac{1}{2}t = \frac{1}{2}$$

$$\frac{9}{2}t = \frac{1}{2}$$

$$\frac{\cancel{2}}{\cancel{9}} \cdot \frac{\cancel{9}}{\cancel{2}}t = \frac{1}{\cancel{2}} \cdot \frac{\cancel{2}}{9}$$

$$t = \frac{1}{9} \text{ hr.}$$

Review Exercises

1. $\dfrac{6}{17} < \dfrac{7}{19}$ 2. $\dfrac{9x \div 3x}{24x^2 \div 3x} = \dfrac{3}{8x}$

3. $\dfrac{-\overset{2}{\cancel{4}}}{\underset{3}{\cancel{9}}} \cdot \dfrac{\overset{7}{\cancel{21}}}{\underset{15}{\cancel{30}}} = \dfrac{-2 \cdot 7}{3 \cdot 15} = -\dfrac{14}{45}$

4. $\dfrac{14a}{15ab^4} \div \dfrac{21}{25ab} = \dfrac{14a}{15ab^4} \cdot \dfrac{25ab}{21}$

$$= \dfrac{2 \cdot \cancel{7} \cdot \cancel{a}}{3 \cdot \cancel{3} \cdot \cancel{a} \cdot \cancel{b} \cdot b \cdot b \cdot b} \cdot \dfrac{\cancel{5} \cdot 5 \cdot a \cdot \cancel{b}}{3 \cdot \cancel{7}}$$

$$= \dfrac{10a}{9b^3}$$

5. $\dfrac{7}{8} + 5\dfrac{2}{3} = \dfrac{7 \cdot 3}{8 \cdot 3} + 5\dfrac{2 \cdot 8}{3 \cdot 8}$

$$= \dfrac{21}{24} + 5\dfrac{16}{24}$$

$$= 5\dfrac{37}{24}$$

$$= 5 + 1\dfrac{13}{24}$$

$$= 6\dfrac{13}{24}$$

Chapter 5 Review Exercises

1. true   2. false   3. true

4. true   5. false   6. false

7. multiplying; dividing

8. Write the numerator and denominator in prime factored form, divide out all common prime factors, then multiply the remaining factors.

9. Write an equivalent multiplication using the reciprocal of the divisor.

10. Find the least common denominator. Rewrite the fractions to have the least common denominator. Add numerators, keep the common denominator.

11. a) $\dfrac{6}{12} = \dfrac{1}{2}$ b) $\dfrac{1}{4}$

12. a)

b)

13. a) $9\overline{)40} = 4\frac{4}{9}$ b) $4\overline{)29} = -7\frac{1}{4}$
$\underline{36}$ $\underline{28}$
$\;\,4$ $\;1$

14. a) $\dfrac{3 \cdot 6 + 2}{3} = \dfrac{18 + 2}{3} = \dfrac{20}{3}$

b) $-\dfrac{2 \cdot 5 + 1}{2} = -\dfrac{10 + 1}{2} = -\dfrac{11}{2}$

15. a) 18   b) undefined   c) 0   d) 1

16. a) $\dfrac{15}{35} = \dfrac{15 \div 5}{35 \div 5} = \dfrac{3}{7}$

b) $-\dfrac{84}{105} = -\dfrac{84 \div 21}{105 \div 21} = -\dfrac{4}{5}$

17. a) $\dfrac{-8m^3 n}{26n^2} = \dfrac{-\cancel{2} \cdot 2 \cdot 2 \cdot m \cdot m \cdot m \cdot \cancel{n}}{\cancel{2} \cdot 13 \cdot n \cdot \cancel{n}} = \dfrac{-4m^3}{13n}$

b) $\dfrac{6xy^2}{20y^5} = \dfrac{\cancel{2} \cdot 3 \cdot x \cdot \cancel{y} \cdot \cancel{y}}{\cancel{2} \cdot 2 \cdot 5 \cdot \cancel{y} \cdot \cancel{y} \cdot y \cdot y \cdot y} = \dfrac{3x}{10y^3}$

18. a) $\dfrac{5}{9} \, ? \, \dfrac{7}{13}$

$\dfrac{5}{9} = \dfrac{5 \cdot 13}{9 \cdot 13} = \dfrac{65}{117}$

$\dfrac{7}{13} = \dfrac{7 \cdot 9}{13 \cdot 9} = \dfrac{63}{117}$

$65 > 63$

so: $\dfrac{5}{9} > \dfrac{7}{13}$

b) $\dfrac{3}{16} \, ? \, \dfrac{5}{24}$

$\dfrac{3}{16} = \dfrac{3 \cdot 3}{16 \cdot 3} = \dfrac{9}{48}$

$\dfrac{5}{24} = \dfrac{5 \cdot 2}{24 \cdot 2} = \dfrac{10}{48}$

$9 < 10$

so: $\dfrac{3}{16} < \dfrac{5}{24}$

c) $\dfrac{10}{16} \, ? \, \dfrac{25}{40}$

$\dfrac{10}{16} = \dfrac{10 \cdot 5}{16 \cdot 5} = \dfrac{50}{80}$

$\dfrac{25}{40} = \dfrac{25 \cdot 2}{40 \cdot 2} = \dfrac{50}{80}$

$50 = 50$

so: $\dfrac{10}{16} = \dfrac{25}{40}$

19. a) $\dfrac{4}{9} \cdot \dfrac{12}{20} = \dfrac{4}{3 \cdot \cancel{3}} \cdot \dfrac{\cancel{3} \cdot \cancel{4}}{\cancel{4} \cdot 5} = \dfrac{4}{15}$

b) $-4\dfrac{1}{2} \cdot 2\dfrac{2}{3} = -\dfrac{9}{2} \cdot \dfrac{8}{3} = -\dfrac{72}{6} = -12$

20. a) $\dfrac{10n}{6p^3} \cdot \dfrac{14np}{8n} = \dfrac{\cancel{2} \cdot 5 \cdot \cancel{n}}{\cancel{2} \cdot 3 \cdot \cancel{p} \cdot p \cdot p} \cdot \dfrac{\cancel{2} \cdot 7 \cdot n \cdot \cancel{p}}{\cancel{2} \cdot 2 \cdot 2 \cdot \cancel{n}}$

$= \dfrac{35n}{12p^2}$

b) $\dfrac{h^3}{4k} \cdot \left(\dfrac{-12k}{15h} \right) = \dfrac{\cancel{h} \cdot h \cdot h}{\cancel{4} \cdot \cancel{k}} \cdot \left(\dfrac{-\cancel{3} \cdot \cancel{4} \cdot \cancel{k}}{\cancel{3} \cdot 5 \cdot \cancel{h}} \right) = -\dfrac{h^2}{5}$

21. a) $\left(\dfrac{1}{2} \right)^6 = \dfrac{1^6}{2^6} = \dfrac{1}{64}$

b) $\left(-\dfrac{3}{5} xy^3 \right)^2 = \left(-\dfrac{3}{5} \right)^2 x^{1 \cdot 2} y^{3 \cdot 2}$

$= \dfrac{9}{25} x^2 y^6$

22. a) $-\dfrac{5}{6} \div \left(\dfrac{-15}{28} \right) = -\dfrac{5}{6} \cdot \left(\dfrac{28}{-15} \right)$

$= -\dfrac{\cancel{5}}{\cancel{2} \cdot 3} \cdot \left(\dfrac{\cancel{2} \cdot 2 \cdot 7}{-3 \cdot \cancel{5}} \right)$

$= \dfrac{14}{9}$

$= 1\dfrac{5}{9}$

b) $7\dfrac{1}{5} \div 2\dfrac{4}{5} = \dfrac{36}{5} \cdot \dfrac{5}{14} = \dfrac{\cancel{2} \cdot 18}{\cancel{5}} \cdot \dfrac{\cancel{5}}{\cancel{2} \cdot 7} = \dfrac{18}{7} = 2\dfrac{4}{7}$

23. a) $\dfrac{20a}{12ab^3} \div \dfrac{10}{8b} = \dfrac{20a}{12ab^3} \cdot \dfrac{8b}{10}$

$= \dfrac{\cancel{2} \cdot \cancel{2} \cdot \cancel{5} \cdot \cancel{a}}{\cancel{2} \cdot \cancel{2} \cdot 3 \cdot \cancel{a} \cdot \cancel{b} \cdot b \cdot b} \cdot \dfrac{\cancel{2} \cdot 2 \cdot 2 \cdot \cancel{b}}{\cancel{2} \cdot \cancel{5}}$

$= \dfrac{4}{3b^2}$

b) $\dfrac{-8x^6 y}{25x^3} \div \dfrac{2xy}{5z^2}$

$= \dfrac{-8x^6 y}{25x^3} \cdot \dfrac{5z^2}{2xy}$

$= \dfrac{-\cancel{2} \cdot 2 \cdot 2 \cdot \cancel{x} \cdot \cancel{x} \cdot \cancel{x} \cdot x \cdot x \cdot \cancel{y}}{\cancel{5} \cdot 5 \cdot \cancel{x} \cdot \cancel{x} \cdot \cancel{x}} \cdot \dfrac{\cancel{5} \cdot z \cdot z}{\cancel{2} \cdot \cancel{x} \cdot \cancel{y}}$

$= \dfrac{-4x^2 z^2}{5}$

24. a) $\sqrt{\dfrac{100}{36}} = \dfrac{\sqrt{100}}{\sqrt{36}} = \dfrac{10}{6} = \dfrac{5}{3}$

b) $\sqrt{\dfrac{50}{2}} = \sqrt{25} = 5$

25. a) $28 = 2^2 \cdot 7$

$24 = 2^3 \cdot 3$

$\text{LCM}(28,24) = 2^3 \cdot 3 \cdot 7 = 8 \cdot 3 \cdot 7 = 168$

b) $15x = 3 \cdot 5 \cdot x$

$20x^2 y = 2^2 \cdot 5 \cdot x^2 \cdot y$

$\text{LCM}(15x, 20x^2 y) = 2^2 \cdot 3 \cdot 5 \cdot x^2 \cdot y$

$= 4 \cdot 3 \cdot 5 \cdot x^2 \cdot y$

$= 60x^2 y$

26. a) $\dfrac{5}{6}+\dfrac{1}{5}=\dfrac{5\cdot5}{6\cdot5}+\dfrac{1\cdot6}{5\cdot6}$

$\qquad\qquad =\dfrac{25}{30}+\dfrac{6}{30}$

$\qquad\qquad =\dfrac{31}{30}$

$\qquad\qquad =1\dfrac{1}{30}$

b) $4\dfrac{5}{8}+6\dfrac{2}{3}=\dfrac{37}{8}+\dfrac{20}{3}$

$\qquad\qquad =\dfrac{37\cdot3}{8\cdot3}+\dfrac{20\cdot8}{3\cdot8}$

$\qquad\qquad =\dfrac{111}{24}+\dfrac{160}{24}$

$\qquad\qquad =\dfrac{271}{24}$

$\qquad\qquad =11\dfrac{7}{24}$

c) $\dfrac{9}{15}-\dfrac{4}{5}=\dfrac{9}{15}-\dfrac{4\cdot3}{5\cdot3}$

$\qquad\qquad =\dfrac{9}{15}-\dfrac{12}{15}$

$\qquad\qquad =-\dfrac{3}{15}$

$\qquad\qquad =-\dfrac{1}{5}$

d) $-5\dfrac{1}{6}-\left(-2\dfrac{1}{3}\right)=-5\dfrac{1}{6}+2\dfrac{1}{3}$

$\qquad\qquad =-\dfrac{31}{6}+\dfrac{7}{3}$

$\qquad\qquad =-\dfrac{31}{6}+\dfrac{7\cdot2}{3\cdot2}$

$\qquad\qquad =-\dfrac{31}{6}+\dfrac{14}{6}$

$\qquad\qquad =-\dfrac{17}{6}$

$\qquad\qquad =-2\dfrac{5}{6}$

27. a) $\dfrac{3n}{8}+\dfrac{n}{8}=\dfrac{3n+n}{8}=\dfrac{4n}{8}=\dfrac{n}{2}$

b) $-\dfrac{2}{9x}+\left(-\dfrac{4}{9x}\right)=\dfrac{-2-4}{9x}=\dfrac{-6}{9x}=-\dfrac{2}{3x}$

c) $\dfrac{7}{6h}-\dfrac{3}{4h}=\dfrac{7\cdot2}{6h\cdot2}-\dfrac{3\cdot3}{4h\cdot3}$

$\qquad\qquad =\dfrac{14}{12h}-\dfrac{9}{12h}$

$\qquad\qquad =\dfrac{14-9}{12h}$

$\qquad\qquad =\dfrac{5}{12h}$

d) $\dfrac{5}{8a}-\dfrac{7}{12}=\dfrac{5\cdot3}{8a\cdot3}-\dfrac{7\cdot2a}{12\cdot2a}$

$\qquad\qquad =\dfrac{15}{24a}-\dfrac{14a}{24a}$

$\qquad\qquad =\dfrac{15-14a}{24a}$

28. a) $3\dfrac{1}{2}+\dfrac{1}{4}\left(\dfrac{2}{3}-\dfrac{1}{6}\right)=3\dfrac{1}{2}+\dfrac{1}{4}\left(\dfrac{4}{6}-\dfrac{1}{6}\right)$

$\qquad\qquad =3\dfrac{1}{2}+\dfrac{1}{4}\cdot\dfrac{3}{6}$

$\qquad\qquad =\dfrac{7}{2}+\dfrac{3}{24}$

$\qquad\qquad =\dfrac{7\cdot4}{2\cdot4}+\dfrac{1}{8}$

$\qquad\qquad =\dfrac{28}{8}+\dfrac{1}{8}$

$\qquad\qquad =\dfrac{29}{8}$

$\qquad\qquad =3\dfrac{5}{8}$

b) $\left(4+\dfrac{1}{12}\right)-8\div\dfrac{3}{4}=4\dfrac{1}{12}-8\cdot\dfrac{4}{3}$

$\qquad\qquad =4\dfrac{1}{12}-\dfrac{32}{3}$

$\qquad\qquad =\dfrac{49}{12}-\dfrac{32\cdot4}{3\cdot4}$

$\qquad\qquad =\dfrac{49}{12}-\dfrac{128}{12}$

$\qquad\qquad =-\dfrac{79}{12}$

$\qquad\qquad =-6\dfrac{7}{12}$

29. a) $\dfrac{3}{5}x^2+9x-\dfrac{1}{2}x^2-2-11x$

$\qquad =\dfrac{3}{5}x^2-\dfrac{1}{2}x^2+9x-11x-2$

$\qquad =\dfrac{6}{10}x^2-\dfrac{5}{10}x^2-2x-2$

$\qquad =\dfrac{1}{10}x^2-2x-2$

b) $\left(\dfrac{1}{4}n^2 - \dfrac{2}{3}n - 3\right) + \left(\dfrac{3}{8}n^2 + 1\right)$

$= \dfrac{1}{4}n^2 + \dfrac{3}{8}n^2 - \dfrac{2}{3}n - 3 + 1$

$= \dfrac{2}{8}n^2 + \dfrac{3}{8}n^2 - \dfrac{2}{3}n - 2$

$= \dfrac{5}{8}n^2 - \dfrac{2}{3}n - 2$

c) $\left(12y^3 - \dfrac{5}{6}y + \dfrac{2}{5}\right) - \left(\dfrac{1}{2}y^3 + 2y^2 + \dfrac{1}{3}\right)$

$= \left(12y^3 - \dfrac{5}{6}y + \dfrac{2}{5}\right) + \left(-\dfrac{1}{2}y^3 - 2y^2 - \dfrac{1}{3}\right)$

$= \dfrac{24}{2}y^3 - \dfrac{1}{2}y^3 - 2y^2 - \dfrac{5}{6}y + \dfrac{6}{15} - \dfrac{5}{15}$

$= \dfrac{23}{2}y^3 - 2y^2 - \dfrac{5}{6}y + \dfrac{1}{15}$

d) $\left(\dfrac{2}{7}b^3\right)\left(-\dfrac{7}{8}b\right) = \dfrac{\cancel{2}}{\cancel{7}} \cdot \left(\dfrac{-\cancel{7}}{\cancel{8}_4}\right)b^{3+1}$

$\qquad\qquad = -\dfrac{1}{4}b^4$

e) $\left(\dfrac{1}{4}x + 2\right)\left(5x - \dfrac{3}{2}\right)$

$= \dfrac{1}{4}x \cdot 5x + \dfrac{1}{4}x \cdot \left(-\dfrac{3}{2}\right) + 2 \cdot 5x + 2 \cdot \left(-\dfrac{3}{2}\right)$

$= \dfrac{5}{4}x^2 - \dfrac{3}{8}x + 10x - 3$

$= \dfrac{5}{4}x^2 - \dfrac{3}{8}x + \dfrac{80}{8}x - 3$

$= \dfrac{5}{4}x^2 + \dfrac{77}{8}x - 3$

30. a)

$$y - 3\dfrac{4}{5} = \dfrac{1}{3}$$

$$y - \dfrac{19}{5} = \dfrac{1}{3}$$

$$15 \cdot \left(y - \dfrac{19}{5}\right) = 15 \cdot \left(\dfrac{1}{3}\right)$$

$$15 \cdot y - 15 \cdot \left(\dfrac{19}{5}\right) = \dfrac{15}{3}$$

$$15y - \dfrac{\overset{3}{\cancel{15}}}{1} \cdot \dfrac{19}{\cancel{5}_1} = 5$$

$$15y - 57 = 5$$

$$\dfrac{+57 \qquad +57}{15y + 0 = 62}$$

$$\dfrac{15y}{15} = \dfrac{62}{15}$$

$$1y = \dfrac{62}{15}$$

$$y = 4\dfrac{2}{15}$$

Check: $\quad 4\dfrac{2}{15} - 3\dfrac{4}{5} = \dfrac{1}{3}$

$$\dfrac{62}{15} - \dfrac{19}{5} = \dfrac{1}{3}$$

$$\dfrac{62}{15} - \dfrac{19 \cdot 3}{5 \cdot 3} = \dfrac{1}{3}$$

$$\dfrac{62}{15} - \dfrac{57}{15} = \dfrac{1}{3}$$

$$\dfrac{5}{15} = \dfrac{1}{3}$$

$$\dfrac{5 \div 5}{15 \div 5} = \dfrac{1}{3}$$

$$\dfrac{1}{3} = \dfrac{1}{3}$$

b)

$$-\dfrac{3}{8}n = \dfrac{4}{3}$$

$$-\dfrac{\cancel{8}}{\cancel{3}} \cdot \left(-\dfrac{\cancel{3}}{\cancel{8}}n\right) = -\dfrac{8}{3} \cdot \dfrac{4}{3}$$

$$n = -\dfrac{32}{9}$$

$$n = -3\dfrac{5}{9}$$

Check: $-\dfrac{3}{8}\cdot\left(-3\dfrac{5}{9}\right)=\dfrac{4}{3}$

$-\dfrac{3}{8}\cdot\left(-\dfrac{32}{9}\right)=\dfrac{4}{3}$

$-\dfrac{\cancel{3}^{1}}{\cancel{8}_{1}}\cdot\left(-\dfrac{\cancel{32}^{4}}{\cancel{9}_{3}}\right)=\dfrac{4}{3}$

$\dfrac{4}{3}=\dfrac{4}{3}$

c) $\dfrac{1}{2}m-5=\dfrac{3}{4}$ Check: $\dfrac{1}{2}\cdot11\dfrac{1}{2}-5=\dfrac{3}{4}$

$4\cdot\left(\dfrac{1}{2}m-5\right)=4\cdot\dfrac{3}{4}$ $\dfrac{1}{2}\cdot\dfrac{23}{2}-5=\dfrac{3}{4}$

$4\cdot\dfrac{1}{2}m-4\cdot5=3$ $\dfrac{23}{4}-\dfrac{20}{4}=\dfrac{3}{4}$

$2m-20=\ \ 3$ $\dfrac{3}{4}=\dfrac{3}{4}$

$\underline{\ +20\qquad +20}$

$2m+0=23$

$\dfrac{2m}{2}=\dfrac{23}{2}$

$m=11\dfrac{1}{2}$

d) $\dfrac{1}{2}\left(n-\dfrac{2}{3}\right)=\dfrac{3}{4}n+2$

$12\cdot\dfrac{1}{2}\left(n-\dfrac{2}{3}\right)=12\cdot\left(\dfrac{3}{4}n+2\right)$

$6\left(n-\dfrac{2}{3}\right)=12\cdot\left(\dfrac{3}{4}n+2\right)$

$6\cdot n-\dfrac{6}{1}\cdot\dfrac{2}{3}=\dfrac{12}{1}\cdot\dfrac{3}{4}n+12\cdot2$

$6n-\dfrac{\cancel{6}^{2}}{1}\cdot\dfrac{2}{\cancel{3}_{1}}=\dfrac{\cancel{12}^{3}}{1}\cdot\dfrac{3}{\cancel{4}_{1}}n+24$

$6n-4=9n+24$

$\underline{-6n\qquad\ -6n}$

$-4=3n+24$

$\underline{-24\qquad\ -24}$

$-28=3n$

$-\dfrac{28}{3}=\dfrac{3n}{3}$

$-9\dfrac{1}{3}=n$

Check: $\dfrac{1}{2}\left(-\dfrac{28}{3}-\dfrac{2}{3}\right)=\dfrac{3}{4}\cdot-\dfrac{28}{3}+2$

$\dfrac{1}{2}\left(-\dfrac{30}{3}\right)=\dfrac{-84}{12}+2$

$-\dfrac{30}{6}=-7+2$

$-5=-5$

31. $\dfrac{\cancel{6}}{\cancel{9}_{3}}\cdot\dfrac{\cancel{5}}{\cancel{10}_{2}}=\dfrac{1}{6}$ of the items are discounted shirts

32. $A=\dfrac{1}{2}bh$

$A=\dfrac{1}{2}\cdot8\cdot2\dfrac{1}{4}$

$A=\dfrac{1}{2}\cdot\dfrac{\cancel{8}^{2}}{1}\cdot\dfrac{9}{\cancel{4}}$

$A=\dfrac{18}{2}$

$A=9\ \text{ft.}^{2}$

33. $r=\dfrac{1}{2}d$

$r=\dfrac{1}{2}\cdot3\dfrac{1}{2}$

$r=\dfrac{1}{2}\cdot\dfrac{7}{2}$

$r=\dfrac{7}{4}$

$r=1\dfrac{3}{4}\ \text{in.}$

34. $d=2r$

$d=2\cdot14\dfrac{1}{5}$

$d=\dfrac{2}{1}\cdot\dfrac{71}{5}$

$d=\dfrac{142}{5}$

$d=28\dfrac{2}{5}\ \text{cm}$

35. $C=\pi d$

$C=\dfrac{22}{7}\cdot2\dfrac{1}{2}$

$C=\dfrac{\cancel{22}^{11}}{7}\cdot\dfrac{5}{\cancel{2}}$

$C=\dfrac{55}{7}$

$C=7\dfrac{6}{7}\ \text{ft.}$

36. $\dfrac{1}{4}\cdot x=4\dfrac{1}{2}$

$\dfrac{\cancel{4}}{\cancel{1}}\cdot\dfrac{\cancel{1}}{\cancel{4}}x=4\dfrac{1}{2}\cdot\dfrac{4}{1}$

$1x=\dfrac{9}{\cancel{2}}\cdot\dfrac{\cancel{4}^{2}}{1}$

$x=18$ There are 18 servings.

37. $C=2\pi r$

$33=\dfrac{2}{1}\cdot\dfrac{22}{7}\cdot r$

$33=\dfrac{44}{7}r$

$\dfrac{7}{\cancel{44}_{4}}\cdot\dfrac{\cancel{33}^{3}}{1}=\dfrac{\cancel{7}}{\cancel{44}}\cdot\dfrac{\cancel{44}}{\cancel{7}}r$

$\dfrac{21}{4}=r$

$5\dfrac{1}{4}\ \text{in.}=r$

38.　$\dfrac{2}{5}+\dfrac{1}{4}+x=1$

$\dfrac{8}{20}+\dfrac{5}{20}+x=\dfrac{20}{20}$

$\dfrac{13}{20}+x=\dfrac{20}{20}$

$\dfrac{13}{20}-\dfrac{13}{20}+x=\dfrac{20}{20}-\dfrac{13}{20}$

$x=\dfrac{7}{20}$ had no opinion

39. Mean:

$\left(18\dfrac{1}{2}+17\dfrac{3}{4}+18\dfrac{3}{4}+19\dfrac{1}{4}+16\dfrac{3}{4}+18\dfrac{1}{4}\right)\div 6$

$=\left(18\dfrac{2}{4}+17\dfrac{3}{4}+18\dfrac{3}{4}+19\dfrac{1}{4}+16\dfrac{3}{4}+18\dfrac{1}{4}\right)\div 6$

$=106\dfrac{13}{4}\cdot\dfrac{1}{6}$

$=\dfrac{437}{4}\cdot\dfrac{1}{6}$

$=\dfrac{437}{24}$ or $18\dfrac{5}{24}$ in.

Median:

$16\dfrac{3}{4},17\dfrac{3}{4},18\dfrac{1}{4},18\dfrac{1}{2},18\dfrac{3}{4},19\dfrac{1}{4}$

$\left(18\dfrac{1}{4}+18\dfrac{1}{2}\right)\div 2$

$=\left(18\dfrac{1}{4}+18\dfrac{2}{4}\right)\div 2$

$=36\dfrac{3}{4}\cdot\dfrac{1}{2}$

$=\dfrac{147}{4}\cdot\dfrac{1}{2}$

$=\dfrac{147}{8}$ or $18\dfrac{3}{8}$ in.

40.　$A=\dfrac{1}{2}h(a+b)$

$A=\dfrac{1}{2}\cdot 7\left(12\dfrac{1}{2}+19\right)$

$A=\dfrac{1}{2}\cdot\dfrac{7}{1}\cdot 31\dfrac{1}{2}$

$A=\dfrac{1}{2}\cdot\dfrac{7}{1}\cdot\dfrac{63}{2}$

$A=\dfrac{441}{4}$

$A=110\dfrac{1}{4}$ cm^2

41. $r=\dfrac{1}{2}\cdot 4$

$r=2$ ft.

$A=\pi r^2$

$A=\dfrac{22}{7}\cdot 2^2$

$A=\dfrac{22}{7}\cdot\dfrac{4}{1}$

$A=\dfrac{88}{7}$

$A=12\dfrac{4}{7}$ ft.^{2}

42. We must subtract the area of the triangle from the area of the trapezoid to find the shaded area.

$A_{\text{triangle}}=\dfrac{1}{2}bh$　　　$A_{\text{trapezoid}}=\dfrac{1}{2}h(a+b)$

$=\dfrac{1}{2}\cdot 11\cdot 5$　　　　$=\dfrac{1}{2}\cdot 12\cdot\left(11\dfrac{1}{4}+30\right)$

$=\dfrac{1}{2}\cdot\dfrac{11}{1}\cdot\dfrac{5}{1}$　　　$=\dfrac{1}{2}\cdot\dfrac{12}{1}\cdot 41\dfrac{1}{4}$

$=\dfrac{55}{2}$　　　　　　$=\dfrac{1}{2}\cdot\dfrac{\overset{3}{\cancel{12}}}{1}\cdot\dfrac{165}{\cancel{4}}$

$=27\dfrac{1}{2}$ cm^2　　　$=\dfrac{495}{2}$

$=247\dfrac{1}{2}$ cm^2

$247\dfrac{1}{2}-27\dfrac{1}{2}=220$ cm^2 is the area of the shaded region.

43.　$\dfrac{1}{4}(n+2)=\dfrac{3}{4}n-\dfrac{3}{5}$

$20\cdot\dfrac{1}{4}(n+2)=20\cdot\left(\dfrac{3}{4}n-\dfrac{3}{5}\right)$

$5(n+2)=\dfrac{20}{1}\cdot\dfrac{3}{4}n-\dfrac{20}{1}\cdot\dfrac{3}{5}$

$5\cdot n+5\cdot 2=\dfrac{\overset{5}{\cancel{20}}}{1}\cdot\dfrac{3}{\cancel{4}_1}n-\dfrac{\overset{4}{\cancel{20}}}{1}\cdot\dfrac{3}{\cancel{5}_1}$

$5n+10=15n-12$

$\underline{-5n\qquad\quad -5n}$

$10=10n-12$

$\underline{+12\qquad\quad +12}$

$22=10n+0$

$\dfrac{22}{10}=\dfrac{10n}{10}$

$\dfrac{22}{10}=1n$

$2\dfrac{1}{5}=n$

44.
$$A = \frac{1}{2}h(a+b)$$
$$19 = \frac{1}{2} \cdot 9 \cdot (2+b)$$
$$19 = \frac{9}{2}(2+b)$$
$$2 \cdot 19 = 2 \cdot \frac{9}{2}(2+b)$$
$$38 = 9(2+b)$$
$$38 = 9 \cdot 2 + 9 \cdot b$$
$$38 = 18 + 9b$$
$$\underline{-18 \quad -18}$$
$$20 = 0 + 9b$$
$$\frac{20}{9} = \frac{9b}{9}$$
$$\frac{20}{9} = b$$
$$2\frac{2}{9} \text{ m} = b$$

45. 1st section of pipe is $2\frac{1}{3}x$

2nd section of pipe is x

$$2\frac{1}{3}x + x = 35$$
$$3\frac{1}{3}x = 35$$
$$\frac{10}{3}x = 35$$
$$3 \cdot \frac{10}{3}x = 3 \cdot 35$$
$$10x = 105$$
$$\frac{10x}{10} = \frac{105}{10}$$
$$x = \frac{105}{10}$$
$$x = 10\frac{1}{2} \text{ ft.}$$

1st section of pipe is

$$2\frac{1}{3}x = 2\frac{1}{3} \cdot 10\frac{1}{2} = \frac{7}{3} \cdot \frac{\overset{7}{\cancel{21}}}{2} = \frac{49}{2} = 24\frac{1}{2} \text{ ft.}$$

2nd section of pipe is $x = 10\frac{1}{2}$ ft.

46. Larger angle $= n$

Smaller angle $= \frac{3}{4}n$

The two angles sum to $180°$.

$$n + \frac{3}{4}n = 180$$
$$\frac{7}{4}n = 180$$
$$\frac{4}{7} \cdot \frac{7}{4}n = \frac{4}{7} \cdot \frac{180}{1}$$
$$\frac{\cancel{4}}{\cancel{7}} \cdot \frac{\cancel{7}}{\cancel{4}}n = \frac{4}{7} \cdot \frac{180}{1}$$
$$n = \frac{720}{7} \text{ or } 102\frac{6}{7}$$

The smaller angle is $\frac{3}{4}n$ which is

$$\frac{3}{4} \cdot 102\frac{6}{7} = \frac{3}{4} \cdot \frac{720}{7} = \frac{3}{\cancel{4}} \cdot \frac{\overset{180}{\cancel{720}}}{7} = \frac{540}{7} \text{ or } 77\frac{1}{7}.$$

Therefore, the larger angle measures $102\frac{6}{7}°$ and

the smaller angle measures $77\frac{1}{7}°$.

47.

Categories	Rate	Time	Distance
Emma	$2\frac{1}{2}$	t	$2\frac{1}{2}t$
Bill	5	t	$5t$

$$2\frac{1}{2}t + 5t = \frac{7}{10}$$
$$7\frac{1}{2}t = \frac{7}{10}$$
$$\frac{15}{2}t = \frac{7}{10}$$
$$20 \cdot \frac{15}{2}t = 20 \cdot \frac{7}{10}$$
$$\frac{\overset{10}{\cancel{20}}}{1} \cdot \frac{15}{\cancel{2}_1}t = \frac{\overset{2}{\cancel{20}}}{1} \cdot \frac{7}{\cancel{10}_1}$$
$$150t = 14$$
$$\frac{150t}{150} = \frac{14}{150}$$
$$t = \frac{14}{150}$$
$$t = \frac{7}{75} \text{ hours}$$

48.

Categories	Rate	Time	Distance
asteroid A	50	t	$50t$
asteroid B	120	t	$120t$

$$50t + 120t = 2000$$
$$170t = 2000$$
$$\frac{170t}{170} = \frac{2000}{170}$$
$$t = 11\frac{13}{17} \text{ sec.}$$

Chapter 5 Practice Test

1. $\dfrac{3}{8}$ 2.

3. a) $\dfrac{3}{4} ? \dfrac{9}{12}$

$$\frac{3}{4} = \frac{3\cdot 3}{4\cdot 3} = \frac{9}{12}$$
$$9 = 9$$
so: $\dfrac{3}{4} = \dfrac{9}{12}$

   b) $\dfrac{14}{15} ? \dfrac{5}{6}$

$$\frac{14}{15} = \frac{14\cdot 2}{15\cdot 2} = \frac{28}{30}$$
$$\frac{5}{6} = \frac{5\cdot 5}{6\cdot 5} = \frac{25}{30}$$
$$28 > 25$$
so: $\dfrac{14}{15} > \dfrac{5}{6}$

4. a) $6\overline{)37} = 6\frac{1}{6}$ (with $\frac{36}{1}$)   b) $-\dfrac{8\cdot 4 + 5}{8} = -\dfrac{32 + 5}{8} = -\dfrac{37}{8}$

5. a) -16 b) undefined   c) 0 d) 1

6. a) $\dfrac{24}{40} = \dfrac{24 \div 8}{40 \div 8} = \dfrac{3}{5}$

   b) $-\dfrac{9x^5 y}{30x^3} = -\dfrac{\cancel{3}\cdot 3\cdot \cancel{x}\cdot \cancel{x}\cdot \cancel{x}\cdot x\cdot x\cdot y}{\cancel{3}\cdot 2\cdot 5\cdot \cancel{x}\cdot \cancel{x}\cdot \cancel{x}}$
   $$= -\frac{3x^2 y}{10}$$

7. a) $-2\dfrac{1}{4}\cdot 5\dfrac{1}{6} = -\dfrac{\overset{3}{\cancel{9}}}{4}\cdot \dfrac{31}{\underset{2}{\cancel{6}}} = -\dfrac{93}{8} = -11\dfrac{5}{8}$

   b) $\dfrac{2a}{7b^3}\cdot \dfrac{14b}{3a} = \dfrac{2\cdot \cancel{a}}{7\cdot \cancel{b}\cdot b\cdot b}\cdot \dfrac{2\cdot 7\cdot \cancel{b}}{3\cdot \cancel{a}}$
   $$= \frac{4}{3b^2}$$

8. a) $\left(\dfrac{2}{5}\right)^2 = \dfrac{2^2}{5^2} = \dfrac{4}{25}$ b) $\left(\dfrac{1}{4}\right)^3 = \dfrac{1^3}{4^3} = \dfrac{1}{64}$

9. a) $3\dfrac{1}{8} \div \dfrac{5}{4} = \dfrac{\overset{5}{\cancel{25}}}{\underset{2}{\cancel{8}}}\cdot \dfrac{\cancel{4}}{\cancel{5}} = \dfrac{5}{2} = 2\dfrac{1}{2}$

   b)
   $$\frac{-2m^4 n}{9n^2} \div \frac{5m}{12n^2} = \frac{-2m^4 n}{9n^2}\cdot \frac{12n^2}{5m}$$
   $$= \frac{-2\cdot \cancel{m}\cdot m\cdot m\cdot m\cdot \cancel{n}}{\cancel{3}\cdot 3\cdot \cancel{n}\cdot \cancel{n}}\cdot \frac{\cancel{3}\cdot 4\cdot \cancel{n}\cdot n}{5\cdot \cancel{m}}$$
   $$= \frac{-8m^3 n}{15}$$

10.
$$18t^3 = 2\cdot 3^2\cdot t^3$$
$$12t^2 u = 2^2\cdot 3\cdot t^2\cdot u$$
$$\text{LCM}(18t^3, 12t^2 u) = 2^2\cdot 3^2\cdot t^3\cdot u = 4\cdot 9t^3 u = 36t^3 u$$

11. a) $\dfrac{3}{10} + \dfrac{1}{4} = \dfrac{3\cdot 2}{10\cdot 2} + \dfrac{1\cdot 5}{4\cdot 5}$
    $$= \frac{6}{20} + \frac{5}{20}$$
    $$= \frac{11}{20}$$

    b) $-3\dfrac{1}{8} - \left(-1\dfrac{1}{2}\right) = -3\dfrac{1}{8} + 1\dfrac{1}{2}$
    $$= -\frac{25}{8} + \frac{3}{2}$$
    $$= -\frac{25}{8} + \frac{12}{8}$$
    $$= -\frac{13}{8}$$
    $$= -1\frac{5}{8}$$

12. a) $-\dfrac{4x}{15} + \left(-\dfrac{x}{15}\right) = \dfrac{-4x - x}{15}$
    $$= \frac{-5x}{15}$$
    $$= -\frac{x}{3}$$

    b) $\dfrac{7}{4a} - \dfrac{5}{6a} = \dfrac{7\cdot 3}{4a\cdot 3} - \dfrac{5\cdot 2}{6a\cdot 2}$
    $$= \frac{21}{12a} - \frac{10}{12a}$$
    $$= \frac{21 - 10}{12a}$$
    $$= \frac{11}{12a}$$

13. $2\dfrac{1}{4}+\dfrac{5}{8}\left(\dfrac{4}{5}-\dfrac{3}{10}\right)=2\dfrac{1}{4}+\dfrac{5}{8}\left(\dfrac{8}{10}-\dfrac{3}{10}\right)$

$$=2\dfrac{1}{4}+\dfrac{\cancel{5}}{8}\left(\dfrac{5}{\cancel{10}_{2}}\right)$$

$$=2\dfrac{4}{16}+\dfrac{5}{16}$$

$$=2\dfrac{9}{16}$$

14. a) $\left(\dfrac{1}{4}n^2-\dfrac{1}{2}n+2\right)-\left(\dfrac{1}{4}n^2+n+5\right)$

$$=\left(\dfrac{1}{4}n^2-\dfrac{1}{2}n+2\right)+\left(-\dfrac{1}{4}n^2-n-5\right)$$

$$=-\dfrac{1}{2}n-\dfrac{2}{2}n+2-5$$

$$=-\dfrac{3}{2}n-3$$

b) $\left(m+\dfrac{3}{4}\right)\left(2m-\dfrac{1}{2}\right)$

$$=m\cdot 2m+m\cdot\left(-\dfrac{1}{2}\right)+\dfrac{3}{4}\cdot 2m+\dfrac{3}{4}\cdot\left(-\dfrac{1}{2}\right)$$

$$=2m^2-\dfrac{1}{2}m+\dfrac{6}{4}m-\dfrac{3}{8}$$

$$=2m^2-\dfrac{1}{2}m+\dfrac{3}{2}m-\dfrac{3}{8}$$

$$=2m^2+\dfrac{2}{2}m-\dfrac{3}{8}$$

$$=2m^2+m-\dfrac{3}{8}$$

15. $$\dfrac{1}{4}m-\dfrac{2}{3}=\dfrac{5}{6}$$

$$12\cdot\left(\dfrac{1}{4}m-\dfrac{2}{3}\right)=12\cdot\left(\dfrac{5}{6}\right)$$

$$\dfrac{12}{1}\cdot\dfrac{1}{4}m-\dfrac{12}{1}\cdot\dfrac{2}{3}=\dfrac{12}{1}\cdot\dfrac{5}{6}$$

$$\dfrac{\overset{3}{\cancel{12}}}{1}\cdot\dfrac{1}{\cancel{4}_1}m-\dfrac{\overset{4}{\cancel{12}}}{1}\cdot\dfrac{2}{\cancel{3}_1}=\dfrac{\overset{2}{\cancel{12}}}{1}\cdot\dfrac{5}{\cancel{6}_1}$$

$$3m-8=10$$

$$\underline{+8\quad +8}$$

$$3m+0=18$$

$$\dfrac{3m}{3}=\dfrac{18}{3}$$

$$1m=6$$

$$m=6$$

Check: $\dfrac{1}{4}\cdot 6-\dfrac{2}{3}=\dfrac{5}{6}$

$$\dfrac{6}{4}-\dfrac{2}{3}=\dfrac{5}{6}$$

$$\dfrac{18}{12}-\dfrac{8}{12}=\dfrac{5}{6}$$

$$\dfrac{10}{12}=\dfrac{5}{6}$$

$$\dfrac{5}{6}=\dfrac{5}{6}$$

16. $\dfrac{\cancel{8}}{4}\cdot\dfrac{5}{\cancel{6}_2}=\dfrac{5}{8}$ of the lots had finished houses by the end of the first year.

17. $A=\dfrac{1}{2}bh$

$$A=\dfrac{1}{2}\cdot 10\cdot 4\dfrac{1}{2}$$

$$A=\dfrac{1}{\cancel{2}}\cdot\dfrac{\overset{5}{\cancel{10}}}{1}\cdot\dfrac{9}{2}$$

$$A=\dfrac{45}{2}$$

$$A=22\dfrac{1}{2}\ \text{m}^2$$

18. $C=\pi d$

$$C=\dfrac{22}{7}\cdot 4\dfrac{3}{4}$$

$$C=\dfrac{\overset{11}{\cancel{22}}}{7}\cdot\dfrac{19}{\cancel{4}_2}$$

$$C=\dfrac{209}{14}$$

$$C=14\dfrac{13}{14}\ \text{in.}$$

19. $4\dfrac{3}{8}\cdot n=35$ 8 pieces can be cut.

$$\dfrac{35}{8}\cdot n=35$$

$$8\cdot\dfrac{35}{8}\cdot n=8\cdot 35$$

$$35n=280$$

$$\dfrac{35n}{35}=\dfrac{280}{35}$$

$$n=8$$

20. First we must find the total that chose A and B.

$$\dfrac{5}{8}+\dfrac{1}{6}=\dfrac{5\cdot 3}{8\cdot 3}+\dfrac{1\cdot 4}{6\cdot 4}=\dfrac{15}{24}+\dfrac{4}{24}=\dfrac{19}{24}\ \text{chose A or B}$$

So, $\dfrac{5}{24}$ chose C.

21. To find the mean we need to sum the heights and divide by the number of heights.

$$\left(69\frac{1}{2}+70+70\frac{3}{4}+71\frac{1}{2}\right)\div 4$$

$$=281\frac{3}{4}\div 4$$

$$=\frac{1127}{4}\cdot\frac{1}{4}$$

$$=\frac{1127}{16}$$

$$=70\frac{7}{16}$$

The mean height is $70\frac{7}{16}$ in.

Since the scores are arranged in order from least to greatest, we find the mean of the middle two scores which are 70 and $70\frac{3}{4}$.

$$\frac{1}{2}\left(70+70\frac{3}{4}\right)=\frac{1}{2}\left(140\frac{3}{4}\right)=\frac{1}{2}\left(\frac{563}{4}\right)=\frac{563}{8}=70\frac{3}{8}$$

The median is $70\frac{3}{8}$ in.

22. $A=\dfrac{1}{2}h(a+b)$

$$A=\frac{1}{2}\cdot 8\left(9+6\frac{1}{2}\right)$$

$$A=\frac{1}{2}\cdot 8\cdot 15\frac{1}{2}$$

$$A=\frac{1}{\cancel{2}}\cdot\frac{\overset{4}{\cancel{8}}}{1}\cdot\frac{31}{\cancel{2}}$$

$$A=62 \text{ in.}^2$$

23. $r=\dfrac{1}{2}d$ $A=\pi r^2$

$$r=\frac{1}{2}\cdot 5 \qquad\qquad A=\frac{22}{7}\cdot\left(2\frac{1}{2}\right)^2$$

$$r=\frac{1}{2}\cdot\frac{5}{1} \qquad\qquad A=\frac{22}{7}\cdot\left(\frac{5}{2}\right)^2$$

$$r=\frac{5}{2} \qquad\qquad\qquad A=\frac{\overset{11}{\cancel{22}}}{7}\cdot\frac{25}{\underset{2}{\cancel{4}}}$$

$$r=2\frac{1}{2} \text{ in.} \qquad\qquad A=\frac{275}{14}$$

$$A=19\frac{9}{14} \text{ in.}^2$$

24. $\dfrac{2}{3}\cdot(n+5)=\dfrac{1}{2}n-\dfrac{3}{4}$

$$12\cdot\frac{2}{3}\cdot(n+5)=12\cdot\left(\frac{1}{2}n-\frac{3}{4}\right)$$

$$8\cdot(n+5)=12\cdot\left(\frac{1}{2}n-\frac{3}{4}\right)$$

$$8\cdot n+8\cdot 5=12\cdot\frac{1}{2}n-12\cdot\frac{3}{4}$$

$$8n+40=6n-9$$

$$\underline{-6n\qquad\quad -6n}$$

$$2n+40=0-9$$

$$\underline{-40\qquad -40}$$

$$2n+0=-49$$

$$\frac{2n}{2}=\frac{-49}{2}$$

$$n=-24\frac{1}{2}$$

25.

Categories	Rate	Time	Distance
Keisha	$5\frac{1}{2}$	t	$5\frac{1}{2}t$
Ken	8	t	$8t$

$$5\frac{1}{2}t+8t=\frac{3}{4}$$

$$13\frac{1}{2}t=\frac{3}{4}$$

$$\frac{27}{2}t=\frac{3}{4}$$

$$4\cdot\frac{27}{2}t=4\cdot\frac{3}{4}$$

$$\frac{\overset{2}{\cancel{4}}}{1}\cdot\frac{27}{\underset{1}{\cancel{2}}}t=\frac{\cancel{4}}{1}\cdot\frac{3}{\cancel{4}}$$

$$54t=3$$

$$\frac{54t}{54}=\frac{3}{54}$$

$$t=\frac{1}{18} \text{ hr}$$

Chapter 5 Cumulative Review Exercises

1. false   2. false   3. false   4. false

5. true   6. true   7. multiply

8. Write the prime factorization of each number in exponential form. Create a factorization for the GCF that contains only those prime factors

common to all factorizations, each raised to the least of its exponents. Then multiply.

9. Mistake: added -9 and -3 to get -6.

10. Find the LCD. Rewrite. Add or subtract the numerators and keep the LCD. Simplify.

11. four million, five hundred eighty-two thousand, six hundred one

12.

$$\overset{\longleftrightarrow}{\underset{-9\ -8\ -7\ -6\ -5\ -4\ -3\ -2\ -1\quad 0}{\quad}}$$

13. 850,000

14. $460 \times 70 = 32{,}200$

15. 1   16. $3600 = 2^4 \cdot 3^2 \cdot 5^2$ 17. $9x$

18. $-(-15) = 15$

19. $\begin{aligned} 5 - 12 + (-17) + 9 &= -7 + (-17) + 9 \\ &= -24 + 9 \\ &= -15 \end{aligned}$

20. $\begin{aligned} 2^3 - 7(-4) &= 8 - 7(-4) \\ &= 8 - (-28) \\ &= 8 + 28 \\ &= 36 \end{aligned}$

21. $\begin{aligned} -3^4 - 5\sqrt{100 - 64} &= -3^4 - 5\sqrt{36} \\ &= -81 - 5 \cdot 6 \\ &= -81 - 30 \\ &= -111 \end{aligned}$

22. $\begin{aligned} 18 - 5(6) \div [9 + 3(-7)] &= 18 - 5(6) \div [9 + (-21)] \\ &= 18 - 5(6) \div (-12) \\ &= 18 - 30 \div (-12) \\ &= 18 - \left(-\dfrac{30}{12}\right) \\ &= 18 + \dfrac{5}{2} \\ &= 18 + 2\dfrac{1}{2} \\ &= 20\dfrac{1}{2} \end{aligned}$

23. $\dfrac{36}{40} = \dfrac{36 \div 4}{40 \div 4} = \dfrac{9}{10}$

24. $\begin{aligned} 4\dfrac{1}{3} \div 5\dfrac{1}{9} &= \dfrac{13}{3} \div \dfrac{46}{9} \\ &= \dfrac{13}{\cancel{3}} \cdot \dfrac{\cancel{9}^3}{46} \\ &= \dfrac{39}{46} \end{aligned}$

25. $\begin{aligned} -\dfrac{5x^3}{9} \cdot \dfrac{27}{40x} &= -\dfrac{\cancel{5} \cdot \cancel{x} \cdot x \cdot x}{\cancel{3} \cdot \cancel{3}} \cdot \dfrac{\cancel{3} \cdot \cancel{3} \cdot 3}{\cancel{5} \cdot 8 \cdot \cancel{x}} \\ &= -\dfrac{3x^2}{8} \end{aligned}$

26. $\begin{aligned} 7\dfrac{1}{6} - 3\dfrac{1}{2} &= 7\dfrac{1}{6} - 3\dfrac{3}{6} \\ &= 6\dfrac{7}{6} - 3\dfrac{3}{6} \\ &= 3\dfrac{4}{6} \\ &= 3\dfrac{2}{3} \end{aligned}$

27. $\begin{aligned} \dfrac{4}{5x} + \dfrac{2}{3} &= \dfrac{4 \cdot 3}{5x \cdot 3} + \dfrac{2 \cdot 5x}{3 \cdot 5x} \\ &= \dfrac{12}{15x} + \dfrac{10x}{15x} \\ &= \dfrac{12 + 10x}{15x} \end{aligned}$

28. $\begin{aligned} \left(\dfrac{5}{6}\right)(-4) - 2(-4)^3 &= \left(\dfrac{5}{6}\right)(-4) - 2(-64) \\ &= -\dfrac{20}{6} - (-128) \\ &= -3\dfrac{1}{3} + 128 \\ &= -3\dfrac{1}{3} + 127\dfrac{3}{3} \\ &= 124\dfrac{2}{3} \end{aligned}$

29. $\begin{aligned} &12b^2 - \dfrac{2}{3}b + 3b^2 - b^3 - \dfrac{1}{4}b \\ &= -b^3 + 12b^2 + 3b^2 - \dfrac{2}{3}b - \dfrac{1}{4}b \\ &= -b^3 + 15b^2 - \dfrac{8}{12}b - \dfrac{3}{12}b \\ &= -b^3 + 15b^2 - \dfrac{11}{12}b \end{aligned}$

30. $\begin{aligned} &(x^4 - 5x^3 - 10x + 18) - (x^4 + 2x^3 - 3x - 8) \\ &= (x^4 - 5x^3 - 10x + 18) + (-x^4 - 2x^3 + 3x + 8) \\ &= -7x^3 - 7x + 26 \end{aligned}$

31. $\begin{aligned} (-8x^3)(-4x^2) &= 32x^{3+2} \\ &= 32x^5 \end{aligned}$

32. $(x - 7)(x + 7) = x^2 - 49$

33. $\dfrac{t^9}{t^5} = t^{9-5} = t^4$

34. $18m^3 + 24m^2 - 30m$

$$= 6m\left(\frac{18m^3 + 24m^2 - 30m}{6m}\right)$$

$$= 6m\left(\frac{18m^3}{6m} + \frac{24m^2}{6m} - \frac{30m}{6m}\right)$$

$$= 6m(3m^2 + 4m - 5)$$

35. $x + 24 = 8$ Check: $-16 + 24 = 8$

     $\underline{-24 \quad -24}$ $8 = 8$

        $x = -16$

36. $\dfrac{2}{3}a = -18$ Check: $\dfrac{2}{3} \cdot (-27) = -18$

$$\frac{\cancel{3}}{\cancel{2}} \cdot \frac{\cancel{2}}{\cancel{3}} a = \frac{3}{\cancel{2}} \cdot \left(\frac{-\cancel{18}^{\,9}}{1}\right) \qquad \frac{-54}{3} = -18$$

$$\qquad\qquad\qquad\qquad\qquad\qquad -18 = -18$$

$$a = -27$$

37.

$$\frac{3}{4}n - 1 = -\frac{2}{5}$$

$$20 \cdot \left(\frac{3}{4}n - 1\right) = 20 \cdot \left(-\frac{2}{5}\right)$$

$$\frac{\overset{5}{\cancel{20}}}{1} \cdot \frac{3}{\underset{1}{\cancel{4}}} n - 20 \cdot 1 = \frac{\overset{4}{\cancel{20}}}{1} \cdot \left(-\frac{2}{\underset{1}{\cancel{5}}}\right)$$

$$15n - 20 = -8$$

$$\underline{\qquad +20 \quad +20}$$

$$15n = 12$$

$$\frac{15n}{15} = \frac{12}{15}$$

$$1n = \frac{12 \div 3}{15 \div 3}$$

$$n = \frac{4}{5}$$

Check: $\dfrac{3}{\cancel{4}} \cdot \dfrac{\cancel{4}}{5} - 1 = -\dfrac{2}{5}$

$$\frac{3}{5} - \frac{5}{5} = -\frac{2}{5}$$

$$-\frac{2}{5} = -\frac{2}{5}$$

38. $3(y - 4) - 8 = 7y - 16$

   $3y - 12 - 8 = 7y - 16$

      $3y - 20 = 7y - 16$

     $\underline{-3y \qquad\quad -3y}$

    $0 - 20 = 4y - 16$

      $\underline{+16 \qquad\quad +16}$

       $-4 = 4y + 0$

       $\dfrac{-4}{4} = \dfrac{4y}{4}$

       $-1 = y$

Check: $3(-1 - 4) - 8 = 7(-1) - 16$

        $3(-5) - 8 = -7 - 16$

         $-15 - 8 = -23$

           $-23 = -23$

39. a) $178{,}900 - 139{,}200 = \$39{,}700$

   b) $178{,}900 - 128{,}300 = \$50{,}600$

   c) $\dfrac{\left(\begin{array}{c} 178{,}900 + 139{,}200 + 135{,}000 \\ +131{,}000 + 128{,}300 \end{array}\right)}{5}$

$$= \frac{712{,}400}{5}$$

$$= \$142{,}480$$

   d) $\$135{,}000$

40. We must find the total of the assets and debts to find the net worth.

$$1260 + 945 + 13{,}190 + (-872)$$

$$+(-57{,}189) + (-3782) + (-12{,}498)$$

$$= 15{,}395 + (-74{,}341)$$

$$= -58{,}946$$

The Krueger family's net worth is $-\$58{,}946$.

41. $h = -16t^2 + h_0$

   $h = -16 \cdot 2^2 + 92$

   $h = -16 \cdot 4 + 92$

   $h = -64 + 92$

   $h = 28$ ft. after 2 seconds

42. a) To find the perimeter, we add all of the sides.

$$P = (2y - 5) + (3y + 4) + (2y - 5) + (3y + 4)$$

$$P = 10y - 2$$

   b) $P = 10y - 2$

      $P = 10 \cdot 8 - 2$

      $P = 80 - 2$

      $P = 78$ ft.

43. $A = \dfrac{1}{2}bh$

$$A = \frac{1}{2} \cdot (h + 3) \cdot h$$

$$A = \frac{1}{2} \cdot (h \cdot h + 3 \cdot h)$$

$$A = \frac{1}{2} \cdot \left(h^2 + 3h\right)$$

$$A = \frac{1}{2} \cdot h^2 + \frac{1}{2} \cdot 3h$$

$$A = \frac{1}{2}h^2 + \frac{3}{2}h$$

44.
$$C = \pi \cdot d$$
$$44 = \frac{22}{7} \cdot d$$
$$\frac{7}{\cancel{22}} \cdot \frac{\cancel{44}^2}{1} = \frac{\cancel{22}}{\cancel{7}} \cdot d \cdot \frac{\cancel{7}}{\cancel{22}}$$
$$14 = 1d$$
$$14 \text{ in.} = d$$

45.
$$V = lwh$$
$$24 = 4 \cdot 3 \cdot h$$
$$24 = 12h$$
$$\frac{24}{12} = \frac{12h}{12}$$
$$2 \text{ ft.} = h$$

46. To find the perimeter of a rectangle, we must use the formula $P = 2l + 2w$. We will let the width be w. The length is "three times the width" so we say $3w$. We also know that the perimeter is 48.
$$P = 2l + 2w$$
$$48 = 2(3w) + 2w$$
$$48 = 6w + 2w$$
$$48 = 8w$$
$$\frac{48}{8} = \frac{8w}{8}$$
$$6 = w$$

The width is 6 ft. and the length $3w$ is $3 \times 6 = 18$ ft.

47. other number $= x$
one number $= 2x + 1$

$$x + (2x + 1) = 25$$
$$3x + 1 = 25$$
$$\underline{-1 \quad -1}$$
$$3x + 0 = 24$$
$$3x = 24$$
$$\frac{3x}{3} = \frac{24}{3}$$
$$1x = 8$$
$$x = 8$$

One number is $2(8) + 1 = 17$. The other number is 8.

48.

Categories	Value	Number	Amount
large box	3	$657 - x$	$3(657 - x)$
small box	2	x	$2x$

$$3(657 - x) + 2x = 1612$$
$$1971 - 3x + 2x = 1612$$
$$1971 - x = 1612$$
$$\underline{-1971 \qquad -1971}$$
$$-x = -359$$
$$x = 359$$

359 small boxes and $657 - 359 = 298$ large boxes of popcorn were sold.

49. Since the angle is on the ground, we conclude that we are dealing with supplementary angles that will total 180°. Let's call the angle made on one side x and the angle on the other side "is 35° less than twice the adjacent angle," so we will call it $2x - 35$.
$$x + (2x - 35) = 180$$
$$3x - 35 = 180$$
$$\underline{+35 \quad +35}$$
$$3x + 0 = 215$$
$$\frac{3x}{3} = \frac{215}{3}$$
$$x = 71\frac{2}{3}°$$

One angle is $71\frac{2}{3}°$ and the other angle is $2x - 35 = 2 \cdot 71\frac{2}{3} - 35 = 108\frac{1}{3}°$

50.

Categories	Rate	Time	Distance
Shelly	$5\frac{1}{2}$	t	$5\frac{1}{2}t$
Ryan	3	t	$3t$

$$5\frac{1}{2}t + 3t = 3\frac{2}{5}$$
$$8\frac{1}{2}t = 3\frac{2}{5}$$
$$\frac{17}{2}t = \frac{17}{5}$$
$$\frac{\cancel{2}}{\cancel{17}} \cdot \frac{\cancel{17}}{\cancel{2}}t = \frac{\cancel{17}}{5} \cdot \frac{2}{\cancel{17}}$$
$$1t = \frac{2}{5}$$
$$t = \frac{2}{5} \text{ hr.}$$

Chapter 6 Decimals

Exercise Set 6.1

1.

$$\begin{array}{c}\text{Hundreds} \\ \text{Tens} \\ \text{Ones} \\ . \\ \text{Tenths} \\ \text{Hundredths} \\ \text{Thousandths} \\ \text{Ten -thousandths}\end{array}$$

3. and

5. $0.2 = \dfrac{2}{10} = \dfrac{2 \div 2}{10 \div 2} = \dfrac{1}{5}$

7. $0.25 = \dfrac{25}{100} = \dfrac{25 \div 25}{100 \div 25} = \dfrac{1}{4}$

9. $0.375 = \dfrac{375}{1000} = \dfrac{375 \div 125}{1000 \div 125} = \dfrac{3}{8}$

11. $0.24 = \dfrac{24}{100} = \dfrac{24 \div 4}{100 \div 4} = \dfrac{6}{25}$

13. $1.5 = 1\dfrac{5}{10} = 1\dfrac{5 \div 5}{10 \div 5} = 1\dfrac{1}{2}$

15. $18.75 = 18\dfrac{75}{100} = 18\dfrac{75 \div 25}{100 \div 25} = 18\dfrac{3}{4}$

17. $9.625 = 9\dfrac{625}{1000} = 9\dfrac{625 \div 125}{1000 \div 125} = 9\dfrac{5}{8}$

19. $7.36 = 7\dfrac{36}{100} = 7\dfrac{36 \div 4}{100 \div 4} = 7\dfrac{9}{25}$

21. $-0.008 = -\dfrac{8}{1000} = -\dfrac{8 \div 8}{1000 \div 8} = -\dfrac{1}{125}$

23. $-13.012 = -13\dfrac{12}{1000} = -13\dfrac{12 \div 4}{1000 \div 4} = -13\dfrac{3}{250}$

25. ninety-seven thousandths

27. two thousand fifteen millionths

29. thirty-one and ninety-eight hundredths

31. five hundred twenty-one and six hundred eight thousandths

33. four thousand one hundred fifty-nine and six tenths

35. negative one hundred seven and ninety-nine hundredths

37. negative fifty thousand ninety-two hundred-thousandths

39.

41.

43.

45.

47.

49.

51. $0.81 > 0.8$ 53. $2.891 > 2.8909$

55. $0.001983 < 0.001985$

57. $0.0245 > 0.00963$

59. $-1.01981 > -1.10981$

61. $-145.7183 < -14.57183$

63. 610.3   65. 610.2832   67. 610

69. 1   71. 1.0   73. 0.951

75. –408.1   77. –408.0626   79. –410

Review Exercises

1. $\dfrac{37}{100} + \dfrac{4}{100} = \dfrac{37 + 4}{100} = \dfrac{41}{100}$

2. $-29 - (-16) = -29 + 16$
$ = -13$

3. $7a^2 - a + 15 - 10a^2 + a - 7a^3$
$= -7a^3 + 7a^2 - 10a^2 - a + a + 15$
$= -7a^3 - 3a^2 + 15$

4. $(5x^2 + 9x - 18) - (x^2 - 12x - 19)$
   $= (5x^2 + 9x - 18) + (-x^2 + 12x + 19)$
   $= 4x^2 + 21x + 1$

5. $x - 19 = -36$ Check: $-17 - 19 = -36$
   $\underline{+19 \quad +19}$ $-36 = -36$
   $x + 0 = -17$
   $\quad x = -17$

6. $P = 2l + 2w$
   $P = 2 \cdot 16 + 2 \cdot 12\dfrac{1}{2}$
   $P = 32 + \dfrac{\cancel{2}}{1} \cdot \dfrac{25}{\cancel{2}}$
   $P = 32 + 25$
   $P = 57$ ft.

Exercise Set 6.2

1. 1. Stack the numbers so that the place values align. (Line up decimal points.)
   2. Add the digits in the corresponding place values.
   3. Place the decimal point in the sum so that it aligns with the decimal points in the addends.

3. 27.31

5. $\quad 34.51$ 7. $\quad 58.9150$ 9. $\quad 312.980$
    $\underline{+125.20}$ $\underline{+0.8361}$ $\quad 6.337$
    $\quad 159.71$ $\quad 59.7511$ $\underline{+14.000}$
                                   $\quad 333.317$

11. $\quad 864.55$ 13. $\quad 809.06$ 15. $\quad 0.10050$
    $\underline{-23.10}$ $\underline{-24.78}$ $\underline{-0.08261}$
    $\quad 841.45$ $\quad 784.28$ $\quad 0.01789$

17. $\quad 25.00$
    $\underline{-14.89}$
    $\quad 10.11$

   Since the two numbers have different signs, we subtract their absolute values and keep the sign of the number with the greater absolute value. Because 25 has the greater absolute value, our sum will be positive.

19. $\quad 0.160$ $-0.16 + (-4.157) = -4.317$
    $\underline{+4.157}$
    $\quad 4.317$

   Because we must add two numbers that have the same sign, we add the absolute values and keep the negative sign.

21. $\quad 1.0000$ $0.0015 - 1 = -0.9985$
    $\underline{-0.0015}$
    $\quad 0.9985$

   Because we are adding numbers with different signs, we subtract and keep the sign of the larger absolute value. Because 1 has the greater absolute value, our sum will be negative.

23. $\quad 30.75$ $-30.75 - 159.27 = -190.02$
    $\underline{+159.27}$
    $\quad 190.02$

   Because we must add two numbers that have the same sign, we add the absolute values and keep the negative sign.

25. $-90 - (-16.75) = -90 + 16.75 = -73.25$

    $\quad 90.00$
    $\underline{-16.75}$
    $\quad 73.25$

   Because we are adding numbers with different signs, we subtract and keep the sign of the larger absolute value. Because 90 has the greater absolute value, our sum will be negative.

27. $-0.008 - (-1.0032) = -0.008 + 1.0032$
    $\qquad\qquad\qquad\quad = 0.9952$

    $\quad 1.0032$
    $\underline{-0.0080}$
    $\quad 0.9952$

   Because we are adding numbers with different signs, we subtract and keep the sign of the larger absolute value. Because 1.0032 has the greater absolute value, our sum will be positive.

29. $5y - 2.81y^2 + 7.1 - 10y^2 - 8.03$
    $= -2.81y^2 - 10y^2 + 5y + 7.1 - 8.03$
    $= -12.81y^2 + 5y - 0.93$

31. $a^3 - 2.5a + 9a^2 + a - 12 + 3.5a^3 + 1.4$
    $= a^3 + 3.5a^3 + 9a^2 - 2.5a + a - 12 + 1.4$
    $= 4.5a^3 + 9a^2 - 1.5a - 10.6$

33. $(x^2 + 5.2x + 3.4) + (9.2x^2 - 6.1x + 5)$
    $= x^2 + 9.2x^2 + 5.2x - 6.1x + 3.4 + 5$
    $= 10.2x^2 - 0.9x + 8.4$

35. $(3.1a^3 - a^2 + 7.5a - 0.01) + (6.91a^3 - 3.91a^2 + 6)$
    $= 3.1a^3 + 6.91a^3 - a^2 - 3.91a^2 + 7.5a - 0.01 + 6$
    $= 10.01a^3 - 4.91a^2 + 7.5a + 5.99$

37. $(1.4n^2 + 8.3n - 0.6) - (n^2 - 2.5n + 3)$

$= (1.4n^2 + 8.3n - 0.6) + (-n^2 + 2.5n - 3)$

$= 1.4n^2 - n^2 + 8.3n + 2.5n - 0.6 - 3$

$= 0.4n^2 + 10.8n - 3.6$

39. $\left(5k^3 + 2.5k^2 - 6.2k - 0.44\right) - \left(2.2k^3 - 6.2k - 0.5\right)$

$= \left(5k^3 + 2.5k^2 - 6.2k - 0.44\right) + \left(-2.2k^3 + 6.2k + 0.5\right)$

$= 5k^3 - 2.2k^3 + 2.5k^2 - 6.2k + 6.2k - 0.44 + 0.5$

$= 2.8k^3 + 2.5k^2 + 0.06$

41. $\left(12.91x^3 - 16.2x^2 - x + 7\right)$

$\quad - \left(10x^3 - 5.1x^2 - 3.6x + 11.45\right)$

$= \left(12.91x^3 - 16.2x^2 - x + 7\right)$

$\quad + \left(-10x^3 + 5.1x^2 + 3.6x - 11.45\right)$

$= 12.91x^3 - 10x^3 - 16.2x^2 + 5.1x^2$

$\quad - x + 3.6x + 7 - 11.45$

$= 2.91x^3 - 11.1x^2 + 2.6x - 4.45$

43. $m + 3.67 = 14.5$ Check: $10.83 + 3.67 \overset{?}{=} 14.5$

$\quad \underline{-3.67 \quad -3.67}$ $\qquad\qquad 14.5 = 14.5$

$\quad m + 0 = 10.83$

$\qquad m = 10.83$

45. $2.19 = 8 + y$ Check: $2.19 \overset{?}{=} 8 + (-5.81)$

$\quad \underline{-8 \quad -8}$ $\qquad\qquad 2.19 = 2.19$

$-5.81 = 0 + y$

$-5.81 = y$

47. $k - 4.8 = -8.02$ Check: $-3.22 - 4.8 \overset{?}{=} -8.02$

$\quad \underline{+4.8 \quad +4.8}$ $\qquad\qquad -8.02 = -8.02$

$\quad k + 0 = -3.22$

$\qquad k = -3.22$

49. balance = initial balance – sum of deductions + transfers

$B = 682.20 - (34.58 + 30.00 + 560.00$

$\qquad + 58.85 + 45.60) + 0$

$B = 682.20 - 729.03 + 0$

$B = -46.83$

Adam's account balance is overdrawn $46.83.

51. balance = payments – charges

$B = 200 - (75 + 15.50 + 485.65 + 45.79)$

$B = 200 - 621.94$

$B = -421.94$

The client owes $421.94.

53. change = amount paid – cost

$C = 20 - (1.59 + 2.49 + 1.79 + 0.79 + 3.45 + 0.51)$

$C = 20 - 10.62$

$C = 9.38$

The change is $9.38.

55. net pay = gross pay – deductions

$N = 2904.17 - (174.25 + 39.64 + 235.62$

$\qquad + 131.56 + 45.00 + 12.74 + 138.15 + 200.00 + 7.32)$

$N = 2904.17 - 984.28$

$N = 1919.89$

The net pay is $1919.89.

57. closing price of stock at the end of the first day + amount stock price changed = closing price of stock at the end of the second day

$35.25 + n = 32.50$

$\underline{-35.25 \qquad -35.25}$

$\quad 0 + n = -2.75$

$\qquad n = -2.75$

The stock price decreased by $2.75.

59. To find the perimeter, we add all of the sides.

$96.1 = 36.8 + 18.5 + d + 18.5$

$96.1 = 73.8 + d$

$\underline{-73.8 \quad -73.8}$

$22.3 = 0 + d$

$22.3 = d$

The missing side is 22.3m.

61. a)

	January 24	January 25
Equation	$43.55 + n = 43.92$ $\underline{-43.55 \quad -43.55}$ $0 + n = 0.37$ $n = 0.37$	$43.92 + n = 43.29$ $\underline{-43.92 \quad -43.92}$ $0 + n = -0.63$ $n = -0.63$
Price change	$0.37	–$0.63
	January 26	January 27
Equation	$43.29 + n = 48.57$ $\underline{-43.29 \quad -43.29}$ $0 + n = 5.28$ $n = 5.28$	$48.57 + n = 50.30$ $\underline{-48.57 \quad -48.57}$ $0 + n = 1.73$ $n = 1.73$
Price change	$5.28	$1.73

b) January 26

Review Exercises

1. $-12(25) = -300$

2. $\dfrac{3}{10} \cdot \dfrac{7}{10} = \dfrac{3 \cdot 7}{10 \cdot 10} = \dfrac{21}{100}$

3. $34\overline{)10404}$

$\underline{102}$
204
$\underline{204}$
0

4. $9\frac{1}{6} \div 3\frac{1}{4} = \frac{55}{6} \div \frac{13}{4}$

$= \frac{55}{\underset{3}{\cancel{6}}} \cdot \frac{\cancel{4}^{2}}{13}$

$= \frac{110}{39}$

$= 2\frac{32}{39}$

5. $\sqrt{121} = 11$

6. $-2x^2 \cdot 9x^5 = -2 \cdot 9x^{2+5}$
$= -18x^7$

7. $(y+3)(y-5)$
$= y \cdot y + y \cdot (-5) + 3 \cdot y + 3 \cdot (-5)$
$= y^2 - 5y + 3y - 15$
$= y^2 - 2y - 15$

8. $-8x = 72$ Check: $-8 \cdot (-9) = 72$

$\frac{-8x}{-8} = \frac{72}{-8}$ $72 = 72$

$1x = -9$

$x = -9$

Exercise Set 6.3

1. Place the decimal point in the product so that the product has the same number of decimal places as the total number of decimal places in the factors.

3. No, because 24.7 is greater than 10. The absolute value of the decimal number needs to be greater than or equal to 1 but less than 10.

5. 0.6
$\underline{\times 0.9}$
0.54

7. 3.65
$\underline{\times 2.9}$
3285
$\underline{730}$
10.585

9. 71.62
$\underline{\times 9.81}$
7162
57296
$\underline{64458}$
702.5922

11. 619.45
$\underline{\times 0.01}$
6.1945

13. 19.6
$\underline{\times 10}$
196

15. $(0.1508)(-100) = -15.08$

17. $152(-0.001) = -0.152$

19. $(-0.029)(-0.15) = 0.00435$

21. $(0.9)^2 = (0.9)(0.9) = 0.81$

23. $(2.1)^3 = (2.1)(2.1)(2.1) = 9.261$

25. $(-0.03)^4 = (-0.03)(-0.03)(-0.03)(-0.03)$
$= 0.00000081$

27. $(-0.4)^3 = (-0.4)(-0.4)(-0.4) = -0.064$

29. $300,000,000$ m/sec.; three hundred million m/sec.

31. $2,140,000$ light-years; two million, one hundred forty thousand light-years

33. $298,000,000$ people; two hundred ninety-eight million people

35. 2.3×10^6 blocks   37. 2.795×10^9 mi.

39. 5.3×10^9 nucleotide pairs

41. $(3.2x^5)(1.8x^3) = 3.2 \cdot (1.8) \cdot x^{5+3}$
$= 5.76x^8$

43. $\left(0.2a^2\right)\left(-0.08a\right) = 0.2 \cdot \left(-0.08\right)a^{2+1}$
$= -0.016a^3$

45. $\left(0.5t^3\right)^2 = \left(0.5\right)^2 t^{3 \cdot 2}$
$= 0.25t^6$

47. $0.5(1.8y^2 + 2.6y + 38)$
$= 0.5 \cdot \left(1.8y^2\right) + 0.5 \cdot \left(2.6y\right) + 0.5 \cdot \left(38\right)$
$= 0.9y^2 + 1.3y + 19$

49. $-6.8\left(0.02a^3 - a + 1.9\right)$
$= -6.8 \cdot \left(0.02a^3\right) - 6.8 \cdot (-a) - 6.8 \cdot \left(1.9\right)$
$= -0.136a^3 + 6.8a - 12.92$

51. $\left(x + 1.12\right)\left(x - 0.2\right)$
$= x \cdot x + x \cdot \left(-0.2\right) + 1.12 \cdot \left(x\right) + 1.12 \cdot \left(-0.2\right)$
$= x^2 - 0.2x + 1.12x - .224$
$= x^2 + 0.92x - 0.224$

53. $(0.4x - 1.22)(16x + 3)$
$= 0.4x \cdot \left(16x\right) + 0.4x \cdot \left(3\right) - 1.22 \cdot \left(16x\right) - 1.22 \cdot \left(3\right)$
$= 6.4x^2 + 1.2x - 19.52x - 3.66$
$= 6.4x^2 - 18.32x - 3.66$

55. $(5.8a + 9)(5.8a - 9) = 33.64a^2 - 81$

57. $(k - 0.1)(14.7k - 2.8)$
$= k \cdot \left(14.7k\right) + k \cdot (-2.8) - 0.1 \cdot \left(14.7k\right) - 0.1 \cdot (-2.8)$
$= 14.7k^2 - 2.8k - 1.47k + 0.28$
$= 14.7k^2 - 4.27k + 0.28$

59. number of gallons · price per gallon = cost
    $32 \cdot 3.69 = 118.08$
    It will cost $118.08.

61. cost = rate · time
    cost = $0.12 \cdot 32$
    cost = $3.84

63. cost = rate · weight
    cost = $0.79 \cdot 4.75$
    cost ≈ $3.75

65. number of therms used · cost per therm = total cost
    $76 \cdot 1.3548 = 102.96$
    Ruth will pay $102.96.

67. tax = dollar value · rate
    tax = $6400 \cdot 0.0285$
    tax = $182.40

69. $V = ir$
    $V = (-3.5)(0.8)$
    $V = -2.8$ volts

71. $d = rt$
    $d = 12.5(4.5)$
    $d = 56.25$ ft.

73. $A = lw$
    $A = (4.8)(4.8)$
    $A = 23.04$ cm^2

75. $A = bh$
    $A = 160.5(78.2)$
    $A = 12551.1$ ft.^{2}

77. $V = lwh$
    $V = 0.18(0.09)(0.5)$
    $V = 0.0081$ m^3

Review Exercises

1. $16\overline{)3280}$ = 205
    $\underline{32}$
    080
    $\underline{80}$
    0

2. $\sqrt{169} = 13$

3. $\dfrac{12x^7}{2x^3} = \dfrac{12}{2}x^{7-3}$
    $= 6x^4$

4. $9t = -81$ Check: $9 \cdot (-9) = -81$
    $\dfrac{9t}{9} = \dfrac{-81}{9}$ $-81 = -81$
    $t = -9$

5. $3x + 5 = 32$ Check: $3 \cdot 9 + 5 = 32$
    $\underline{-5 \quad -5}$ $27 + 5 = 32$
    $3x + 0 = 27$ $32 = 32$
    $3x = 27$
    $\dfrac{3x}{3} = \dfrac{27}{3}$
    $1x = 9$
    $x = 9$

6. $-2a + 9 = 3a - 6$ Check: $-2 \cdot 3 + 9 = 3 \cdot 3 - 6$
    $\underline{+2a \qquad +2a}$ $-6 + 9 = 9 - 6$
    $0 + 9 = 5a - 6$ $3 = 3$
    $\underline{+6 \qquad +6}$
    $15 = 5a + 0$
    $15 = 5a$
    $\dfrac{15}{5} = \dfrac{5a}{5}$
    $3 = 1a$
    $3 = a$

7. $r = \dfrac{1}{2}d$ $C = \pi \cdot d$
    $r = \dfrac{1}{2} \cdot \dfrac{15}{1}$ $C = \dfrac{22}{7} \cdot 15$
    $r = \dfrac{15}{2}$ $C = \dfrac{330}{7}$
    $r = 7\dfrac{1}{2}$ in. $C = 47\dfrac{1}{7}$ in.

Exercise Set 6.4

1. 6.2

3. To write a fraction as a decimal number, divide the denominator into the numerator.

5. $12\overline{)286.20}$ = 23.85
    $\underline{24}$
    46
    $\underline{36}$
    102
    $\underline{96}$
    60
    $\underline{60}$
    0

7. $20\overline{)0.480}$ = 0.024
    $\underline{40}$
    80
    $\underline{80}$
    0

9. $5640 \div 9.4 = 56400 \div 94$ so, $94\overline{)56400}$ = 600
    $\underline{564}$
    000

11. $0.288 \div 0.06 = 28.8 \div 6$ so, $6\overline{)28.8}$

$$\begin{array}{r} 4.8 \\ 6\overline{)28.8} \\ \underline{24} \\ 48 \\ \underline{48} \\ 0 \end{array}$$

so, $85\overline{)90.10}$

$$\begin{array}{r} 1.06 \\ 85\overline{)90.10} \\ \underline{85} \\ 5\,10 \\ \underline{5\,10} \\ 0 \end{array}$$

13. $6.4 \div 0.16 = 640 \div 16$ so, $16\overline{)640}$

$$\begin{array}{r} 40 \\ 16\overline{)640} \\ \underline{64} \\ 00 \end{array}$$

so, $-6.4 \div 0.16 = -40$

25. $\dfrac{3}{5} = 5\overline{)3.0}$

$$\begin{array}{r} 0.6 \\ 5\overline{)3.0} \\ \underline{3\,0} \\ 0 \end{array}$$

27. $\dfrac{9}{20} = 20\overline{)9.00}$

$$\begin{array}{r} 0.45 \\ 20\overline{)9.00} \\ \underline{8\,0} \\ 1\,00 \\ \underline{1\,00} \\ 0 \end{array}$$

15. $1 \div 0.008 = 1000 \div 8$ so, $8\overline{)1000}$

$$\begin{array}{r} 125 \\ 8\overline{)1000} \\ \underline{8} \\ 20 \\ \underline{16} \\ 40 \\ \underline{40} \\ 0 \end{array}$$

29. $\dfrac{7}{16} = 16\overline{)7.0000}$

$$\begin{array}{r} 0.4375 \\ 16\overline{)7.0000} \\ \underline{6\,4} \\ 60 \\ \underline{48} \\ 120 \\ \underline{112} \\ 80 \\ \underline{80} \\ 0 \end{array}$$

so, $-\dfrac{7}{16} = -0.4375$

31. $\dfrac{13}{30} = 30\overline{)13.000} = 0.4\overline{3}$

$$\begin{array}{r} 0.4333 \\ 30\overline{)13.000} \\ \underline{12\,0} \\ 1\,00 \\ \underline{90} \\ 100 \\ \underline{90} \end{array}$$

17.

$$\begin{array}{r} 0.0206 \\ 1000\overline{)20.6000} \\ \underline{20\,00} \\ 6000 \\ \underline{6000} \\ 0 \end{array}$$

so, $20.6 \div (-1000) = -0.0206$

19. $-8.145 \div (-0.01) = -814.5 \div (-1) = 814.5$

21. $19.6 \div 0.11 = 1960 \div 11$

$$\begin{array}{r} 178.1818 \\ 11\overline{)1960.0000} \\ \underline{11} \\ 86 \\ \underline{77} \\ 90 \\ \underline{88} \\ 20 \\ \underline{11} \\ 90 \\ \underline{88} \\ 20 \\ \underline{11} \end{array}$$

so, $19.6 \div 0.11 = 178.\overline{18}$

23. $-0.0901 \div (-0.085) = 90.1 \div 85$

33. $13\dfrac{1}{4} = 13 + \dfrac{1}{4}$ so, $4\overline{)1.00}$

$$\begin{array}{r} 0.25 \\ 4\overline{)1.00} \\ \underline{8} \\ 20 \\ \underline{20} \\ 0 \end{array}$$

so, $13 + 0.25 = 13.25$

35. $-17\dfrac{5}{8} = -\left(17 + \dfrac{5}{8}\right)$ so, $8\overline{)5.000}$

$$\begin{array}{r} 0.625 \\ 8\overline{)5.000} \\ \underline{4\,8} \\ 20 \\ \underline{16} \\ 40 \\ \underline{40} \\ 0 \end{array}$$

so, $-(17 + 0.625) = -17.625$

37. $104\dfrac{2}{3} = 104 + \dfrac{2}{3}$ so, $3\overline{)2.00} = 0.\overline{6}$

$$\begin{array}{r} 0.66 \\ 3\overline{)2.00} \\ \underline{1\,8} \\ 20 \\ \underline{18} \\ 2 \end{array}$$

so, $104 + 0.\overline{6} = 104.\overline{6}$

39. $-216\frac{4}{7} = -\left(216 + \frac{4}{7}\right)$

so, $7\overline{)4.00000000} = 0.\overline{571428}$ with quotient 0.57142857

$\quad\underline{3\,5}$

$\quad 50$

$\quad\underline{49}$

$\quad 10$

$\quad\underline{7}$

$\quad 30$

$\quad\underline{28}$

$\quad 20$

$\quad\underline{14}$

$\quad 60$

$\quad\underline{56}$

$\quad 40$

$\quad\underline{35}$

$\quad 50$

$\quad\underline{49}$

$\quad 1$

so, $-(216 + 0.\overline{571428}) = -216.\overline{571428}$

41. $\sqrt{0.0016} = 0.04$ because $(0.04)^2 = 0.0016$

43. $\sqrt{0.25} = 0.5$ because $(0.5)^2 = 0.25$

45. $\sqrt{24} \approx 4.89897 \approx 4.90$

47. $\sqrt{200} \approx 14.142 \approx 14.14$

49. $\sqrt{1.69} = 1.3$ because $(1.3)^2 = 1.69$

51. $\sqrt{0.009} \approx 0.09486 \approx 0.09$

53. $10.2x^6 \div 0.4x^4 = \dfrac{10.2}{0.4}x^{6-4}$

$\qquad\qquad\qquad = 25.5x^2$

55. $\dfrac{-0.96m^3n}{0.15m} = \dfrac{-0.96}{0.15}m^{3-1}n$

$\qquad\qquad\quad = -6.4m^2n$

57. $\dfrac{-3.03a^5bc}{-20.2a^4b} = \dfrac{-3.03}{-20.2}a^{5-4}b^{1-1}c$

$\qquad\qquad\qquad = 0.15ac$

59. $2.1b = 12.642$ 　　Check: $2.1\cdot(6.02) = 12.642$

$\dfrac{2.1b}{2.1} = \dfrac{12.642}{2.1}$ 　　　　　　$12.642 = 12.642$

$1b = 6.02$

$b = 6.02$

61. $-0.88h = 1.408$ 　Check: $-0.88(-1.6) = 1.408$

$\dfrac{-0.88h}{-0.88} = \dfrac{1.408}{-0.88}$ 　　　　　$1.408 = 1.408$

$1h = -1.6$

$h = -1.6$

63. $-2.28 = -3.8n$ 　Check: $-2.28 = -3.8(0.6)$

$\dfrac{-2.28}{-3.8} = \dfrac{-3.8n}{-3.8}$ 　　　　$-2.28 = -2.28$

$0.6 = 1n$

$0.6 = n$

65. $-28.7 = 8.2x$ 　Check: $-28.7 = 8.2(-3.5)$

$\dfrac{-28.7}{8.2} = \dfrac{8.2x}{8.2}$ 　　　　$-28.7 = -28.7$

$-3.5 = 1x$

$-3.5 = x$

67. To calculate the payments, divide $1839.96 by 3. $1839.96 \div 3 = 613.32$ so, each payment will be $613.32.

69. To calculate the bursts of radiation, divide 324 by 8. $324 \div 8 = 40.5$ so, each burst of radiation will be 40.5 rads.

71. $V = ir$

$12 = 0.03r$

$\dfrac{12}{0.03} = \dfrac{0.03r}{0.03}$

$400\Omega = r$

73. $d = rt$

$600 = 350t$

$\dfrac{600}{350} = \dfrac{350t}{350}$

$1.71\text{ hr.} \approx t$

75. $d = 2r$ 　　　　$C = \pi d$

$d = 2\cdot 1087.5$ 　　$C = 2175\pi$

$d = 2175\text{ mi.}$ 　　$C = 2175(3.14)$

$\qquad\qquad\qquad C \approx 6829.5\text{ mi.}$

77. $C = \pi d$

$21 = \pi d$

$21 \approx 3.14d$

$\dfrac{21}{3.14} \approx \dfrac{3.14d}{3.14}$

$6.69\text{ mi.} \approx d$

79. $C = \pi d$

$6.5 \approx 3.14d$

$\dfrac{6.5}{3.14} \approx \dfrac{3.14d}{3.14}$

$2.07 \approx 1d$

$2.07 \approx d$

so, $r \approx 1.04\text{ ft.}$

Review Exercises

1. $16 - 9[12 + (8 - 13)] \div 3 = 16 - 9[12 + (-5)] \div 3$
    $= 16 - 9 \cdot 7 \div 3$
    $= 16 - 63 \div 3$
    $= 16 - 21$
    $= -5$

2. $\dfrac{3}{5}(30) = \dfrac{3}{5} \cdot \dfrac{30}{1}$
    $= \dfrac{90}{5}$
    $= 18$

3. $\dfrac{1}{2} \cdot -10 \cdot 6^2 = \dfrac{1}{2} \cdot \dfrac{-10}{1} \cdot \dfrac{36}{1}$
    $= \dfrac{-360}{2}$
    $= -180$

4. $A = \dfrac{1}{2}bh$
    $A = \dfrac{1}{2} \cdot 14 \cdot 8$
    $A = \dfrac{1}{\cancel{2}_1} \cdot \dfrac{\cancel{14}^{7}}{1} \cdot \dfrac{8}{1}$
    $A = 56$
    $A = 56 \text{ in.}^2$

5. $A = \dfrac{1}{2}h(a + b)$
    $A = \dfrac{1}{2} \cdot 20 \cdot (46 + 22)$
    $A = \dfrac{1}{2} \cdot 20 \cdot 68$
    $A = \dfrac{1360}{2}$
    $A = 680 \text{ cm}^2$

Exercise Set 6.5

1. Writing the decimals as fractions is better in this case because the decimal equivalent of $\dfrac{2}{3}$ is $0.\overline{6}$, which would have to be rounded for the calculation and rounding would yield an approximation.

3. Cylinder: $V = \pi r^2 h$, Cone: $V = \dfrac{1}{3}\pi r^2 h$. The volume of a cone is one-third of the volume of a cylinder with the same radius and height.

5. $0.64 + 2.5(0.8) = 0.64 + 2$
    $= 2.64$

7. $-5.1(3.4 - 5.7) = -5.1(-2.3)$
    $= 11.73$

9. $88.2 - 2.2(0.45 + 20.1) = 88.2 - 2.2(20.55)$
    $= 88.2 - 45.21$
    $= 42.99$

11. $(0.4)^2 - 2.8 \div 0.2(1.6) = 0.16 - 2.8 \div 0.2(1.6)$
    $= 0.16 - 14(1.6)$
    $= 0.16 - 22.4$
    $= -22.24$

13. $10.7 - 18\sqrt{1.96} + (74.6 - 88.1)$
    $= 10.7 - 18\sqrt{1.96} + (-13.5)$
    $= 10.7 - 18(1.4) + (-13.5)$
    $= 10.7 - 25.2 + (-13.5)$
    $= -28$

15. $\sqrt{0.0081} + 120.8 \div 4(2.5 - 6.4)$
    $= 0.09 + 120.8 \div 4(-3.9)$
    $= 0.09 + 30.2(-3.9)$
    $= 0.09 + (-117.78)$
    $= -117.69$

17. $[-5.53 \div (0.68 + 0.9)] + 0.7\sqrt{2.25}$
    $= [-5.53 \div 1.58] + 0.7(1.5)$
    $= -3.5 + 1.05$
    $= -2.45$

19. $\dfrac{3}{5}(-0.85) = 0.6(-0.85)$
    $= -0.51$

21. $\dfrac{1}{4} + \dfrac{2}{3}(0.06) = \dfrac{1}{4} + \dfrac{\cancel{2}}{\cancel{3}_1} \cdot \dfrac{\cancel{6}^{\,2^{\,1}}}{\cancel{100}_{\cancel{50}\,25}}$
    $= \dfrac{1}{4} + \dfrac{1}{25}$
    $= \dfrac{1 \cdot 25}{4 \cdot 25} + \dfrac{1 \cdot 4}{25 \cdot 4}$
    $= \dfrac{25}{100} + \dfrac{4}{100}$
    $= \dfrac{29}{100}$
    $= 0.29$

23. $-2\dfrac{4}{9}(1.8) - \dfrac{3}{8} = -\dfrac{\cancel{22}^{\,11}}{\cancel{9}} \cdot \dfrac{\cancel{18}^{\,2}}{\cancel{10}_5} - \dfrac{3}{8}$
    $= -\dfrac{22}{5} - \dfrac{3}{8}$
    $= -\dfrac{22 \cdot 8}{5 \cdot 8} - \dfrac{3 \cdot 5}{8 \cdot 5}$
    $= -\dfrac{176}{40} - \dfrac{15}{40}$
    $= -\dfrac{191}{40}$
    $= -4.775$

25. $\dfrac{2}{5} \div (-0.8) + \left(\dfrac{1}{3}\right)^2 = \dfrac{2}{5} \div (-0.8) + \dfrac{1}{9}$

$= \dfrac{2}{5} \div -\dfrac{8}{10} + \dfrac{1}{9}$

$= \dfrac{\cancel{2}}{\cancel{5}} \cdot \left(-\dfrac{\cancel{10}^{\,2}}{\cancel{8}_{\,4}}\right) + \dfrac{1}{9}$

$= -\dfrac{\cancel{2}^{\,1}}{\cancel{4}_{\,2}} + \dfrac{1}{9}$

$= -\dfrac{1}{2} + \dfrac{1}{9}$

$= -\dfrac{1 \cdot 9}{2 \cdot 9} + \dfrac{1 \cdot 2}{9 \cdot 2}$

$= -\dfrac{9}{18} + \dfrac{2}{18}$

$= -\dfrac{7}{18}$

$= -0.3\overline{8}$

27. a) Mean:

$\dfrac{(15 + 38 + 12 + 26)}{4} = \dfrac{91}{4} = 22.75$ minutes

Median: 12, 15, 26, 38
The median is the mean of the middle two values which are 15 and 26.

$\dfrac{15 + 26}{2} = \dfrac{41}{2} = 20.5$ minutes

b) You must divide the Phoenix call into two parts since it was included in two time frames.

Total cost $= \left(\begin{array}{c}\text{cost of call} \\ \text{to Denver}\end{array}\right) + \left(\begin{array}{c}\text{cost of call} \\ \text{to Austin}\end{array}\right)$
$\qquad + \left(\begin{array}{c}\text{cost of call} \\ \text{to Phoenix}\end{array}\right) + \left(\begin{array}{c}\text{cost of call} \\ \text{to Portland}\end{array}\right)$

$C = \left[0.12(15)\right] + \left[0.09(38)\right]$
$\qquad + \left[0.09(3) + 0.12(9)\right] + \left[0.10(26)\right]$
$C = 1.8 + 3.42 + 0.27 + 1.08 + 2.60$
$C = 9.17$

The total bill is $9.17.

29. Mean:

$\dfrac{\left(\begin{array}{c}73 + 73 + 69 + 69 + 76 + 74 + 66 \\ + 70 + 74 + 73 + 74 + 67 + 69\end{array}\right)}{13} = \dfrac{927}{13} \approx 71.3$ in.

Median: 66, 67, 69, 69, 69, 70, 73, 73, 73, 74, 74, 74, 76
The median is the middle height which is 73 inches.

31. Mean:

$\dfrac{\left(\begin{array}{c}47.03 + 46.39 + 46.47 + 46.32 \\ + 47.66 + 47.68 + 47.91 + 47.5\end{array}\right)}{8} = \dfrac{376.96}{8} = \$47.12$

Median: 46.32, 46.39, 46.47, 47.03, 47.5, 47.66, 47.68, 47.91
The median is the mean of the two middle scores which are 47.03 and 47.5.

$\dfrac{47.03 + 47.5}{2} = \dfrac{94.53}{2} \approx \$47.27$

33. Begin by finding the actual mileage Aimee traveled: ending odometer reading – beginning odometer reading.

$76264.1 - 75618.4 = 645.7$ miles She will be reimbursed $0.35 for every mile:

$0.35(645.7) = 225.995 \approx 226.00$ On Tuesday, Aimee will be paid for lunch and dinner ($12 + 15 = \$27$). On Wednesday, she will be paid for breakfast, lunch and dinner ($10 + 12 + 15 = \$37$). And on Thursday, she will again receive payment for breakfast, lunch, and dinner ($37). Her total reimbursement = mileage + meals

$R = 226.00 + 27.00 + 37.00 + 37.00$
$R = \$327.00$

35. Find the total of Chan's bill and divide by the three equal payments.

$1645.95 + 2(14.95) + 83.79$
$= 1645.95 + 29.90 + 83.79$
$= 1759.64$

The total cost is $1759.64. The payments are $1759.64 \div 3 \approx \$586.55$.

37. GPA is the ratio of total grade points to total credit hours. To calculate the total grade points, we must multiply the credits for each course by the value of the letter grade received, then add all those results.

	Credit	Grade	Grade Points
Math 101	3.0	B	(3)(3) = 9
Eng 101	3.0	C	(3)(2) = 6
His 102	3.0	B+	(3)(3.5) = 10.5
Chm 101	4.0	A	(4)(4) = 16

$\text{GPA} = \dfrac{\text{total grade points}}{\text{total credits}}$

$\text{GPA} = \dfrac{41.5}{13} = 3.192$

The student's GPA is 3.192.

39. GPA is the ratio of total grade points to total credit hours. To calculate the total grade points, we must multiply the credits for each course by

the value of the letter grade received, then add all those results.

	Credit	Grade	Grade Points
Math 102	5.0	B	$(5)(3) = 15$
Eng 100	3.0	B+	$(3)(3.5) = 10.5$
Psy 101	3.0	D+	$(3)(1.5) = 4.5$
Bio 101	4.0	C	$(4)(2) = 8$
Col 101	2.0	A	$(2)(4) = 8$

$$\text{GPA} = \frac{\text{total grade points}}{\text{total credits}}$$

$$\text{GPA} = \frac{46}{17} = 2.706$$

The student's GPA is 2.667.

41. $\frac{1}{2}mv^2 = \frac{1}{2}(1.25)(-8)^2$

$\qquad = \frac{1}{2}(1.25)(64)$

$\qquad = 0.5(1.25)(64)$

$\qquad = 40$

43. $\dfrac{w}{h^2} \cdot 705 = \dfrac{150}{67^2} \cdot 705$

$\qquad = \dfrac{150}{4489} \cdot 705$

$\qquad = \dfrac{150}{4489} \cdot \dfrac{705}{1}$

$\qquad = \dfrac{105750}{4489}$

$\qquad \approx 23.6$

45. $mc^2 = \left(2.5 \times 10^6\right)\left(3 \times 10^8\right)^2$

$\qquad = \left(2.5 \times 10^6\right)\left(3^2 \times 10^{8 \cdot 2}\right)$

$\qquad = \left(2.5 \times 10^6\right)\left(9 \times 10^{16}\right)$

$\qquad = (2.5)(9) \times 10^{6+16}$

$\qquad = 22.5 \times 10^{22}$

$\qquad = 2.25 \times 10^{23}$

47. $vt + \frac{1}{2}at^2 = 20(0.6) + \frac{1}{2}(-12.5)(0.6)^2$

$\qquad = 20(0.6) + \frac{1}{2}(-12.5)(0.36)$

$\qquad = 12 + 0.5(-12.5)(0.36)$

$\qquad = 12 + (-6.25)(0.36)$

$\qquad = 12 + (-2.25)$

$\qquad = 9.75$

49. $\left(1 + \dfrac{r}{n}\right)^{nt} = \left(1 + \dfrac{0.12}{4}\right)^{4(0.5)}$

$\qquad = (1 + 0.03)^{4(0.5)}$

$\qquad = (1.03)^{4(0.5)}$

$\qquad = (1.03)^2$

$\qquad = 1.0609$

51. $A = \frac{1}{2}bh$

$\quad A = \frac{1}{2} \cdot 9 \cdot 12.8$

$\quad A = \frac{1}{2} \cdot \frac{9}{1} \cdot \frac{12.8}{1}$

$\quad A = \dfrac{115.2}{2}$

$\quad A = 57.6 \text{ cm}^2$

53. $A = \frac{1}{2}h(a+b)$

$\quad A = \frac{1}{2}(1.08)(1.12 + 1.28)$

$\quad A = \frac{1}{2}(1.08)(2.4)$

$\quad A = \frac{1}{2} \cdot \frac{1.08}{1} \cdot \frac{2.4}{1}$

$\quad A = \dfrac{2.592}{2}$

$\quad A = 1.296 \text{ m}^2$

55. $r = \frac{1}{2}d$

$\quad r = \frac{1}{2} \cdot 10.5$

$\quad r = \frac{1}{2} \cdot \frac{10.5}{1}$

$\quad r = \dfrac{10.5}{2}$

$\quad r = 5.25$

$\quad A = \pi r^2$

$\quad A = \pi(5.25)^2$

$\quad A = 27.6\pi$

$\quad A \approx 27.6(3.14)$

$\quad A \approx 86.5 \text{ in.}^2$

57. $r = \frac{1}{2}d$

$\quad r = \frac{1}{2}(2.4)$

$\quad r = \frac{1}{2} \cdot \frac{2.4}{1}$

$\quad r = \dfrac{2.4}{2}$

$\quad r = 1.2 \text{ cm}$

$\quad V = \pi r^2 h$

$\quad V = \pi \cdot (1.2)^2(0.2)$

$\quad V = \pi(1.44)(0.2)$

$\quad V = 0.288\pi$

$\quad V \approx 0.288(3.14)$

$\quad V \approx 0.9 \text{ cm}^3$

59. $V = \frac{1}{3}lwh$

$\quad V = \frac{1}{3} \cdot 75 \cdot 75 \cdot 90$

$\quad V = \frac{1}{\cancel{3}} \cdot \frac{75}{1} \cdot \frac{75}{1} \cdot \frac{\cancel{90}^{30}}{1}$

$\quad V = 168,750 \text{ ft.}^3$

61. $V = \frac{1}{3}\pi r^2 h$

$V = \frac{1}{3} \cdot \pi \cdot 4^2 \cdot (5.5)$

$V = \frac{1}{3} \cdot \pi \cdot 16 \cdot (5.5)$

$V = \frac{1}{3} \cdot \frac{16}{1} \cdot \frac{5.5}{1} \pi$

$V = \frac{88}{3} \pi$

$V \approx \frac{88}{3}(3.14)$

$V \approx 92.10\overline{6} \text{ in.}^3$

63. $V = \frac{4}{3}\pi r^3$

$V = \frac{4}{3} \cdot \pi \cdot (3884.3)^3$

$V = \frac{4}{3} \cdot \pi \cdot (5.86 \times 10^{10})$

$V \approx \frac{4}{3}(3.14)(5.86 \times 10^{10})$

$V \approx 2.45 \times 10^{11} \text{ mi.}^3$

65. The composite area is a triangle and half of a circle. We must find the area of the triangle and add the area of the half circle. Note that the diameter of the circle is 1.8 m so the radius is 0.9 m.

$A = \text{area of triangle} + \text{area of half circle}$

$A = \frac{1}{2}bh + \frac{1}{2}\pi r^2$

$A = \frac{1}{2}(1.6)(1.8) + \frac{1}{2}\pi(0.9)^2$

$A = 0.5(1.6)(1.8) + 0.5\pi(0.81)$

$A = 1.44 + 0.405\pi$

$A = 1.44 + 0.405(3.14)$

$A \approx 1.44 + 1.2717$

$A \approx 2.71 \text{ cm}^2$

67. Subtract the area of the circle from the area of the square. Note that the diameter of the circle is 25 cm so the radius is 12.5 cm

$A = \text{area of square} - \text{area of circle}$

$A = bh - \pi r^2$

$A = 25(25) - \pi(12.5)^2$

$A = 25(25) - 156.25\pi$

$A \approx 625 - 156.25(3.14)$

$A \approx 625 - 490.625$

$A \approx 134.375 \text{ cm}^2$

69. We need to add the volume of the sphere to the volume of the cylinder. The diameter of the sphere is 64 ft., so the radius is 32 ft. The diameter of the cylinder is 20 ft., so its radius is 10 ft. $V = \text{volume of sphere} + \text{volume of cylinder}$

$V = \frac{4}{3}\pi r^3 + \pi r^2 h$

$V = \frac{4}{3}\pi(32)^3 + \pi(10)^2 45.5$

$V = \frac{4}{3}\pi(32,768) + \pi(100)(45.5)$

$V = \frac{4}{3} \cdot \pi \cdot \frac{32,768}{1} + 4550\pi$

$V = \frac{131,072}{3}\pi + \frac{13,650}{3}\pi$

$V = \frac{144,722}{3}\pi$

$V \approx \frac{144,722}{3} \cdot \frac{3.14}{1}$

$V \approx 151,475.69 \text{ ft.}^3$

71. Find the volume of the pyramid and subtract the volume of the inner chamber.

$V = \text{volume of pyramid} - \text{volume of inner chamber}$

$V = \frac{1}{3}lwh - lwh$

$V = \frac{1}{3}(230)(230)(146.5) - (5.8)(5.2)(10.8)$

$V = \frac{1}{3} \cdot \frac{230}{1} \cdot \frac{230}{1} \cdot \frac{146.5}{1} - 325.728$

$V = \frac{7,749,850}{3} - 325.728$

$V \approx 2,583,283.33 - 325.728$

$V \approx 2,582,957.6 \text{ m}^3$

73. a) $V = lwh$

$V = 0.03(0.03)(0.03)$

$V = 0.000027 \text{ in.}^3$

b) $0.000027 \cdot x = 1$

$\frac{0.000027x}{0.000027} = \frac{1}{0.000027}$

$x = 37037.037$

There are approximately 37,037 grains of salt in 1 cubic inch.

Review Exercises

1. $4(x-6)+1 = 9x-2-8x$

$4x-24+1 = x-2$

$4x-23 = x-2$

$\underline{-x \qquad\quad -x}$

$3x-23 = 0-2$

$\underline{+23 \qquad +23}$

$3x+0 = 21$

$\dfrac{3x}{3} = \dfrac{21}{3}$

$x = 7$

Check: $4(7-6)+1 = 9\cdot 7-2-8\cdot 7$

$4\cdot 1+1 = 63-2-56$

$4+1 = 61-56$

$5 = 5$

2. $\dfrac{1}{3}y-\dfrac{3}{4} = 6$ Check: $\dfrac{1}{3}\cdot 20\dfrac{1}{4}-\dfrac{3}{4} = 6$

$\dfrac{1}{3}y-\dfrac{3}{4}+\dfrac{3}{4} = 6+\dfrac{3}{4}$ $\dfrac{1}{3}\cdot\dfrac{81}{4}-\dfrac{3}{4} = 6$

$\dfrac{1}{3}y+0 = 6\dfrac{3}{4}$ $\dfrac{81}{12}-\dfrac{9}{12} = 6$

$\dfrac{1}{3}y = \dfrac{27}{4}$ $\dfrac{72}{12} = 6$

$\dfrac{\cancel{3}}{1}\cdot\dfrac{1}{\cancel{3}}y = \dfrac{27}{4}\cdot\dfrac{3}{1}$ $6 = 6$

$1y = \dfrac{81}{4}$

$y = 20\dfrac{1}{4}$

3. Find the area of the rectangle and add the area of the triangle. The height of the triangle is $18-12 = 6$ ft. A = area of rectangle + area of triangle

$A = lw+\dfrac{1}{2}bh$

$A = 20\cdot 12+\dfrac{1}{2}\cdot 20\cdot 6$

$A = 240+\dfrac{1}{\cancel{2}}\cdot\dfrac{\cancel{20}^{10}}{1}\cdot\dfrac{6}{1}$

$A = 240+60$

$A = 300$ ft.^{2}

4. Find the area of the larger rectangle and subtract the area of the smaller rectangle.

$A = lw-lw$

$A = 48\cdot 24-24\cdot 18$

$A = 1152-432$

$A = 720$ cm^2

5. $\dfrac{1}{2}(x+12) = \dfrac{3}{4}x-9$

$\dfrac{1}{2}x+\dfrac{1}{\cancel{2}}\cdot\dfrac{\cancel{12}^{6}}{1} = \dfrac{3}{4}x-9$

$\dfrac{1}{2}x+6 = \dfrac{3}{4}x-9$

$\dfrac{1}{2}x-\dfrac{1}{2}x+6 = \dfrac{3}{4}x-\dfrac{1}{2}x-9$

$0+6 = \dfrac{3}{4}x-\dfrac{2}{4}x-9$

$6 = \dfrac{1}{4}x-9$

$6+9 = \dfrac{1}{4}x-9+9$

$15 = \dfrac{1}{4}x+0$

$\dfrac{4}{1}\cdot 15 = \dfrac{1}{4}x\cdot\dfrac{4}{1}$

$60 = x$

6.

Categories	Value	Number	Amount
$10 bills	10	$x+4$	$10(x+4)$
$5 bills	5	x	$5x$

$10(x+4)+5x = 160$

$10x+40+5x = 160$

$15x+40 = 160$

$\underline{-40 \quad -40}$

$15x+0 = 120$

$\dfrac{15x}{15} = \dfrac{120}{15}$

$x = 8$

There are 8 five-dollar bills and $8+4 = 12$ ten dollar bills.

Exercise Set 6.6

1. Multiply both sides of the equation by an appropriate power of 10 as determined by the decimal number with the most decimal places.

3. A right triangle has one angle that measures 90°.

5. The hypotenuse is the longest side in a right triangle.

7.
$$4.5n + 7 = 7.9$$
$$10(4.5n + 7) = 10(7.9)$$
$$10 \cdot (4.5n) + 10 \cdot (7) = 79$$
$$45n + 70 = 79$$
$$\underline{-70 = -70}$$
$$45n + 0 = 9$$
$$\frac{45n}{45} = \frac{9}{45}$$
$$1n = 0.2$$
$$n = 0.2$$

Check: $4.5(0.2) + 7 = 7.9$
$$0.9 + 7 = 7.9$$
$$7.9 = 7.9$$

9.
$$15.5y + 11.8 = 21.1$$
$$10(15.5y + 11.8) = 10(21.1)$$
$$10 \cdot (15.5y) + 10 \cdot (11.8) = 10 \cdot 21.1$$
$$155y + 118 = 211$$
$$\underline{-118 = -118}$$
$$155y + 0 = 93$$
$$\frac{155y}{155} = \frac{93}{155}$$
$$1y = 0.6$$
$$y = 0.6$$

Check: $15.5(0.6) + 11.8 = 21.1$
$$9.3 + 11.8 = 21.1$$
$$21.1 = 21.1$$

11.
$$16.7 - 3.5t = 17.12$$
$$100(16.7 - 3.5t) = 100(17.12)$$
$$100 \cdot (16.7) - 100 \cdot (3.5t) = 100 \cdot (17.12)$$
$$1670 - 350t = 1712$$
$$\underline{-1670 \qquad = -1670}$$
$$0 - 350t = 42$$
$$\frac{-350t}{-350} = \frac{42}{-350}$$
$$1t = -0.12$$
$$t = -0.12$$

Check: $16.7 - 3.5(-0.12) = 17.12$
$$16.7 - (-0.42) = 17.12$$
$$16.7 + 0.42 = 17.12$$
$$17.12 = 17.12$$

13.
$$0.62k - 12.01 = 0.17k - 14.8$$
$$100(0.62k - 12.01) = 100(0.17k - 14.8)$$
$$100 \cdot (0.62k) - 100 \cdot (12.01) = 100 \cdot (0.17k) - 100 \cdot (14.8)$$
$$62k - 1201 = 17k - 1480$$
$$\underline{-17k \qquad = -17k}$$
$$45k - 1201 = -1480$$
$$\underline{+1201 = +1201}$$
$$45k + 0 = -279$$
$$\frac{45k}{45} = \frac{-279}{45}$$
$$1k = -6.2$$
$$k = -6.2$$

Check: $0.62(-6.2) - 12.01 = 0.17(-6.2) - 14.8$
$$-3.844 - 12.01 = -1.054 - 14.8$$
$$-15.854 = -15.854$$

15.
$$0.8n + 1.22 = 0.408 - 0.6n$$
$$1000(0.8n + 1.22) = 1000(0.408 - 0.6n)$$
$$800n + 1220 = 408 - 600n$$
$$\underline{+600n \qquad = \qquad +600n}$$
$$1400n + 1220 = 408 + 0$$
$$\underline{-1220 = -1220}$$
$$1400n + 0 = -812$$
$$\frac{1400n}{1400} = \frac{-812}{1400}$$
$$1n = -0.58$$
$$n = -0.58$$

Check: $0.8(-0.58) + 1.22 = 0.408 - 0.6(-0.58)$
$$-0.464 + 1.22 = 0.408 - (-0.348)$$
$$0.756 = 0.408 + 0.348$$
$$0.756 = 0.756$$

17.
$$4.1 - 1.96x = 4.2 - 1.99x$$
$$100(4.1 - 1.96x) = 100(4.2 - 1.99x)$$
$$100 \cdot (4.1) - 100 \cdot (1.96x) = 100 \cdot (4.2) - 100 \cdot (1.99x)$$
$$410 - 196x = 420 - 199x$$
$$\underline{+199x = \qquad +199x}$$
$$410 + 3x = 420 + 0$$
$$\underline{-410 \qquad = -410}$$
$$0 + 3x = 10$$
$$\frac{3x}{3} = \frac{10}{3}$$
$$x = 3.\overline{3}$$

Check: $4.1 - 1.96(3.\overline{3}) \approx 4.2 - 1.99(3.\overline{3})$
$$4.1 - 6.468 \approx 4.2 - 6.567$$
$$-2.368 \approx -2.367$$

19.
$$0.4(8+t)=5t+0.9$$
$$0.4\cdot 8+0.4t=5t+0.9$$
$$3.2+0.4t=5t+0.9$$
$$10\left(3.2+0.4t\right)=10\left(5t+0.9\right)$$
$$10\cdot\left(3.2\right)+10\cdot\left(0.4t\right)=10\cdot\left(5t\right)+10\cdot\left(0.9\right)$$
$$32+4t=50t+9$$
$$\underline{-4t}=\underline{-4t}$$
$$32+0=46t+9$$
$$\underline{-9}\quad=\quad\underline{-9}$$
$$23=46t+0$$
$$\frac{23}{46}=\frac{46t}{46}$$
$$0.5=1t$$
$$0.5=t$$

Check: $0.4(8+0.5)=5\cdot 0.5+0.9$
$$0.4(8.5)=2.5+0.9$$
$$3.4=3.4$$

21.
$$1.5\left(x+8\right)=3.2+0.7x$$
$$1.5\cdot x+1.5\cdot\left(8\right)=3.2+0.7x$$
$$1.5x+12=3.2+0.7x$$
$$10\left(1.5x+12\right)=10\left(3.2+0.7x\right)$$
$$10\cdot\left(1.5x\right)+10\cdot\left(12\right)=10\cdot\left(3.2\right)+10\cdot\left(0.7x\right)$$
$$15x+120=32+7x$$
$$\underline{-7x}\quad=\quad\underline{-7x}$$
$$8x+120=32+0x$$
$$\underline{-120}=\underline{-120}$$
$$8x+0=-88$$
$$\frac{8x}{8}=\frac{-88}{8}$$
$$1x=-11$$
$$x=-11$$

Check: $1.5\left(-11+8\right)=3.2+0.7\left(-11\right)$
$$1.5\left(-3\right)=3.2-7.7$$
$$-4.5=-4.5$$

23.
$$20(0.2n+0.28)=3.98-(0.3-3.92n)$$
$$20\cdot\left(0.2n\right)+20\cdot\left(0.28\right)=3.98-0.3+3.92n$$
$$4n+5.6=3.68+3.92n$$
$$100\left(4n+5.6\right)=100\left(3.68+3.92n\right)$$
$$100\cdot\left(4n\right)+100\cdot\left(5.6\right)=100\cdot\left(3.68\right)+100\cdot\left(3.92n\right)$$
$$400n+560=368+392n$$
$$\underline{-392n}\quad=\quad\underline{-392n}$$
$$8n+560=368+0$$
$$\underline{-560}=\underline{-560}$$
$$8n+0=-192$$
$$\frac{8n}{8}=\frac{-192}{8}$$
$$n=-24$$

Check:
$$20(0.2(-24)+0.28)=3.98-(0.3-3.92(-24))$$
$$20(-4.8+0.28)=3.98-(0.3-(-94.08))$$
$$20(-4.52)=3.98-(0.3+94.08)$$
$$-90.4=3.98-94.38$$
$$-90.4=-90.4$$

25.
$$3.2p+7=9.56$$
$$\underline{-7}=\underline{-7}$$
$$3.2p+0=2.56$$
$$100\cdot\left(3.2p\right)=100\cdot\left(2.56\right)$$
$$320p=256$$
$$\frac{320p}{320}=\frac{256}{320}$$
$$1p=0.8$$
$$p=0.8$$

27.
$$18.75+3.5t=1.5-t$$
$$100\left(18.75+3.5t\right)=100\left(1.5-t\right)$$
$$100\cdot\left(18.75\right)+100\cdot\left(3.5t\right)=100\cdot\left(1.5\right)-100\cdot\left(t\right)$$
$$1875+350t=150-100t$$
$$\underline{+100t}=\quad\underline{+100t}$$
$$1875+450t=150+0$$
$$\underline{-1875}\quad=\underline{-1875}$$
$$0+450t=-1725$$
$$\frac{450t}{450}=\frac{-1725}{450}$$
$$1t=-3.8\overline{3}$$
$$t=-3.8\overline{3}$$

29.

$$0.2y - 0.48 = 0.1(2.76 - y)$$
$$0.2y - 0.48 = 0.1 \cdot (2.76) - 0.1 \cdot y$$
$$0.2y - 0.48 = 0.276 - 0.1y$$
$$1000(0.2y - 0.48) = 1000(0.276 - 0.1y)$$
$$1000 \cdot (0.2y) - 1000 \cdot (0.48) = 1000 \cdot (0.276) - 1000 \cdot (0.1y)$$
$$200y - 480 = 276 - 100y$$
$$\underline{+100y \qquad = \qquad +100y}$$
$$300y - 480 = 276 + 0$$
$$\underline{+480 = +480}$$
$$300y + 0 = 756$$
$$\frac{300y}{300} = \frac{756}{300}$$
$$y = 2.52$$

31.

$$A = \pi r^2$$
$$7850 \approx 3.14r^2$$
$$\frac{7850}{3.14} \approx \frac{3.14r^2}{3.14}$$
$$2500 \approx r^2$$
$$\sqrt{2500} \approx r$$
$$50 \approx r$$

Since the radius is 50 km, the diameter is ≈ 100 km.

33. We must use the Pythagorean Theorem.

$$a^2 + b^2 = c^2$$
$$4^2 + 3^2 = c^2$$
$$16 + 9 = c^2$$
$$25 = c^2$$
$$\sqrt{25} = c$$
$$5 = c$$

The distance is 5 miles.

35. We must use the Pythagorean Theorem.

$$a^2 + b^2 = c^2$$
$$6^2 + 8^2 = c^2$$
$$36 + 64 = c^2$$
$$100 = c^2$$
$$\sqrt{100} = c$$
$$10 = c$$

The distance is 10 in.

37. We must use the Pythagorean theorem.

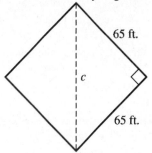

$$a^2 + b^2 = c^2$$
$$65^2 + 65^2 = c^2$$
$$4225 + 4225 = c^2$$
$$8450 = c^2$$
$$\sqrt{8450} = c^2$$
$$91.9 \text{ ft.} \approx c$$

39. We must use the Pythagorean theorem. One leg of the triangle is 7 ft. and the hypotenuse is 25 ft. We must find the other leg.

$$a^2 + b^2 = c^2$$
$$7^2 + b^2 = 25^2$$
$$49 + b^2 = 625$$
$$b^2 = 576$$
$$b = \sqrt{576}$$
$$b = 24$$

The ladder reaches 24 ft.

41. We must use the Pythagorean theorem. One leg of the triangle is 3.5 ft. and the hypotenuse is 14.5 ft. We must find the other leg.

$$a^2 + b^2 = c^2$$
$$3.5^2 + b^2 = 14.5^2$$
$$12.25 + b^2 = 210.25$$
$$b^2 = 198$$
$$b = \sqrt{198}$$
$$b \approx 14.1$$

The support beam is ≈ 14.1 ft.

43. We must use the Pythagorean theorem.

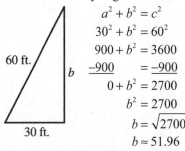

$$a^2 + b^2 = c^2$$
$$30^2 + b^2 = 60^2$$
$$900 + b^2 = 3600$$
$$\underline{-900 \qquad = -900}$$
$$0 + b^2 = 2700$$
$$b^2 = 2700$$
$$b = \sqrt{2700}$$
$$b \approx 51.96$$

a) Since the bottom of the ladder is 8 feet above the ground, the highest point the ladder can reach is $51.96 + 8 = 59.96$ ft.

b) If each story is 10.5 feet then the bottom of the seventh floor will be $6 \times 10.5 = 63$ ft. high. The ladder could not reach a person on the seventh floor.

45.

Categories	Selling price	Number of boxes sold	Income
Paperback	0.50	$n + 14$	$0.50(n + 14)$
Hardcover	1.50	n	$1.50n$

47.

Categories	Selling price	Number of T-shirts sold	Income
Children's T-shirts	8.99	$45 - n$	$8.99(45 - n)$
Adult T-shirts	14.99	n	$14.99n$

49.

Categories	Value	Number	Amount
quarters	0.25	$x + 12$	$0.25(x+12)$
half dollars	0.50	x	$0.50x$

$$0.25(x + 12) + 0.50x = 6.75$$
$$0.25 \cdot (x) + 0.25 \cdot (12) + 0.50x = 6.75$$
$$0.25x + 3 + 0.50x = 6.75$$
$$0.75x + 3 = 6.75$$
$$100(0.75x + 3) = 100(6.75)$$
$$100 \cdot (0.75x) + 100 \cdot (3) = 100 \cdot (6.75)$$
$$75x + 300 = 675$$
$$\underline{-300 = -300}$$
$$75x + 0 = 375$$
$$\frac{75x}{75} = \frac{375}{75}$$
$$x = 5$$

Latisha has 5 half dollars and $5 + 12 = 17$ quarters.

51.

Categories	Value	Number	Amount
12	1.50	x	$1.5x$
16	2.00	$65 - x$	$2(65 - x)$

$$1.5x + 2(65 - x) = 109$$
$$1.5x + 130 - 2x = 109$$
$$-0.5x + 130 = 109$$
$$10(-0.5x + 130) = 10(109)$$
$$-5x + 1300 = 1090$$
$$\underline{-1300 = -1300}$$
$$-5x + 0 = -210$$
$$\frac{-5x}{-5} = \frac{-210}{-5}$$
$$x = 42$$

Bernice sold 42 12-oz. drinks and $65 - 42 = 23$ 16-oz. drinks.

Review Exercises

1. twenty-four billion, nine hundred fifteen million, two hundred four.

2. 46,300,000

3. $4\dfrac{5}{8} + 8\dfrac{5}{6} = 4\dfrac{5 \cdot 3}{8 \cdot 3} + 8\dfrac{5 \cdot 4}{6 \cdot 4}$

$$= 4\dfrac{15}{24} + 8\dfrac{20}{24}$$
$$= 12\dfrac{35}{24}$$
$$= 12 + 1\dfrac{11}{24}$$
$$= 13\dfrac{11}{24}$$

4. $12\dfrac{1}{4} - 7\dfrac{4}{5} = 12\dfrac{1 \cdot 5}{4 \cdot 5} - 7\dfrac{4 \cdot 4}{5 \cdot 4}$

$$= 12\dfrac{5}{20} - 7\dfrac{16}{20}$$
$$= 11\dfrac{25}{20} - 7\dfrac{16}{20}$$
$$= 4\dfrac{9}{20}$$

5. $\dfrac{28x^9}{7x^3} = \dfrac{28}{7}x^{9-3}$

$$= 4x^6$$

6. Mistake: divided by 3 rather than by –3.

$$-3x - 18 = -24$$
$$\underline{+18 = +18}$$
$$-3x = -6$$
$$\frac{-3x}{-3} = \frac{-6}{-3}$$
$$x = 2$$

Chapter 6 Review Exercises

1. true 2. true 3. true

4. false 5. false 6. true

7. 1. Write all the digits to the left of the decimal point as the integer part of a mixed number.
2. Write all digits to the right of the decimal point in the numerator of the fraction.
3. Write the denominator indicated by the last place value.
4. Simplify to lowest terms.

8. 1. Stack the numbers so that the place values align. (Line up the decimal points.)
2. Add the digits in the corresponding place values.
3. Place the decimal point in the sum so that it aligns with the decimal points in the addends.

9. Decimal point in the product was placed incorrectly. Correct: 0.0312.

10. denominator; numerator

11. a) $0.8 = \frac{8}{10} = \frac{8 \div 2}{10 \div 2} = \frac{4}{5}$

b) $0.024 = \frac{24 \div 8}{1000 \div 8} = \frac{3}{125}$

c) $-2.65 = -2\frac{65 \div 5}{100 \div 5} = -2\frac{13}{20}$

12. a) twenty-four and thirty-nine hundredths
b) five hundred eighty-one and four hundred fifty-nine thousandths
c) two thousand nine hundred seventeen ten-thousandths

13. a) $2.001 > 2.0009$ b) $-0.016 < -0.008$

14. a) 32 b) 31.8 c) 31.81 d) 31.806

15. a) 24.810
54.200
$\underline{+9.005}$
88.015

b) 9.005
$\underline{-1.089}$
7.916

c) $6.7 + (-12.9) = -6.2$ Because we are adding numbers with different signs, we subtract and keep the sign of the larger absolute value.

d) $(-20) - (-13.88) = (-20) + 13.88 = -6.12$ Because we are adding numbers with different signs, we subtract and keep the sign of the larger absolute value.

e) $-1.082 + 29.1 - 4.9 = -5.982 + 29.1 = 23.118$

f) $14.1 - (-6.39) = 14.1 + 6.39 = 20.49$

16. a) $9y^4 + 8y - 12.2y^2 + 2.1y^4 - 16y + 3.8$
$= 9y^4 + 2.1y^4 - 12.2y^2 + 8y - 16y + 3.8$
$= 11.1y^4 - 12.2y^2 - 8y + 3.8$

b) $(0.3a^2 + 8.1a + 6) + (5.1a^2 - 2.9a + 4.33)$
$= 0.3a^2 + 5.1a^2 + 8.1a - 2.9a + 6 + 4.33$
$= 5.4a^2 + 5.2a + 10.33$

c) $(2.6x^3 - 98.1x^2 + 4) - (5.3x^3 - 2.1x + 8.2)$
$= (2.6x^3 - 98.1x^2 + 4) + (-5.3x^3 + 2.1x - 8.2)$
$= 2.6x^3 - 5.3x^3 - 98.1x^2 + 2.1x + 4 - 8.2$
$= -2.7x^3 - 98.1x^2 + 2.1x - 4.2$

17. a) 3.44
$\underline{\times 5.1}$
344
$\underline{17200}$
17.544

b) –9.67
$\underline{\times 1.5}$
4835
$\underline{9670}$
–14.505

c) $(1.2)^2 = (1.2)(1.2) = 1.44$

d) $(-0.2)^3 = (-0.2)(-0.2)(-0.2) = -0.008$

e) $(2.5h)(0.01hk^3) = (2.5)(0.01)h^{1+1}k^3$
$= 0.025h^2k^3$

f) $(-0.9x^3y)(4.2x^5y) = (-0.9)(4.2)x^{3+5}y^{1+1}$
$= -3.78x^8y^2$

g) $(-0.3x^3)^3 = (-0.3)^3 x^{3\cdot3}$
$= -0.027x^9$

h) $(1.2y^3)^2 = (1.2)^2 y^{3\cdot2}$
$= 1.44y^6$

i) $0.2(0.6n^3 - 1.4n^2 + n - 18)$
$= 0.2\cdot(0.6n^3) - 0.2\cdot(1.4n^2) + 0.2\cdot(n) - 0.2\cdot(18)$
$= 0.12n^3 - 0.28n^2 + 0.2n - 3.6$

j) $-1.1(2.8a - 4.9b) = -1.1\cdot\left(2.8a\right) - 1.1\cdot\left(-4.9a\right)$
$= -3.08a - (-5.39b)$
$= -3.08a + 5.39b$

k) $(4.5a - 6.1)(2a - 0.5)$
$= 4.5a\left(2a\right) + 4.5a(-0.5) - 6.1(2a) - 6.1(-0.5)$
$= 9a^2 - 2.25a - 12.2a - (-3.05)$
$= 9a^2 - 14.45a + 3.05$

l) $(1.3x + 4.2)(1.3x - 4.2) = 1.69x^2 - 17.64$

18. a) 3,580,000,000   b) −420,000

19. a) 9.2×10^9 b) -1.03×10^5

20. a)
$$\begin{array}{r} 5.25 \\ 4\overline{)21.00} \\ \underline{20} \\ 1\,0 \\ \underline{8} \\ 20 \\ \underline{20} \\ 0 \end{array}$$
    b)
$$\begin{array}{r} 0.4 \\ 9\overline{)3.6} \\ \underline{3\,6} \\ 0 \end{array}$$

c) $17.304 \div 2.06 = 1730.4 \div 206$ so,
$$\begin{array}{r} 8.4 \\ 206\overline{)1730.4} \\ \underline{1648} \\ 82\,4 \\ \underline{82\,4} \\ 0 \end{array}$$

d) $-54 \div (-0.09) = 5400 \div 9$ so,
$$\begin{array}{r} 600 \\ 9\overline{)5400} \\ \underline{54} \\ 000 \end{array}$$

e) $3.18x^7 \div 0.4x^5 = \dfrac{3.18x^7}{0.4x^5} = \dfrac{3.18}{0.4}x^{7-5} = 7.95x^2$

f) $\dfrac{-2.38a^4bc}{6.8ab} = -\dfrac{-2.38}{6.8}a^{4-1}b^{1-1}c$
$= -0.35a^3c$

21. a) $\dfrac{3}{5} = $
$$\begin{array}{r} 0.6 \\ 5\overline{)3.0} \\ \underline{3\,0} \\ 0 \end{array}$$
    b) $\dfrac{1}{6} = $
$$\begin{array}{r} 0.16\overline{6} \\ 6\overline{)1.000} \\ \underline{6} \\ 40 \\ \underline{36} \\ 40 \\ \underline{36} \end{array}$$ $= 0.1\overline{6}$

c) $6\dfrac{1}{4} = 6 + \dfrac{1}{4}$ so,
$$\begin{array}{r} 0.25 \\ 4\overline{)1.00} \\ \underline{8} \\ 20 \\ \underline{20} \\ 0 \end{array}$$ so, $6 + 0.25 = 6.25$

d) $-4\dfrac{2}{11} = -\left(4 + \dfrac{2}{11}\right)$ so,
$$\begin{array}{r} 0.181 \\ 11\overline{)2.000} \\ \underline{1\,1} \\ 90 \\ \underline{88} \\ 20 \\ \underline{11} \\ 90 \\ \underline{88} \end{array}$$ $= 0.\overline{18}$

so, $-4\dfrac{2}{11} = -(4 + 0.\overline{18}) = -4.\overline{18}$

22. a) $\sqrt{0.64} = 0.8$ because $(0.8)^2 = 0.64$

b) $\sqrt{1.21} = 1.1$ because $(1.1)^2 = 1.21$

23. a) $\sqrt{148} \approx 12.17$ b) $\sqrt{1.6} \approx 1.26$

24. a) $4.59 - 2.8\cdot 6.7 + \sqrt{1.44} = 4.59 - 2.8\cdot 6.7 + 1.2$
$= 4.59 - 18.76 + 1.2$
$= -12.97$

b) $3.6 + (2.5)^2 - 1.5(9 - 0.4)$
$= 3.6 + (2.5)^2 - 1.5(8.6)$
$= 3.6 + 6.25 - 1.5(8.6)$
$= 3.6 + 6.25 - 12.9$
$= -3.05$

c) $\dfrac{3}{4}(0.86) - \dfrac{3}{8} = 0.75(0.86) - 0.375$
$= 0.645 - 0.375$
$= 0.27$

d) $\dfrac{1}{6}(0.24) + \dfrac{2}{3}(-3.3) = \dfrac{1}{6}\left(\dfrac{0.24}{1}\right) + \dfrac{2}{3}\left(\dfrac{-3.3}{1}\right)$
$= \dfrac{0.24}{6} + \dfrac{-6.6}{3}$
$= 0.04 + (-2.2)$
$= -2.16$

25. $P + Pr = 200 + 200\left(0.03\right)$
$= 200 + 6$
$= 206$

26. $\dfrac{1}{2}at^2 = \dfrac{1}{2}(9.8)(2)^2$

$\quad\quad\quad = \dfrac{1}{2}(9.8)(4)$

$\quad\quad\quad = 19.6$

27. The radius is half of the diameter. Since the diameter is 15 cm, the radius is 7.5 cm.

$\begin{array}{ll} C = \pi d & A = \pi r^2 \\ C = 15\pi & A = \pi(7.5)^2 \\ C \approx 15(3.14) & A = 56.25\pi \\ C \approx 47.1 \text{ cm} & A \approx 176.625 \text{ cm}^2 \end{array}$

28. a) We need to add the area of the rectangle to the area of the half circle. The diameter of the half circle is $14 - 3 - 3 = 8$ in., so the radius is 4 in.

$A = bh + \dfrac{1}{2}\pi r^2$

$A = 14(1.5) + \dfrac{1}{2}\pi(4^2)$

$A = 21 + \dfrac{1}{\cancel{2}}\pi \cdot \dfrac{\cancel{16}^{\,8}}{1}$

$A = 21 + 8\pi$

$A \approx 21 + 8(3.14)$

$A \approx 21 + 25.12$

$A \approx 46.12 \text{ in.}^2$

b) We must find the area of the parallelogram and add it to the area of the triangle. The base of the triangle is $0.65 - 0.45 = 0.2$ m.

$A = bh + \dfrac{1}{2}bh$

$A = 0.65(0.2) + \dfrac{1}{2}(0.2)(0.2)$

$A = 0.13 + \dfrac{1}{2}\left(\dfrac{0.2}{1}\right)\left(\dfrac{0.2}{1}\right)$

$A = 0.13 + \dfrac{0.04}{2}$

$A = 0.13 + 0.02$

$A = 0.15 \text{ m}^2$

29. a) We must subtract the area of the small circle from the area of the larger circle. Since the diameter of the small circle is 3 in., the radius is 1.5 in. The diameter of the large circle is 12 in., so the radius is 6 in. $A = \pi r^2 - \pi r^2$

$\quad\quad A = \pi(6^2) - \pi(1.5^2)$

$\quad\quad A = 36\pi - 2.25\pi$

$\quad\quad A = 33.75\pi$

$\quad\quad A \approx 105.975 \text{ in.}^2$

b) We must subtract the area of the half circle from the area of the trapezoid. Since the diameter

of the half circle is 48 in., the radius is 24 in. The height of the trapezoid is $15 + 24 = 39$ in.

$A = \dfrac{1}{2}h(a + b) - \dfrac{1}{2}\pi r^2$

$A = \dfrac{1}{2}(39)(72.5 + 48) - \dfrac{1}{2}\pi(24)^2$

$A = \dfrac{1}{2}(39)(120.5) - \dfrac{1}{\cancel{2}}\cdot \cancel{576}^{\,288}\,\pi$

$A = \dfrac{1}{2}\left(\dfrac{39}{1}\right)\left(\dfrac{120.5}{1}\right) - 288\pi$

$A = \dfrac{4699.5}{2} - 288\pi$

$A = 2349.75 - 288\pi$

$A \approx 2349.75 - 288(3.14)$

$A \approx 2349.75 - 904.32$

$A \approx 1445.43 \text{ cm}^2$

30. a) $V = \pi r^2 h$

$\quad V = \pi(3^2)(1.5)$

$\quad V = \pi(9)(1.5)$

$\quad V = 13.5\pi$

$\quad V \approx 42.39 \text{ in.}^3$

b) $V = \dfrac{1}{3}\pi r^2 h$

$\quad V = \dfrac{1}{3}\pi(2^2)(4.8)$

$\quad V = \dfrac{1}{3}\left(\dfrac{4}{1}\right)\left(\dfrac{4.8}{1}\right)\pi$

$\quad V = \dfrac{19.2}{3}\pi$

$\quad V = 6.4\pi$

$\quad V \approx 6.4(3.14)$

$\quad V \approx 20.096 \text{ cm}^3$

c) $V = \dfrac{4}{3}\pi r^3$

$\quad V = \dfrac{4}{3}\pi(6^3)$

$\quad V = \dfrac{4}{3}\left(\dfrac{216}{1}\right)\pi$

$\quad V = \dfrac{864}{3}\pi$

$\quad V = 288\pi$

$\quad V \approx 288(3.14)$

$\quad V \approx 904.32 \text{ m}^3$

d) $V = \dfrac{1}{3}lwh$

$\quad V = \dfrac{1}{3}(42)(42)(38.6)$

$\quad V = \dfrac{1}{3}\left(\dfrac{42}{1}\right)\left(\dfrac{42}{1}\right)\left(\dfrac{38.6}{1}\right)$

$\quad V = \dfrac{68{,}090.4}{3}$

$\quad V = 22{,}696.8 \text{ m}^3$

e) We must find the volume of the rectangle and add the volume of the cylinder. Since the diameter of the circle is 3 in., the radius is 1.5 in.

$V = lwh + \pi r^2 h$

$V = 4.5(2)(3.5) + \pi(1.5^2)(1.2)$

$V = 31.5 + \pi(2.25)(1.2)$

$V = 31.5 + 2.7\pi$

$V \approx 31.5 + 2.7(3.14)$

$V \approx 31.5 + 8.478$

$V \approx 39.978 \text{ in.}^3$

31. We must find the volume of the smaller cylinder and subtract it from volume of the larger cylinder. Since the diameter of the smaller cylinder is 0.4 in., the radius is 0.2 in. And the diameter of the larger cylinder is 0.8 in., so its radius is 0.4 in.

$V = \pi r^2 h - \pi r^2 h$

$V = \pi(0.4^2)(20) - \pi(0.2^2)(20)$

$V = \pi(0.16)(20) - \pi(0.04)(20)$

$V = 3.2\pi - 0.8\pi$

$V = 2.4\pi$

$V \approx 2.4(3.14)$

$V \approx 7.536 \text{ in.}^3$

32. a) $x + 0.56 = 2.1$

$\underline{-0.56 \quad -0.56}$

$\qquad x = 1.54$

Check: $1.54 + 0.56 = 2.1$

$\qquad\qquad 2.1 = 2.1$

b) $y - 0.58 = -1.22$

$\underline{+0.58 \quad +0.58}$

$\qquad y = -0.64$

Check: $-0.64 - 0.58 = -1.22$

$\qquad\qquad -1.22 = -1.22$

c) $0.8y = 2.4$ 　　　Check: $0.8 \cdot 3 = 2.4$

$\dfrac{0.8y}{0.8} = \dfrac{2.4}{0.8}$ 　　　　　$2.4 = 2.4$

$1y = 3$

$y = 3$

d) $-1.6n = 0.032$

$\dfrac{-1.6n}{-1.6} = \dfrac{0.032}{-1.6}$

$1n = -0.02$

$n = -0.02$

Check: $-1.6 \cdot (-0.02) = 0.032$

$\qquad\qquad 0.032 = 0.032$

e) $-0.562b = 0$

$\dfrac{-0.562b}{-0.562} = \dfrac{0}{-0.562}$

$1b = 0$

$b = 0$

Check: $-0.562 \cdot 0 = 0$

$\qquad\qquad 0 = 0$

f) $\qquad 4.5k - 2.61 = 0.99$

$100(4.5k - 2.61) = 100(0.99)$

$100 \cdot (4.5k) - 100 \cdot (2.61) = 100 \cdot (0.99)$

$450k - 261 = 99$

$\underline{+261 = +261}$

$450k + 0 = 360$

$\dfrac{450k}{450} = \dfrac{360}{450}$

$k = 0.8$

Check: $4.5 \cdot (0.8) - 2.61 = 0.99$

$\qquad\qquad 3.6 - 2.61 = 0.99$

$\qquad\qquad 0.99 = 0.99$

g) $\qquad 51.2 = 0.4a + 51.62$

$100(51.2) = 100(0.4a + 51.62)$

$100 \cdot (51.2) = 100 \cdot (0.4a) + 100 \cdot (51.62)$

$5120 = 40a + 5162$

$\underline{-5162 = \qquad -5162}$

$-42 = 40a + 0$

$\dfrac{-42}{40} = \dfrac{40a}{40}$

$-1.05 = a$

Check: $51.2 = 0.4 \cdot (-1.05) + 51.62$

$\qquad\quad 51.2 = -0.42 + 51.62$

$\qquad\quad 51.2 = 51.2$

h)

$\qquad 2.1x - 12.6 = 1.9x - 10.98$

$100(2.1x - 12.6) = 100(1.9x - 10.98)$

$100 \cdot (2.1x) - 100 \cdot (12.6) = 100 \cdot (1.9x) - 100 \cdot (10.98)$

$210x - 1260 = 190x - 1098$

$\underline{-190x \qquad = -190x}$

$20x - 1260 = 0 - 1098$

$\underline{+1260 = \quad +1260}$

$20x + 0 = 162$

$\dfrac{20x}{20} = \dfrac{162}{20}$

$x = 8.1$

Check: $2.1 \cdot (8.1) - 12.6 = 1.9 \cdot (8.1) - 10.98$

$$17.01 - 12.6 = 15.39 - 10.98$$
$$4.41 = 4.41$$

i)
$$3.98 - 19.6x = 3.2(x - 4.1)$$
$$3.98 - 19.6x = 3.2 \cdot x - 3.2 \cdot (4.1)$$
$$3.98 - 19.6x = 3.2x - 13.12$$
$$100 \cdot (3.98 - 19.6x) = 100 \cdot (3.2x - 13.12)$$
$$398 - 1960x = 320x - 1312$$
$$\underline{+1312} \quad\quad = \quad\quad \underline{+1312}$$
$$1710 - 1960x = 320x + 0$$
$$\underline{+1960x} = \underline{+1960x}$$
$$1710 + 0 = 2280x$$
$$\frac{1710}{2280} = \frac{2280x}{2280}$$
$$0.75 = x$$

Check: $3.98 - 19.6(0.75) = 3.2(0.75 - 4.1)$

$$3.98 - 14.7 = 2.4 - 13.12$$
$$-10.72 = -10.72$$

j) $3.6n - (n - 13.2) = 10(0.4n + 0.62)$

$$3.6n - n + 13.2 = 4n + 6.2$$
$$2.6n + 13.2 = 4n + 6.2$$
$$10(2.6n + 13.2) = 10(4n + 6.2)$$
$$26n + 132 = 40n + 62$$
$$\underline{-26n} \quad\quad = \underline{-26n}$$
$$0 + 132 = 14n + 62$$
$$\underline{-62} = \quad\quad \underline{-62}$$
$$70 = 14n + 0$$
$$\frac{70}{14} = \frac{14n}{14}$$
$$5 = n$$

Check: $3.6 \cdot 5 - (5 - 13.2) = 10(0.4 \cdot 5 + 0.62)$

$$3.6 \cdot (5) - (-8.2) = 10(2 + 0.62)$$
$$18 - (-8.2) = 10 \cdot (2.62)$$
$$18 + 8.2 = 26.2$$
$$26.2 = 26.2$$

33. $150.20 - 84.61 - 245.16 - 67.54$
$-584.95 + 1545.24 = 713.18$

Sonja's balance is $713.18.

34. Total cost = number of pounds purchased
    · price per pound

$$C = 3.4(5.99)$$
$$C \approx 20.37$$

The flounder cost $20.37.

35. Since there are 12 months in a year, we must multiply $12 \times 5 = 60$ payments. Divide the total balance by the number of payments.
$2854.80 \div 60 = \$47.58$
The payments will be $47.58.

36. a) Mean:
$$\frac{\left(\begin{array}{c}139.87 + 140.78 + 137.96 \\ +155.85 + 161.89 + 158.35\end{array}\right)}{6} = \frac{894.7}{6} \approx \$149.12$$

b) Median: 137.96, 139.87, 140.78, 155.85, 158.35, 161.89
$$\frac{140.78 + 155.85}{2} = \frac{296.63}{2} \approx \$148.32$$

37. GPA is the ratio of total grade points to total credit hours. To calculate the total grade points, we must multiply the credits for each course by the value of the letter grade received, then add all those results.

	Credit	Grade	Grade Points
ENG 101	3.0	C	$(3)(2) = 6$
BIO 101	4.0	D	$(4)(1) = 4$
MAT 102	3.0	B	$(3)(3) = 9$
PSY 101	3.0	C	$(3)(2) = 6$

$$\text{GPA} = \frac{\text{total grade points}}{\text{total credits}}$$
$$\text{GPA} = \frac{25}{13} = 1.923$$

The student's GPA is 1.923.

38. $d = rt$
$$115 = 65t$$
$$\frac{115}{65} = \frac{65t}{65}$$
$$1.8 \approx t$$

It will take about 1.8 hours.

39. We must use the Pythagorean theorem.

$$a^2 + b^2 = c^2$$
$$3^2 + b^2 = 12^2$$
$$9 + b^2 = 144$$
$$\underline{-9} \quad\quad = \underline{-9}$$
$$0 + b^2 = 135$$
$$b = \sqrt{135}$$
$$b \approx 11.62$$

The top of the board from the base of the tree is 11.62 ft.

40. $72 = 45 + 4.5x$

$10 \cdot 72 = 10\left(45 + 4.5x\right)$

$720 = 10 \cdot \left(45\right) + 10 \cdot \left(4.5x\right)$

$720 = 450 + 45x$

$\underline{-450 = -450}$

$270 = 0 + 45x$

$\dfrac{270}{45} = \dfrac{45x}{45}$

$6 = x$

Brittany can afford to hire the electrician for 6 periods of 15 minutes for a total of 90 minutes.

41.

Categories	Value	Number	Amount
nickels	0.05	x	$0.05x$
dimes	0.10	$x + 9$	$0.10(x+9)$

$0.05x + 0.10(x + 9) = 2.40$

$0.05x + 0.10x + 0.90 = 2.40$

$0.15x + 0.90 = 2.40$

$100\left(0.15x + 0.90\right) = 100\left(2.40\right)$

$100 \cdot \left(0.15x\right) + 100 \cdot \left(0.90\right) = 100 \cdot \left(2.40\right)$

$15x + 90 = 240$

$\underline{-90 = -90}$

$15x + 0 = 150$

$\dfrac{15x}{15} = \dfrac{150}{15}$

$x = 10$

Elissa has 10 nickels and $10 + 9 = 19$ dimes.

42.

Categories	Value	Number	Amount
Large	14.50	x	$14.50x$
Small	9.50	$24 - x$	$9.50(24 - x)$

$14.5x + 9.5(24 - x) = 273$

$14.5x + 228 - 9.5x = 273$

$5x + 228 = 273$

$\underline{-228 = -228}$

$5x + 0 = 45$

$\dfrac{5x}{5} = \dfrac{45}{5}$

$x = 9$

Conrad sold 9 larger size figurines and $24 - 9 = 15$ smaller size figurines.

Chapter 6 Practice Test

1. fifty-six and seven hundred eighty-nine thousandths

2. $0.68 = \dfrac{68 \div 4}{100 \div 4} = \dfrac{17}{25}$

3. $-0.0059 < -0.0058$

4. a) 2   b) 2.1   c) 2.09   d) 2.092

5. $\begin{array}{r} 4.591 \\ 34.600 \\ +2.800 \\ \hline 41.991 \end{array}$

6. $\begin{array}{r} 9.005 \\ -1.089 \\ \hline 7.916 \end{array}$

7. $\begin{array}{r} -8.61 \\ \times 4.5 \\ \hline 4305 \\ 34440 \\ \hline -38.745 \end{array}$

8. $16\overline{)4.8}$ with quotient 0.3; $\begin{array}{r} 0.3 \\ 16\overline{)4.8} \\ \underline{4\,8} \\ 0 \end{array}$

9. $(-55.384) \div (-9.2) = (-553.84) \div (-92)$ so,

$\begin{array}{r} 6.02 \\ 92\overline{)553.84} \\ \underline{552} \\ 184 \\ \underline{184} \\ 0 \end{array}$

10. $\sqrt{0.81} = 0.9$ because $(0.9)^2 = 0.81$

11. $3\dfrac{5}{6} = 3 + \dfrac{5}{6}$ so, $\begin{array}{r} 0.83 \\ 6\overline{)5.00} \\ \underline{48} \\ 20 \\ \underline{18} \\ 2 \end{array}$ $= 0.8\overline{3}$ so, $3 + 0.8\overline{3} = 3.8\overline{3}$

12. 2,970,000   13. -3.56×10^7

14. $3.2 + (0.12)^2 - 0.6(2.4 - 5.6)$

$= 3.2 + (0.12)^2 - 0.6(-3.2)$

$= 3.2 + 0.0144 - 0.6(-3.2)$

$= 3.2 + 0.0144 - (-1.92)$

$= 3.2 + 0.0144 + 1.92$

$= 5.1344$

15. $\frac{1}{2}mv^2 = \frac{1}{2}(12.8)(0.2^2)$

$= \frac{1}{2}(12.8)(0.04)$

$= \frac{1}{2}\left(\frac{12.8}{1}\right)\left(\frac{0.04}{1}\right)$

$= \frac{0.512}{2}$

$= 0.256$

16. $(2.6x^3 - 98.1x^2 + 4) - (5.3x^3 - 2.1x + 8.2)$

$= (2.6x^3 - 98.1x^2 + 4) + (-5.3x^3 + 2.1x - 8.2)$

$= 2.6x^3 - 5.3x^3 - 98.1x^2 + 2.1x + 4 - 8.2$

$= -2.7x^3 - 98.1x^2 + 2.1x - 4.2$

17. $2.5x^3 \cdot 0.6x = 1.5x^{3+1}$

$= 1.5x^4$

18. $(4.5a - 6.1)(2a - 0.5)$

$= 4.5a(2a) + 4.5a(-0.5) - 6.1(2a) - 6.1(-0.5)$

$= 9a^2 - 2.25a - 12.2a - (-3.05)$

$= 9a^2 - 14.45a + 3.05$

19.

$6.5t - 12.8 = 13.85$

$100(6.5t - 12.8) = 100(13.85)$

$100 \cdot (6.5t) - 100 \cdot (12.8) = 100 \cdot (13.85)$

$650t - 1280 = 1385$

$\underline{+1280 = +1280}$

$650t + 0 = 2665$

$\frac{650t}{650} = \frac{2665}{650}$

$t = 4.1$

Check: $6.5 \cdot (4.1) - 12.8 = 13.85$

$26.65 - 12.8 = 13.85$

$13.85 = 13.85$

20.

$1.8(k - 4) = -3.5k - 4.02$

$1.8k - 7.2 = -3.5k - 4.02$

$100(1.8k - 7.2) = 100(-3.5k - 4.02)$

$100 \cdot (1.8k) - 100 \cdot (7.2) = -100 \cdot (3.5k) - 100 \cdot (4.02)$

$180k - 720 = -350k - 402$

$\underline{+350k \qquad = +350k}$

$530k - 720 = 0 - 402$

$\underline{+720 = \quad +720}$

$530k + 0 = 318$

$\frac{530k}{530} = \frac{318}{530}$

$k = 0.6$

Check: $1.8(0.6 - 4) = -3.5 \cdot (0.6) - 4.02$

$1.8 \cdot (-3.4) = -2.1 - 4.02$

$-6.12 = -6.12$

21. $C = 20.00 - (7.95 + 2.95 + 1.50 + 0.57)$

$C = 20.00 - 12.97$

$C = \$7.03$

Dedra will get $7.03 in change.

22. Total cost = number of pounds purchased

$\cdot$ price per pound

$C = 5(7.99)$

$C = 39.95$

The shrimp cost $39.95.

23. We must find the total bill and then divide the total by 3 payments.

$385.80 + 2(59.90) + 25.28$

$= 385.80 + 119.80 + 25.28$

$= \$530.88$

$530.88 \div 3 \approx \$176.96$ payment amount

24. a) Mean:

$$\frac{(31.23 + 30.52 + 30.90 + 32.04)}{4} = \frac{124.69}{4} \approx \$31.17$$

b) Median: 30.52, 30.90, 31.23, 32.04

$$\frac{30.90 + 31.23}{2} = \frac{62.13}{2} \approx \$31.07$$

25. If the radius is 12 ft., then the diameter is 24 ft.

a) $C = \pi d$

$C = 24\pi$

$C \approx 24(3.14)$

$C \approx 75.36$ ft.

b) $A = \pi r^2$

$A = \pi \cdot 12^2$

$A = 144\pi$

$A \approx 144(3.14)$

$A \approx 452.16$ ft.^{2}

26. $V = \frac{4}{3}\pi r^3$

$V = \frac{4}{3}\pi \cdot 2^3$

$V = \frac{4}{3}\cdot \pi \cdot 8$

$V = \frac{32}{3}\pi$

$V \approx \frac{32}{3}(3.14)$

$V \approx \frac{32}{3}\left(\frac{3.14}{1}\right)$

$V \approx \frac{100.48}{3}$

$V \approx 33.49\overline{3}$ cm^3

27. $V = \frac{1}{3}lwh$

$V = \frac{1}{3}(12)(12)(16.2)$

$V = \frac{1}{\cancel{3}}\left(\frac{\cancel{12}^{4}}{1}\right)(12)(16.2)$

$V = 4(12)(16.2)$

$V = 48(16.2)$

$V = 777.6$ ft.^{3}

28. We must add the area of the triangle to the area of the rectangle. The base of the triangle is $56.5 - 35 = 21.5$ ft. And the height of the triangle is $22 - 12 = 10$ ft.

$A = \frac{1}{2}bh + bh$

$A = \frac{1}{2}(21.5)(10) + 56.5(12)$

$A = \frac{1}{\cancel{2}}\left(\frac{21.5}{1}\right)\left(\frac{\cancel{10}^{5}}{1}\right) + 678$

$A = 107.5 + 678$

$A = 785.5$ ft.^{2}

29. We must use the Pythagorean theorem.

$a^2 + b^2 = c^2$

$5^2 + b^2 = 20^2$

$25 + b^2 = 400$

$\underline{-25 \qquad = -25}$

$b^2 = 375$

$b = \sqrt{375}$

$b \approx 19.36$ ft.

The ladder will reach 19.36 ft.

30.

Categories	Value	Number	Amount
round	14	x	$14x$
sheet	12.50	$12 - x$	$12.50(12 - x)$

$14x + 12.5(12 - x) = 162$

$14x + 150 - 12.5x = 162$

$1.5x + 150 = 162$

$\underline{-150 = -150}$

$1.5x + 0 = 12$

$\frac{1.5x}{1.5} = \frac{12}{1.5}$

$x = 8$

Yolanda sold 8 round-layered cakes and $12 - 8 = 4$ sheet cakes.

Chapter 6 Cumulative Review Exercises

1. true

2. false

3. true

4. true

5. false

6. false

7. undefined

8. coefficients; variables

9. $2000 \div 50$: cancel 1 zero $200 \div 5 = 40$

10. 6

11. Missing a term. Should have $5x$ term from combining $2x + 3x$. The correct answer is $x^2 + 5x + 6$.

12. $2^3 \cdot 3 \cdot 5 \cdot 7$

13.

14. Change the divisor to its reciprocal, then multiply.

15. $56 = 2^3 \cdot 7$

$42 = 2 \cdot 3 \cdot 7$

$LCM(56, 42) = 2^3 \cdot 3 \cdot 7 = 8 \cdot 3 \cdot 7 = 168$

16. twenty-nine and six thousand eighty-one ten-thousandths

17. 2.02

18. 768,000,000 : seven hundred sixty-eight million

19. $\{9 - 4[6 + (-16)]\} \div 7 = \{9 - 4 \cdot (-10)\} \div 7$

$= \{9 - (-40)\} \div 7$

$= \{9 + 40\} \div 7$

$= 49 \div 7$

$= 7$

20. $(-2)^5 - 4\sqrt{16 + 9} = (-2)^5 - 4\sqrt{25}$

$= (-2)^5 - 4 \cdot 5$

$= -32 - 20$

$= -52$

21. $-\dfrac{48}{60} = -\dfrac{48 \div 12}{60 \div 12} = -\dfrac{4}{5}$

22. $\dfrac{18x^2 y}{24xy^4} = \dfrac{18 \div 6}{24 \div 6} x^{2-1} y^{1-4} = \dfrac{3}{4} xy^{-3} = \dfrac{3x}{4y^3}$

23. $-\dfrac{1}{9} \cdot 5\dfrac{1}{6} = -\dfrac{1}{9} \cdot \dfrac{31}{6} = -\dfrac{31}{54}$

24. $(0.5)^2 = (0.5)(0.5) = 0.25$

25. $-\dfrac{12n^2}{13m} \div \dfrac{-15n}{26m} = -\dfrac{12n^2}{13m} \cdot \dfrac{26m}{-15n}$

$= -\dfrac{2 \cdot 2 \cdot \cancel{3} \cdot \cancel{n} \cdot n}{\cancel{13} \cdot \cancel{m}} \cdot \dfrac{2 \cdot \cancel{13} \cdot \cancel{m}}{-\cancel{3} \cdot 5 \cdot \cancel{n}}$

$= \dfrac{8n}{5}$

$= \dfrac{8}{5}n$

26. $4\dfrac{5}{8} + 7\dfrac{1}{2} = 4\dfrac{5}{8} + 7\dfrac{1 \cdot 4}{2 \cdot 4}$

$= 4\dfrac{5}{8} + 7\dfrac{4}{8}$

$= 11\dfrac{9}{8}$

$= 11 + 1\dfrac{1}{8}$

$= 12\dfrac{1}{8}$

27. $\dfrac{x}{6} + \dfrac{3}{4} = \dfrac{x \cdot 2}{6 \cdot 2} + \dfrac{3 \cdot 3}{4 \cdot 3}$

$= \dfrac{2x}{12} + \dfrac{9}{12}$

$= \dfrac{2x + 9}{12}$

28. $0.48 + 0.6\left(4\dfrac{1}{3}\right) + \sqrt{0.81} = 0.48 + 0.6\left(4\dfrac{1}{3}\right) + 0.9$

$= 0.48 + \dfrac{0.6}{1} \cdot \dfrac{13}{3} + 0.9$

$= 0.48 + \dfrac{7.8}{3} + 0.9$

$= 0.48 + 2.6 + 0.9$

$= 3.98$

29. $2.6x^3 - \dfrac{1}{4}x^2 + 3.8 - x^3 + x^2 - \dfrac{1}{4}$

$= 2.6x^3 - x^3 - \dfrac{1}{4}x^2 + x^2 + 3.8 - \dfrac{1}{4}$

$= 1.6x^3 - \dfrac{1}{4}x^2 + \dfrac{4}{4}x^2 + 3.8 - 0.25$

$= 1.6x^3 + \dfrac{3}{4}x^2 + 3.55$

30. $(14m^3 - 5.7m^2 + 7) - (9.1m^3 + m^2 - 11.6)$

$= (14m^3 - 5.7m^2 + 7) + (-9.1m^3 - m^2 + 11.6)$

$= 14m^3 - 9.1m^3 - 5.7m^2 - m^2 + 7 + 11.6$

$= 4.9m^3 - 6.7m^2 + 18.6$

31. $4\left(\dfrac{3}{4}\right)^2 - (-0.2)^3 = 4 \cdot \dfrac{9}{16} - (-0.008)$

$= \dfrac{\cancel{4}}{1} \cdot \dfrac{9}{\cancel{16}_4} + 0.008$

$= \dfrac{9}{4} + 0.008$

$= 2.25 + 0.008$

$= 2.258$

32. $(4y - 3)(y + 2) = 4y \cdot y + 4y \cdot 2 - 3 \cdot y - 3 \cdot 2$

$= 4y^2 + 8y - 3y - 6$

$= 4y^2 + 5y - 6$

33. $4x^3 \cdot ? = -48x^5 y$

$? = -48x^5 y \div 4x^3$

$? = (-48 \div 4)x^{5-3} y$

$? = -12x^2 y$

34. $30n^5 + 15n^2 - 25n$

$= 5n\left(\dfrac{30n^5 + 15n^2 - 25n}{5n}\right)$

$= 5n\left(\dfrac{30n^5}{5n} + \dfrac{15n^2}{5n} - \dfrac{25n}{5n}\right)$

$= 5n(6n^4 + 3n - 5)$

35. $y - 15.2 = 9.5$

$\underline{+15.2 \quad +15.2}$

$y = 24.7$

Check: $24.7 - 15.2 = 9.5$

$9.5 = 9.5$

36. $-0.06k = 0.408$ \quad Check: $-0.06 \cdot -6.8 = 0.408$

$\dfrac{-0.06k}{-0.06} = \dfrac{0.408}{-0.06}$ \qquad\qquad $0.408 = 0.408$

$1k = -6.8$

$k = -6.8$

37. $\dfrac{2}{3}x + \dfrac{1}{4} = \dfrac{5}{6}$ Check: $\dfrac{2}{3} \cdot \dfrac{7}{\cancel{8}_4} + \dfrac{1}{4} = \dfrac{5}{6}$

$\dfrac{2}{3}x + \dfrac{1}{4} - \dfrac{1}{4} = \dfrac{5}{6} - \dfrac{1}{4}$ $\dfrac{7}{12} + \dfrac{1 \cdot 3}{4 \cdot 3} = \dfrac{5}{6}$

$\dfrac{2}{3}x + 0 = \dfrac{10}{12} - \dfrac{3}{12}$ $\dfrac{7}{12} + \dfrac{3}{12} = \dfrac{5}{6}$

$\dfrac{2}{3}x = \dfrac{7}{12}$ $\dfrac{10}{12} = \dfrac{5}{6}$

$\dfrac{\cancel{3}}{\cancel{2}} \cdot \dfrac{\cancel{2}}{\cancel{3}}x = \dfrac{7}{12} \cdot \dfrac{3}{2}$ $\dfrac{5}{6} = \dfrac{5}{6}$

$1x = \dfrac{21}{24}$

$x = \dfrac{7}{8}$

38. $2n + 8 = 5(n - 3) + 2$

$2n + 8 = 5n - 15 + 2$

$2n + 8 = 5n - 13$

$\underline{-2n \qquad = -2n}$

$0 + 8 = 3n - 13$

$\underline{+13 = \qquad +13}$

$21 = 3n + 0$

$\dfrac{21}{3} = \dfrac{3n}{3}$

$7 = n$

Check: $2 \cdot 7 + 8 = 5(7 - 3) + 2$

$14 + 8 = 5 \cdot 4 + 2$

$22 = 20 + 2$

$22 = 22$

39. $24 + x = -8$

$\underline{-24 \qquad = -24}$

$0 + x = -32$

$x = -32°$

The temperature dropped 32°F.

40. $N = R - C$

$N = 2,609,400 - 1,208,500$

$N = \$1,400,900$

The net was a profit at $1,400,900.

41. a) $V = lwh$

$V = (2w - 1)w \cdot 4w$

$V = 4w^2(2w - 1)$

$V = 4w^2 \cdot 2w - 4w^2 \cdot 1$

$V = 8w^3 - 4w^2$

b) $V = 8 \cdot 6^3 - 4 \cdot 6^2$

$V = 8 \cdot 216 - 4 \cdot 36$

$V = 1728 - 144$

$V = 1584 \text{ cm}^3$

42. $SA = 2lw + 2lh + 2wh$

$52 = 2(2)(3) + 2(2)h + 2(3)h$

$52 = 12 + 4h + 6h$

$52 = 12 + 10h$

$\underline{-12 = -12}$

$40 = 0 + 10h$

$\dfrac{40}{10} = \dfrac{10h}{10}$

$4 \text{ ft.} = h$

43. $5 - 2(x - 9) = 7 + 6x$

$5 - 2x + 18 = 7 + 6x$

$23 - 2x = 7 + 6x$

$\underline{+2x = \qquad +2x}$

$23 + 0 = 7 + 8x$

$\underline{-7 \qquad = -7}$

$16 = 0 + 8x$

$\dfrac{16}{8} = \dfrac{8x}{8}$

$2 = x$

44. An isosceles triangle has two equal sides. Let's call the base b. Since the two equal sides are "12 ft. longer than twice the base," we can say the equal sides are $2b + 12$. To find the perimeter of any figure, we add the sides. We also know the total perimeter is 89 ft.

$P = b + (2b + 12) + (2b + 12)$

$89 = 5b + 24$

$\underline{-24 = \qquad -24}$

$65 = 5b + 0$

$\dfrac{65}{5} = \dfrac{5b}{5}$

$13 = b$

The base of the triangle is 13 ft. and the two equal-length sides are $2b + 12 = 26 + 12 = 38$ ft. each.

45. $\dfrac{3}{4} \cdot \dfrac{2}{3} = \dfrac{6}{12} = \dfrac{1}{2}$ of these employees have taken supplies.

46. $\dfrac{1}{6} + \dfrac{3}{5} = \dfrac{1 \cdot 5}{6 \cdot 5} + \dfrac{3 \cdot 6}{5 \cdot 6} = \dfrac{5}{30} + \dfrac{18}{30} = \dfrac{23}{30}$ said yes or no.

$\dfrac{7}{30}$ had no opinion.

47. We must divide her salary by the 12 months.

$23,272 \div 12 = 1939.\overline{33}$ Her salary is $1939.33 per month.

48. We must add the area of the rectangle to the area of the half circle. Since the diameter of the circle is 10 in, so the radius is 5 in.

$$A = lw + \frac{1}{2}\pi r^2$$

$$A = (10 \cdot 4.2) + \left(\frac{1}{2} \cdot \pi \cdot 5^2\right)$$

$$A = 42 + \left(\frac{1}{2} \cdot \pi \cdot \frac{25}{1}\right)$$

$$A = 42 + \frac{25}{2}\pi$$

$$A = 42 + 12.5\pi$$

$$A \approx 42 + 12.5(3.14)$$

$$A \approx 42 + 39.25$$

$$A \approx 81.25 \text{ in.}^2$$

49. We must use the Pythagorean theorem.

$$a^2 + b^2 = c^2$$
$$6^2 + b^2 = 20^2$$
$$36 + b^2 = 400$$
$$\underline{-36 \qquad = -36}$$
$$b^2 = 364$$
$$b = \sqrt{364}$$
$$b \approx 19.08 \text{ ft.}$$

The top of the ladder is 19.08 ft. from the base of the building.

50.

Categories	Value	Number	Amount
larger size	12.50	x	$12.5x$
smaller size	8.50	$20 - x$	$8.5(20 - x)$

$$12.5x + 8.5(20 - x) = 206$$
$$12.5x + 170 - 8.5x = 206$$
$$4x + 170 = 206$$
$$\underline{-170 = -170}$$
$$4x + 0 = 36$$
$$\frac{4x}{4} = \frac{36}{4}$$
$$x = 9$$

Cedrick sold 9 larger size bowls and $20 - 9 = 11$ smaller size bowls.

Chapter 7 Ratios, Proportions, and Measurement

Exercise Set 7.1

1. A ratio is a comparison between two quantities using a quotient.

3. Divide the denominator into the numerator.

5. A unit price is a unit ratio of price to quantity.

7. total attendance = males + females = 62 + 70 = 132

   a) $\dfrac{62}{132} = \dfrac{31}{66}$ 　　b) $\dfrac{70}{132} = \dfrac{35}{66}$

   c) $\dfrac{62}{70} = \dfrac{31}{35}$ 　　d) $\dfrac{70}{62} = \dfrac{35}{31}$

9. total atoms = calcium atoms + nitrogen atoms = 3 + 2 = 5

   a) $\dfrac{3}{2}$ 　 b) $\dfrac{2}{3}$ 　 c) $\dfrac{3}{5}$ 　 d) $\dfrac{2}{5}$

11. $\dfrac{14}{16} = \dfrac{7}{8}$

13. $\dfrac{3\frac{1}{2}}{2\frac{1}{4}} = 3\frac{1}{2} \div 2\frac{1}{4}$

    $= \dfrac{7}{2} \div \dfrac{9}{4}$

    $= \dfrac{7}{\cancel{2}} \cdot \dfrac{\cancel{4}^2}{9}$

    $= \dfrac{14}{9}$

15. total mail = bills + letters + advertisements + credit card offers = 7 + 3 + 17 + 3 = 30

    a) $\dfrac{7}{30}$ 　 b) $\dfrac{17}{30}$ 　 c) $\dfrac{3}{3} = 1$ 　 d) $\dfrac{10}{20} = \dfrac{1}{2}$

17. a) $\dfrac{128}{172} = \dfrac{128 \div 4}{172 \div 4} = \dfrac{32}{43}$

    b) $\dfrac{172}{642} = \dfrac{172 \div 2}{642 \div 2} = \dfrac{86}{321}$

    c) We must first find the total population of Rhode Island. 53 + 172 + 85 + 642 + 128 = 1080

    $\dfrac{128}{1080} = \dfrac{128 \div 8}{1080 \div 8} = \dfrac{16}{135}$

    d) $\dfrac{642}{1080} = \dfrac{642 \div 6}{1080 \div 6} = \dfrac{107}{180}$

19. There are 4 Kings in a standard deck of 52 cards.

    $P = \dfrac{4}{52} = \dfrac{1}{13}$

21. There are 4 Aces and 4 Kings in a standard deck of 52 cards. $P = \dfrac{8}{52} = \dfrac{2}{13}$

23. There are 2 favorable outcomes out of six possible outcomes. $P = \dfrac{2}{6} = \dfrac{1}{3}$

25. There are no 7's on a normal six-sided die.

    $P = \dfrac{0}{6} = 0$

27. The total number of entries is 12 + 14 = 26. There are 14 favorable outcomes out of 26 possible outcomes. $P = \dfrac{14}{26} = \dfrac{7}{13}$

29. total candies = green + red + brown + blue = 42 + 28 + 30 + 25 = 125.

    a) There are 25 favorable outcomes out of 125 possible outcomes. $P = \dfrac{25}{125} = \dfrac{1}{5}$

    b) There are 28 favorable outcomes out of 125 possible outcomes. $P = \dfrac{28}{125}$

    c) There are 42 + 30 = 72 favorable outcomes out of 125 possible outcomes. $P = \dfrac{72}{125}$

    d) There are 42 + 30 + 25 = 97 favorable outcomes out of 125 possible outcomes. $P = \dfrac{97}{125}$

31. Debt-to-income ratio = $\dfrac{4798}{29,850} \approx 0.16$. This means that out of every dollar of gross income, $0.16 is paid towards debt.

33. Payment-to-income ratio = $\dfrac{945}{3650} \approx 0.26$. This means that out of each dollar of gross income, $0.26 is paid to the mortgage.

35. Students-to-faculty ratio $= \dfrac{12,480}{740} \approx 16.9$. This means that there are 16.9 (or about 17) students for each faculty member.

37. Selling price to annual earnings $= \dfrac{54\frac{1}{8}}{4.48}$

$$\dfrac{54\frac{1}{8}}{4.48} = \dfrac{54.125}{4.48} \approx 12.08$$

This means that the stock is selling at $12.08 for every dollar of annual earnings.

39. $r = \dfrac{d}{t}$

$r = \dfrac{306.9}{4.5}$

$r = 68.2$ mph

41. $u = \dfrac{p}{q}$

$u = \dfrac{\$2.76}{23 \text{ min.}}$

$u = \dfrac{276 \text{ cents}}{23 \text{ min.}}$

$u = 12\cent / \text{min.}$

43. $u = \dfrac{p}{q}$

$u = \dfrac{\$3.75}{\frac{1}{2} \text{ lb.}}$

$u = \$7.50 / \text{lb.}$

45. Let F represent the unit price of the 10.5-oz. can of soup and T represent the unit price of the 16-oz. can of soup.

$F = \dfrac{\$1.79}{10.5 \text{ oz.}}$ $T = \dfrac{\$2.19}{16 \text{ oz.}}$

$F \approx \$0.17 / \text{oz.}$ $T \approx \$0.14 / \text{oz.}$

Because $0.14 is the smaller unit price, the 16-oz. can of soup is the better buy.

47. Let x represent the unit price of the bag of 24 diapers and y represent the unit price of the bag of 40 diapers.

$x = \dfrac{\$10.99}{24 \text{ diapers}}$ $y = \dfrac{\$15.99}{40 \text{ diapers}}$

$x \approx \$0.46 / \text{diaper}$ $y \approx \$0.40 / \text{diaper}$

Because $0.40 is the smaller unit price, the bag of 40 diapers is the better buy.

49. Buying 2 of the 15.5-oz. boxes means that we are buying $2(15.5) = 31$ oz. of raisin bran for $5.00. Let N represent the unit price of the name brand box of raisin bran and S represent the unit price of the store brand box of raisin bran.

$N = \dfrac{\$3.45}{20 \text{ oz.}}$ $S = \dfrac{\$5.00}{31 \text{ oz.}}$

$N \approx \$0.17 / \text{oz.}$ $S \approx \$0.16 / \text{oz.}$

Because $0.16 is the smaller unit price, the 2 boxes of 15.5 oz. is a better buy.

51. a)

	Violent Crimes	Aggravated Assault
Ratio per 10,000 households	$\dfrac{226}{10,000}$ $= \dfrac{226 \div 2}{10,000 \div 2}$ $= \dfrac{113}{5000}$	$\dfrac{46}{10,000}$ $= \dfrac{46 \div 2}{10,000 \div 2}$ $= \dfrac{23}{5000}$
Ratio per 1000 households	$\dfrac{226}{10,000}$ $= \dfrac{226 \div 10}{10,000 \div 10}$ $= \dfrac{22.6}{1000}$	$\dfrac{46}{10,000}$ $= \dfrac{46 \div 10}{10,000 \div 10}$ $= \dfrac{4.6}{1000}$
Ratio per 100 households	$\dfrac{226}{10,000}$ $= \dfrac{226 \div 100}{10,000 \div 100}$ $= \dfrac{2.26}{100}$	$\dfrac{46}{10,000}$ $= \dfrac{46 \div 100}{10,000 \div 100}$ $= \dfrac{0.46}{100}$

	Simple Assault	Household Burglary
Ratio per 10,000 households	$\dfrac{114}{10,000}$ $= \dfrac{114 \div 2}{10,000 \div 2}$ $= \dfrac{57}{5000}$	$\dfrac{246}{10,000}$ $= \dfrac{246 \div 2}{10,000 \div 2}$ $= \dfrac{123}{5000}$
Ratio per 1000 households	$\dfrac{114}{10,000}$ $= \dfrac{114 \div 10}{10,000 \div 10}$ $= \dfrac{11.4}{1000}$	$\dfrac{246}{10,000}$ $= \dfrac{246 \div 10}{10,000 \div 10}$ $= \dfrac{24.6}{1000}$

Ratio per 100 households	$\dfrac{114}{10,000}$	$\dfrac{246}{10,000}$
	$= \dfrac{114 \div 100}{10,000 \div 100}$	$= \dfrac{246 \div 100}{10,000 \div 100}$
	$= \dfrac{1.14}{100}$	$= \dfrac{2.46}{100}$

	Motor Vehicle Theft
Ratio per 10,000 households	$\dfrac{67}{10,000}$
Ratio per 1000 households	$\dfrac{67}{10,000}$ $= \dfrac{67 \div 10}{10,000 \div 10}$ $= \dfrac{6.7}{1000}$
Ratio per 100 households	$\dfrac{67}{10,000}$ $= \dfrac{67 \div 100}{10,000 \div 100}$ $= \dfrac{0.67}{100}$

b) 23 out of 5000 households were victims of aggravated assault

c) 22.6   d) 2.46

Review Exercises

1. $5\dfrac{1}{4} \cdot \dfrac{6}{7} = \dfrac{\overset{3}{\cancel{21}}}{\underset{2}{\cancel{4}}} \cdot \dfrac{\cancel{6}^{3}}{\cancel{7}_{1}}$

    $= \dfrac{9}{2}$

    $= 4\dfrac{1}{2}$

2.
$$
\begin{array}{r}
12.5 \\
\times 9.6 \\
\hline
750 \\
1125 \\
\hline
120.00
\end{array}
$$

3. $4.5\left(2\dfrac{3}{4}\right) = 4.5(2.75)$

    $= 12.375$

4. $\dfrac{2}{3}y = 5$ Check: $\dfrac{2}{3}\left(7\dfrac{1}{2}\right) = 5$

   $\dfrac{3}{2} \cdot \dfrac{2}{3}y = \dfrac{5}{1} \cdot \dfrac{3}{2}$ $\dfrac{\overset{1}{\cancel{2}}}{\underset{1}{\cancel{3}}} \cdot \dfrac{\cancel{15}^{5}}{\cancel{2}_{1}} = 5$

   $1y = \dfrac{15}{2}$ $\dfrac{5}{1} = 5$

   $y = 7\dfrac{1}{2}$ $5 = 5$

---

5. $10.4x = 89.44$ Check: $10.4(8.6) = 89.44$

   $\dfrac{10.4x}{10.4} = \dfrac{89.44}{10.4}$ $89.44 = 89.44$

   $1x = 8.6$

   $x = 8.6$

Exercise Set 7.2

1. A proportion is an equation in the form $\dfrac{a}{b} = \dfrac{c}{d}$ where $b \neq 0$ and $d \neq 0$.

3. 1. Calculate the cross-products.

   2. Set the cross products equal to one another.

   3. Use the multiplication/division principle of equality to isolate the variable.

5. Compare the cross products, if the cross products are equal, then the ratios are proportional.

   $12 \cdot 3 = 36$
   $4 \cdot 9 = 36$

   Yes: Because the cross products are equal, these ratios are proportional.

7. Compare the cross products, if the cross products are equal, then the ratios are proportional.

   $(2.4) \cdot 15 = 36$
   $4 \cdot 7 = 28$

   No: Because the cross products are not equal, these ratios are not proportional.

9. Compare the cross products, if the cross products are equal, then the ratios are proportional.

   $(3.2) \cdot 15 = 48$
   $24 \cdot (2.5) = 60$

   No: Because the cross products are not equal, these ratios are not proportional.

11. Compare the cross products, if the cross products are equal, then the ratios are proportional.

    $42 \cdot (9.5) = 399$
    $14 \cdot (28.5) = 399$

    Yes: Because the cross products are equal, these ratios are proportional.

13. Compare the cross products, if the cross products are equal, then the ratios are proportional.

$$42.5 \cdot \left(16.8\right) = 714$$
$$40.2 \cdot \left(17.5\right) = 703.5$$

No: Because the cross products are not equal, these ratios are not proportional.

15. Compare the cross products, if the cross products are equal, then the ratios are proportional.

$$10\frac{1}{2} \cdot \frac{4}{5} = \frac{21}{2} \cdot \frac{\cancel{4}^2}{5} = \frac{42}{5} = 8\frac{2}{5}$$

$$\frac{7}{10} \cdot \frac{12}{1} = \frac{84}{10} = 8\frac{4}{10} = 8\frac{2}{5}$$

Yes: Because the cross products are equal, these ratios are proportional.

17. Compare the cross products, if the cross products are equal, then the ratios are proportional.

$$7\frac{1}{3} \cdot 2\frac{1}{3} = 17\frac{1}{9}$$

$$3\frac{1}{2} \cdot 5\frac{1}{2} = 19\frac{1}{4}$$

No: Because the cross products are not equal, these ratios are not proportional.

19.
$$\frac{x}{8} = \frac{20}{32}$$
$$32 \cdot x = 8 \cdot 20$$
$$32x = 160$$
$$\frac{32x}{32} = \frac{160}{32}$$
$$x = 5$$

21.
$$\frac{-3}{8} = \frac{n}{20}$$
$$20 \cdot (-3) = 8 \cdot n$$
$$-60 = 8n$$
$$\frac{-60}{8} = \frac{8n}{8}$$
$$-7.5 = n$$

23.
$$\frac{2}{n} = \frac{7}{21}$$
$$21 \cdot 2 = n \cdot 7$$
$$42 = 7n$$
$$\frac{42}{7} = \frac{7n}{7}$$
$$6 = n$$

25.
$$-\frac{4}{28} = \frac{5}{m}$$
$$\frac{-4}{28} = \frac{5}{m}$$
$$m \cdot \left(-4\right) = 28 \cdot 5$$
$$-4m = 140$$
$$\frac{-4m}{-4} = \frac{140}{-4}$$
$$m = -35$$

27.
$$\frac{18}{h} = \frac{21.6}{30}$$
$$30 \cdot 18 = h \cdot \left(21.6\right)$$
$$540 = 21.6h$$
$$\frac{540}{21.6} = \frac{21.6h}{21.6}$$
$$25 = h$$

29.
$$\frac{-14}{b} = \frac{3.5}{6.25}$$
$$6.25 \cdot \left(-14\right) = b \cdot 3.5$$
$$-87.5 = 3.5b$$
$$\frac{-87.5}{3.5} = \frac{3.5b}{3.5}$$
$$-25 = b$$

31.
$$\frac{4\frac{3}{4}}{5\frac{1}{2}} = \frac{d}{16\frac{1}{2}}$$
$$16\frac{1}{2} \cdot 4\frac{3}{4} = 5\frac{1}{2} \cdot d$$
$$\frac{33}{2} \cdot \frac{19}{4} = \frac{11}{2}d$$
$$\frac{627}{8} = \frac{11}{2}d$$
$$\frac{\cancel{2}}{\cancel{11}} \cdot \frac{\cancel{627}^{57}}{\cancel{8}_4} = \frac{\cancel{11}}{\cancel{2}}d \cdot \frac{\cancel{2}}{\cancel{11}}$$
$$\frac{57}{4} = 1d$$
$$14\frac{1}{4} = d$$

33.
$$\frac{-9.5}{22} = \frac{-6\frac{1}{3}}{t}$$
$$t \cdot (-9.5) = 22 \cdot \left(-6\frac{1}{3}\right)$$
$$-9.5t = \frac{22}{1} \cdot \left(\frac{-19}{3}\right)$$
$$-\frac{95}{10}t = \frac{-418}{3}$$
$$-\frac{\cancel{10}}{\cancel{95}} \cdot \left(-\frac{\cancel{95}}{\cancel{10}}\right)t = \frac{-418}{3} \cdot \left(-\frac{10}{95}\right)$$
$$1t = \frac{4180}{285}$$
$$t = 14\frac{2}{3}$$

35. 358.4 mi. using 16.4 gal. translates to $\dfrac{358.4 \text{ mi.}}{16.4 \text{ gal.}}$

and 750 miles using ? gal. translates to $\dfrac{750 \text{ mi.}}{x \text{ gal.}}$

So, $\dfrac{358.4 \text{ mi.}}{16.4 \text{ gal.}} = \dfrac{750 \text{ mi.}}{x \text{ gal.}}$

$$x \cdot 358.4 = 16.4 \cdot \left(750\right)$$
$$358.4x = 12{,}300$$
$$x \approx 34.3 \text{ gal.}$$

It will take about 34.3 gallons to drive 750 miles.

37. 12.5 lb. turkey has 30 servings translates to

$\dfrac{12.5 \text{ lb.}}{30 \text{ servings}}$ and 15 lb. will have ? servings

translates to $\dfrac{15 \text{ lb.}}{x \text{ servings}}$.

So, $\dfrac{12.5 \text{ lb.}}{30 \text{ servings}} = \dfrac{15 \text{ lb.}}{x \text{ servings}}$

$$x \cdot (12.5) = 30 \cdot (15)$$
$$12.5x = 450$$
$$\dfrac{12.5x}{12.5} = \dfrac{450}{12.5}$$
$$x = 36 \text{ servings}$$

A 15 lb. turkey will have 36 servings.

39. 16 oz. of yogurt in 5 days translates to $\dfrac{16 \text{ oz.}}{5 \text{ days}}$

and ? oz. in 365.25 days translates to

$\dfrac{x \text{ oz.}}{365.25 \text{ days}}$

So, $\dfrac{16 \text{ oz.}}{5 \text{ days}} = \dfrac{x \text{ oz.}}{365.25 \text{ days}}$

$$365.25 \cdot (16) = 5 \cdot x$$
$$5844 = 5x$$
$$\dfrac{5844}{5} = \dfrac{5x}{5}$$
$$1168.8 = x$$

Nathan will consume 1168.8 oz. of yogurt in a year.

41. 1 ½ cups of bran flakes for 8 muffins translates to

$\dfrac{1\frac{1}{2} \text{ cups}}{8 \text{ muffins}}$ and ? cups of bran flakes for a dozen

muffins translates to $\dfrac{x \text{ cups}}{12 \text{ muffins}}$

So, $\dfrac{1.5 \text{ cups}}{8 \text{ muffins}} = \dfrac{x \text{ cups}}{12 \text{ muffins}}$

$$12 \cdot 1.5 = 8 \cdot x$$
$$18 = 8x$$
$$\dfrac{18}{8} = \dfrac{8x}{8}$$
$$2\frac{1}{4} = x$$

It will take $2\frac{1}{4}$ cups of bran flakes for a dozen muffins.

43. 42 ½ ft. wall measures 8 ½ in. on a scale drawing

translates to $\dfrac{42\frac{1}{2} \text{ ft.}}{8\frac{1}{2} \text{ in.}}$ and 16 ¼ ft. wall measures

? in. on a scale drawing translates to $\dfrac{16\frac{1}{4} \text{ ft.}}{x \text{ in.}}$.

So, $\dfrac{42\frac{1}{2} \text{ ft.}}{8\frac{1}{2} \text{ in.}} = \dfrac{16\frac{1}{4} \text{ ft.}}{x \text{ in.}}$

$$x \cdot 42\frac{1}{2} = 8\frac{1}{2} \cdot 16\frac{1}{4}$$
$$\frac{85}{2}x = \frac{17}{2} \cdot \frac{65}{4}$$
$$\frac{85}{2}x = \frac{1105}{8}$$
$$\frac{\cancel{2}}{\cancel{85}} \cdot \frac{\cancel{85}}{\cancel{2}}x = \frac{^{221}\cancel{1105}}{_4\cancel{8}} \cdot \frac{\cancel{2}}{\cancel{85}_{17}}$$
$$1x = \frac{221}{68}$$
$$x = 3\frac{1}{4} \text{ in.}$$

A 16 ¼ ft. wall measure 3 ¼ in. on a scale drawing.

45. 18 feet every 30 days translates to $\dfrac{18 \text{ ft.}}{30 \text{ days}}$

980 ft. in ? days translates to $\dfrac{980 \text{ ft.}}{x \text{ days}}$.

So, $\dfrac{18 \text{ ft.}}{30 \text{ days}} = \dfrac{980 \text{ ft.}}{x \text{ days}}$

$$x \cdot 18 = 30 \cdot 980$$
$$18x = 29,400$$
$$\dfrac{18x}{18} = \dfrac{29,400}{18}$$
$$1x = 1633\frac{6}{18}$$
$$x = 1633\frac{1}{3} \text{ days}$$

It will take 1633 1/3 days to complete 980 feet.

47. | Smaller triangle | | Larger triangle |
|---|---|---|
| 12.5 cm. | ↔ | 16 cm. |
| 7 cm. | ↔ | c |

$$\frac{12.5}{16} = \frac{7}{c}$$

$$c \cdot (12.5) = 16 \cdot 7$$

$$12.5c = 112$$

$$\frac{12.5c}{12.5} = \frac{112}{12.5}$$

$$c = 8.96 \text{ cm}$$

The missing length is 8.96 cm.

49.

Larger trapezoid		Smaller trapezoid
$4\frac{1}{2}$ in.	$\leftrightarrow$	$3\frac{1}{4}$ in.
5 in.	$\leftrightarrow$	a
$4\frac{1}{2}$ in.	$\leftrightarrow$	b
$6\frac{3}{8}$ in.	$\leftrightarrow$	c

To find a:

$$\frac{4\frac{1}{2}}{3\frac{1}{4}} = \frac{5}{a}$$

$$a \cdot 4\frac{1}{2} = 3\frac{1}{4} \cdot 5$$

$$\frac{9}{2}a = \frac{13}{4} \cdot 5$$

$$\frac{9}{2}a = \frac{65}{4}$$

$$\frac{2}{9} \cdot \frac{9}{2}a = \frac{65}{4} \cdot \frac{2}{9}$$

$$1a = \frac{65}{18}$$

$$a = 3\frac{11}{18} \text{ in.}$$

It isn't necessary to write a proportion to find b. Notice that the two sides in the larger trapezoid measure 4 ½ in. This means that the corresponding sides must match in the smaller trapezoid. Because 4 ½ in. corresponds to 3 ¼ in. we can conclude that $b = 3$ ¼ in.

To find c:

$$\frac{4\frac{1}{2}}{3\frac{1}{4}} = \frac{6\frac{3}{8}}{c}$$

$$c \cdot 4\frac{1}{2} = 3\frac{1}{4} \cdot 6\frac{3}{8}$$

$$\frac{9}{2}c = \frac{13}{4} \cdot \frac{51}{8}$$

$$\frac{9}{2}c = \frac{663}{32}$$

$$\frac{2}{9} \cdot \frac{9}{2}c = \frac{663}{32} \cdot \frac{2}{9}$$

$$1c = \frac{221}{48}$$

$$c = 4\frac{29}{48} \text{ in.}$$

51.

Smaller triangle		Larger triangle
3 m	$\leftrightarrow$	16.5 m
1.8 m	$\leftrightarrow$	x m

$$\frac{3}{16.5} = \frac{1.8}{x}$$

$$x \cdot 3 = 16.5 \cdot (1.8)$$

$$3x = 29.7$$

$$\frac{3x}{3} = \frac{29.7}{3}$$

$$x = 9.9 \text{ m}$$

The house is 9.9 m tall.

53.

Smaller triangle		Larger triangle
$5\frac{1}{2}$ ft.	$\leftrightarrow$	x
6 ft.	$\leftrightarrow$	$524\frac{2}{3}$ ft.

$$\frac{5\frac{1}{2}}{x} = \frac{6}{524\frac{2}{3}}$$

$$524\frac{2}{3} \cdot 5\frac{1}{2} = x \cdot 6$$

$$\frac{\cancel{1574}^{787}}{3} \cdot \frac{11}{\cancel{2}} = 6x$$

$$\frac{8657}{3} = 6x$$

$$\frac{1}{6} \cdot \frac{8657}{3} = \cancel{6}x \cdot \frac{1}{\cancel{6}}$$

$$\frac{8657}{18} = 1x$$

$$480\frac{17}{18} = x$$

The pyramid is $480\frac{17}{18}$ ft. tall.

55.
Smaller triangle		Larger triangle
1.2 m	↔	1384 m
10.4 m	↔	y

$$\frac{1.2}{1384} = \frac{10.4}{y}$$
$$y \cdot (1.2) = 1384 \cdot (10.4)$$
$$1.2y = 14,393.6$$
$$\frac{1.2y}{1.2} = \frac{14,393.6}{1.2}$$
$$1y = 11,994.\overline{6}$$
$$y = 11,994.\overline{6} \text{ m}$$

The width of the Grand Canyon is $11,994.\overline{6}$ m.

Review Exercises

1. $\dfrac{\cancel{4}}{9} \cdot \dfrac{5}{\cancel{12}_3} = \dfrac{5}{27}$

2. $9\frac{1}{3}(144) = \dfrac{28}{\cancel{3}} \cdot \dfrac{\cancel{144}^{48}}{1}$
   $= 1344$

3. $60\frac{3}{4} \div 3 = \dfrac{243}{4} \div \dfrac{3}{1}$
   $= \dfrac{243}{4} \cdot \dfrac{1}{3}$
   $= \dfrac{243}{12}$
   $= 20\frac{1}{4}$

4.
$$\begin{array}{r}
5.875 \\
12\overline{)70.500} \\
\underline{60} \\
105 \\
\underline{96} \\
90 \\
\underline{84} \\
60 \\
\underline{60} \\
0
\end{array}$$

5. $59.2 \cdot \dfrac{1}{2} \cdot \dfrac{1}{2} \cdot \dfrac{1}{4} = \dfrac{59.2}{16}$
   $= 3.7$

Exercise Set 7.3

1. A unit fraction is a fraction with a ratio equivalent to 1.

3. 12; 3; 5280

5. 8; 2; 2; 4

7. 52 yd. $= \dfrac{52 \text{ yd.}}{1} \cdot \dfrac{3 \text{ ft.}}{1 \text{ yd.}} = 156$ ft.

9. 90 in. $= \dfrac{90 \text{ in.}}{1} \cdot \dfrac{1 \text{ ft.}}{12 \text{ in.}} = \dfrac{90}{12}$ ft. $= 7.5$ ft.

11. 6.5 mi. $= \dfrac{6.5 \text{ mi.}}{1} \cdot \dfrac{5280 \text{ ft.}}{1 \text{ mi.}} = 34,320$ ft.

13. $5\frac{1}{2}$ ft. $= \dfrac{11 \text{ ft.}}{2} \cdot \dfrac{12 \text{ in.}}{1 \text{ ft.}} = \dfrac{132}{2}$ in. $= 66$ in.

15. 20.8 mi. $= \dfrac{20.8 \text{ mi.}}{1} \cdot \dfrac{5280 \text{ ft.}}{1 \text{ mi.}} \cdot \dfrac{1 \text{ yd.}}{3 \text{ ft.}}$
    $= \dfrac{109,824}{3}$ yd.
    $= 36,608$ yd.

17. 99 in. $= \dfrac{99 \text{ in.}}{1} \cdot \dfrac{1 \text{ ft.}}{12 \text{ in.}} \cdot \dfrac{1 \text{ yd.}}{3 \text{ ft.}}$
    $= \dfrac{99}{36}$ yd.
    $= 2.75$ yd.

19. 30 yd.$^2 = \dfrac{30 \text{ yd.}^2}{1} \cdot \dfrac{9 \text{ ft.}^2}{1 \text{ yd.}^2} = 270$ ft.^{2}

21. $10\dfrac{1}{2}$ yd.^{2} $= \dfrac{21 \text{ yd.}^2}{2} \cdot \dfrac{9 \text{ ft.}^2}{1 \text{ yd.}^2} \cdot \dfrac{144 \text{ in.}^2}{1 \text{ ft.}^2}$

$= \dfrac{27{,}216}{2}$ in.^{2}

$= 13{,}608$ in.^{2}

23. We must first convert the width from inches to feet. Since 1 ft. = 12 in., $36 \div 12 = 3$ ft.

$A = lw$

$A = 150(3)$

$A = 450$ ft.^{2}

25. $A = lw$ $\qquad$ 168.75 ft.^{2}

$A = (13.5)(12.5)$ $\qquad = \dfrac{168.75 \text{ ft.}^2}{1} \cdot \dfrac{1 \text{ yd.}^2}{9 \text{ ft.}^2}$

$A = 168.75$ ft.^{2} $\qquad = \dfrac{168.75}{9}$ yd.^{2}

$\qquad = 18.75$ yd.^{2}

27. 9 pt. $= \dfrac{9 \text{ pt.}}{1} \cdot \dfrac{2 \text{ c.}}{1 \text{ pt.}} = 18$ c.

29. 2.5 c. $= \dfrac{2.5 \text{ c.}}{1} \cdot \dfrac{8 \text{ oz.}}{1 \text{ c.}} = 20$ oz.

31. 240 oz. $= \dfrac{240 \text{ oz.}}{1} \cdot \dfrac{1 \text{ c.}}{8 \text{ oz.}} \cdot \dfrac{1 \text{ pt.}}{2 \text{ c.}}$

$= \dfrac{240}{16}$ pt.

$= 15$ pt.

33. $8\dfrac{1}{2}$ pt. $= \dfrac{17 \text{ pt.}}{2} \cdot \dfrac{1 \text{ qt.}}{2 \text{ pt.}} = \dfrac{17}{4}$ qt. $= 4.25$ qt.

35. 0.4 gal. $= \dfrac{0.4 \text{ gal.}}{1} \cdot \dfrac{4 \text{ qt.}}{1 \text{ gal.}} \cdot \dfrac{2 \text{ pt.}}{1 \text{ qt.}} \cdot \dfrac{2 \text{ c.}}{1 \text{ pt.}} \cdot \dfrac{8 \text{ oz.}}{1 \text{ c.}}$

$= 51.2$ oz.

37. 40.8 pt. $= \dfrac{40.8 \text{ pt.}}{1} \cdot \dfrac{1 \text{ qt.}}{2 \text{ pt.}} \cdot \dfrac{1 \text{ gal.}}{4 \text{ qt.}}$

$= \dfrac{40.8}{8}$ gal.

$= 5.1$ gal.

39. $12\dfrac{1}{4}$ lb. $= \dfrac{49 \text{ lb.}}{4} \cdot \dfrac{16 \text{ oz.}}{1 \text{ lb.}} = 196$ oz.

41. 180 oz. $= \dfrac{180 \text{ oz.}}{1} \cdot \dfrac{1 \text{ lb.}}{16 \text{ oz.}} = \dfrac{180}{16}$ lb. $= 11.25$ lb.

43. 1200 lb. $= \dfrac{1200 \text{ lb.}}{1} \cdot \dfrac{1 \text{ T}}{2000 \text{ lb.}} = \dfrac{1200}{2000}$ T $= 0.6$ T

45. 4.2 T $= \dfrac{4.2 \text{ T}}{1} \cdot \dfrac{2000 \text{ lb.}}{1 \text{ T}} = 8400$ lb.

47. 150 min. $= \dfrac{150 \text{ min.}}{1} \cdot \dfrac{1 \text{ hr.}}{60 \text{ min.}}$

$= \dfrac{150}{60}$ hr.

$= 2.5$ hr.

49. 90 sec. $= \dfrac{90 \text{ sec.}}{1} \cdot \dfrac{1 \text{ min.}}{60 \text{ sec.}} = \dfrac{90}{60}$ min. $= 1.5$ min.

51. $10\dfrac{1}{2}$ min. $= \dfrac{21 \text{ min.}}{2} \cdot \dfrac{60 \text{ sec.}}{1 \text{ min.}}$

$= \dfrac{1260}{2}$ sec.

$= 630$ sec.

53. $2\dfrac{1}{4}$ hr. $= \dfrac{9 \text{ hr.}}{4} \cdot \dfrac{60 \text{ min.}}{1 \text{ hr.}} = \dfrac{540}{4}$ min. $= 135$ min.

55. 0.2 hr. $= \dfrac{0.2 \text{ hr.}}{1} \cdot \dfrac{60 \text{ min.}}{1 \text{ hr.}} \cdot \dfrac{60 \text{ sec.}}{1 \text{ min.}} = 720$ sec.

57. 30 d. $= \dfrac{30 \text{ d.}}{1} \cdot \dfrac{1 \text{ yr.}}{365.25 \text{ d.}}$

$= \dfrac{30}{365.25}$ yr.

≈ 0.082 yr.

59. 1 d. $= \dfrac{1 \text{ d.}}{1} \cdot \dfrac{24 \text{ hr.}}{1 \text{ d.}} \cdot \dfrac{60 \text{ min.}}{1 \text{ hr.}} = 1440$ min.

61. 72 yr. $= \dfrac{72 \text{ yr.}}{1} \cdot \dfrac{365.25 \text{ d.}}{1 \text{ yr.}} \cdot \dfrac{24 \text{ hr.}}{1 \text{ d.}} \cdot \dfrac{60 \text{ min.}}{1 \text{ hr.}}$

$= 37{,}869{,}120$ min.

63. 1088 ft./sec. $= \dfrac{1088 \text{ ft.}}{1 \text{ sec.}} \cdot \dfrac{1 \text{ mi.}}{5280 \text{ ft.}} \cdot \dfrac{60 \text{ sec.}}{1 \text{ min.}}$

$= \dfrac{65{,}280}{5280}$ mi./min.

$= 12.36$ mi./min.

65. 80 ft./sec. $= \dfrac{80 \text{ ft.}}{1 \text{ sec.}} \cdot \dfrac{1 \text{ mi.}}{5280 \text{ ft.}} \cdot \dfrac{60 \text{ sec.}}{1 \text{ min.}} \cdot \dfrac{60 \text{ min.}}{1 \text{ hr.}}$

$= \dfrac{288,000}{5280}$ mi./hr.

$= 54.\overline{54}$ mi./hr.

No; the driver and the officer are not in agreement because the officer's estimate converts to a speed of $54.\overline{54}$ mi./hr., not 40 mi./hr.

67. 36,687.6 ft./sec.

$= \dfrac{36,687.6 \text{ ft.}}{1 \text{ sec.}} \cdot \dfrac{1 \text{ mi.}}{5280 \text{ ft.}} \cdot \dfrac{60 \text{ sec.}}{1 \text{ min.}} \cdot \dfrac{60 \text{ min.}}{1 \text{ hr.}}$

$= \dfrac{132,075,360}{5280}$ mi./hr.

$= 25,014.\overline{27}$ mi./hr.

69. We will use the formula $d = rt$, but first we must convert mi./hr. to ft./sec.

60 mi./hr. $= \dfrac{60 \text{ mi.}}{1 \text{ hr.}} \cdot \dfrac{5280 \text{ ft.}}{1 \text{ mi.}} \cdot \dfrac{1 \text{ hr.}}{60 \text{ min.}} \cdot \dfrac{1 \text{ min.}}{60 \text{ sec.}}$

$= \dfrac{316,800}{3600}$ ft./sec.

$= 88$ ft./sec.

$d = rt$

$500 \text{ft.} = \left(88 \text{ ft./sec.}\right)t$

$5.\overline{681} \text{ sec.} \approx t$

Review Exercises

1. $8.4(100) = 840$ **2.** $0.95(1000) = 950$

3. $(0.01)(32.9) = 0.329$

4. $45.6 \div 10 = 4.56$

5. $9 \div 1000 = 0.009$

6. $4800 \div 1000 = 4.8$

Exercise Set 7.4

1. A base unit is a basic unit; other units are named relative to it.

3. meter

5. Mass is a measure of the amount of matter that makes up an object.

7. Because cm is 2 units to the right of m, we move the decimal point 2 places to the right in 4.5.

4.5 m = 450 cm

9. Because m is 3 units to the right of km, we move the decimal point 3 places to the right in 0.07.

0.07 km = 70 m

11. Because km is 3 units to the left of m, we move the decimal point 3 places to the left in 3800.

3800 m = 3.8 km

13. Because m is 3 units to the left of mm, we move the decimal point 3 places to the left in 9500.

9500 mm = 9.5 m

15. Because dam is 2 units to the left of dm, we move the decimal point 2 places to the left in 1540.

1540 dm = 15.4 dam

17. Because hm is 2 units to the left of m, we move the decimal point 2 places to the left in 112.

112 m = 1.12 hm

19. Because ml is 3 units to the right of L, we move the decimal point 3 places to the right in 0.12.

0.12 L = 120 ml

21. Because L is 3 units to the right of kl, we move the decimal point 3 places to the right in 0.005.

0.005 kl = 5 L

23. Because dal is 3 units to the left of cl, we move the decimal point 3 places to the left in 9.5.

9.5 cl = 0.0095 dal

25. $1 \text{ cm}^3 = 1 \text{ ml}$: Because ml is 3 units to the right of l, we move the decimal point 3 places to the right in 0.08.

0.08 l = 80 cc

27. Since 1 cc = 1 ml, 12 cc = 12 ml.

29. $1 \text{ cm}^3 = 1 \text{ ml}$: Because ml is 3 units to the right of l, we move the decimal point 3 places to the right in 0.4.

0.4 l = 400 cc

31. Because mg is 3 units to the right of g, we move the decimal point 3 places to the right in 12.

12 g = 12,000 mg

33. Because g is 3 units to the right of kg, we move the decimal point 3 places to the right in 0.009.

0.009 kg = 9 g

35. Because cg is 3 units to the right of dag, we move the decimal point 3 places to the right in 0.05.

    0.05 dag = 50 cg

37. $3600 \text{ kg} = \dfrac{3600 \ \cancel{kg}}{1} \cdot \dfrac{1 \text{ t}}{1000 \ \cancel{kg}} = \dfrac{3600}{1000} = 3.6 \text{ t}$

39. $0.106 \text{ t} = \dfrac{0.106 \ \cancel{t}}{1} \cdot \dfrac{1000 \text{ kg}}{1 \ \cancel{t}} = 106 \text{ kg}$

41. Since kg is 1 unit to the left of hg and t is 3 units to the left of kg. Therefore t is a total of 4 units to the left of hg, so we move the decimal point 4 places to the left in 65,000.

    65,000 hg = 6.5 t

43. $1.8 \text{ m}^2 = \dfrac{1.8 \ \cancel{m^2}}{1} \cdot \dfrac{10,000 \text{ cm}^2}{1 \ \cancel{m^2}} = 18,000 \text{ cm}^2$

45. $2200 \text{ cm}^2 = \dfrac{2200 \ \cancel{cm^2}}{1} \cdot \dfrac{1 \text{ m}^2}{10,000 \ \cancel{cm^2}}$
    $= \dfrac{2200}{10,000} \text{ m}^2$
    $= 0.22 \text{ m}^2$

47. The mm is the 1st unit to the right of cm. With the area units we move the decimal point twice the number of places; so that means the decimal point move to the right 2 places.

    $0.05 \text{ cm}^2 = 5 \text{ mm}^2$

49. The hm is the third unit to the left of dm. With area units, we move the decimal point twice the number of places, so that means the decimal point moves to the left 6 places.

    $44,300 \text{ dm}^2 = 0.0443 \text{ hm}^2$

Review Exercises

1. $6\dfrac{1}{2} \div 3 \cdot 10 = \dfrac{13}{2} \div \dfrac{3}{1} \cdot \dfrac{10}{1}$
   $= \dfrac{13}{\cancel{2}} \cdot \dfrac{1}{3} \cdot \dfrac{\cancel{10}^5}{1}$
   $= \dfrac{65}{3}$
   $= 21\dfrac{2}{3}$

2. $8.4(12) \div 100 = 100.8 \div 100$
   $= 1.008$

3. $6.5x^2 - 8x + 14.4 - 10.2x - 20$
   $= 6.5x^2 - 8x - 10.2x + 14.4 - 20$
   $= 6.5x^2 - 18.2x - 5.6$

4. $100(4.81y^2 - 2.8y + 7)$
   $= 100 \cdot (4.81y^2) - 100 \cdot (2.8y) + 100 \cdot (7)$
   $= 481y^2 - 280y + 700$

5. $$7x - 3.98 = 2.4x + 8.9$$
   $$100(7x - 3.98) = 100(2.4x + 8.9)$$
   $$100 \cdot (7x) - 100 \cdot (3.98) = 100 \cdot (2.4x) + 100 \cdot (8.9)$$
   $$700x - 398 = 240x + 890$$
   $$\underline{-240x \qquad\quad = -240x}$$
   $$460x - 398 = 0 + 890$$
   $$\underline{+398 = \quad +398}$$
   $$460x + 0 = 1288$$
   $$\dfrac{460x}{460} = \dfrac{1288}{460}$$
   $$x = 2.8$$

   Check: $7(2.8) - 3.98 = 2.4(2.8) + 8.9$
   $$19.6 - 3.98 = 6.72 + 8.9$$
   $$15.62 = 15.65$$

Exercise Set 7.5

1. 2.54 cm

3. $F = \dfrac{9}{5}C + 32$

5. $100 \text{ yd.} \approx \dfrac{100 \ \cancel{yd.}}{1} \cdot \dfrac{0.914 \text{ m.}}{1 \ \cancel{yd.}} \approx 91.4 \text{ m}$

7. $26.2 \text{ mi.} \approx \dfrac{26.2 \ \cancel{mi.}}{1} \cdot \dfrac{1.609 \text{ km}}{1 \ \cancel{mi.}} \approx 42.2 \text{ km}$

9. $\dfrac{3}{8} \text{ in.} = \dfrac{\frac{3}{8} \ \cancel{in.}}{1} \cdot \dfrac{2.54 \text{ cm}}{1 \ \cancel{in.}} = 0.9525 \text{ cm} = 9.525 \text{ mm}$

11. $400 \text{ m} \approx \dfrac{400 \ \cancel{m}}{1} \cdot \dfrac{1.094 \text{ yd.}}{1 \ \cancel{m.}} \approx 437.6 \text{ yd.}$

13. $40 \text{ km} \approx \dfrac{40 \ \cancel{km}}{1} \cdot \dfrac{0.621 \text{ mi.}}{1 \ \cancel{km}} \approx 24.84 \text{ mi.}$

15. $45 \text{ cm} = \dfrac{45 \ \cancel{cm}}{1} \cdot \dfrac{1 \text{ in.}}{2.54 \ \cancel{cm}} = \dfrac{45}{2.54} \text{ in.} = 17.72 \text{ in.}$

17. $16 \text{ oz.} \approx \dfrac{16 \text{ oz.}}{1} \cdot \dfrac{1 \text{ c.}}{8 \text{ oz.}} \cdot \dfrac{1 \text{ pt.}}{2 \text{ c.}} \cdot \dfrac{1 \text{ qt.}}{2 \text{ pt.}} \cdot \dfrac{0.946 \text{ l}}{1 \text{ qt.}}$

$\approx \dfrac{15.136}{32} \text{ l}$

$\approx 0.473 \text{ l}$

19. $2\dfrac{1}{2} \text{ c.} \approx \dfrac{2\frac{1}{2} \text{ c.}}{1} \cdot \dfrac{1 \text{ pt.}}{2 \text{ c.}} \cdot \dfrac{1 \text{ qt.}}{2 \text{ pt.}} \cdot \dfrac{0.946 \text{ l}}{1 \text{ qt.}} \cdot \dfrac{1000 \text{ ml}}{1 \text{ l}}$

$\approx \dfrac{2365}{4} \text{ l}$

$\approx 591.25 \text{ ml}$

21. $3 \text{ l} \approx \dfrac{3 \text{ l}}{1} \cdot \dfrac{1.057 \text{ qt.}}{1 \text{ l}} \cdot \dfrac{1 \text{ gal.}}{4 \text{ qt.}}$

$\approx \dfrac{3.171}{4} \text{ gal.}$

$\approx 0.79 \text{ gal.}$

23. $20 \text{ ml} = 0.02 \text{ l}$

$\approx \dfrac{0.02 \text{ l}}{1} \cdot \dfrac{1.057 \text{ qt.}}{1 \text{ l}} \cdot \dfrac{2 \text{ pt.}}{1 \text{ qt.}} \cdot \dfrac{2 \text{ c.}}{1 \text{ pt.}} \cdot \dfrac{8 \text{ oz.}}{1 \text{ c.}}$

$\approx 0.68 \text{ oz.}$

25. $135 \text{ lb.} \approx \dfrac{135 \text{ lb.}}{1} \cdot \dfrac{0.454 \text{ kg}}{1 \text{ lb.}} \approx 61.29 \text{ kg}$

27. $12 \text{ oz.} \approx \dfrac{12 \text{ oz.}}{1} \cdot \dfrac{1 \text{ lb.}}{16 \text{ oz.}} \cdot \dfrac{0.454 \text{ kg}}{1 \text{ lb.}} \cdot \dfrac{1000 \text{ g}}{1 \text{ kg}}$

$\approx \dfrac{5448}{16} \text{ kg}$

$\approx 340.5 \text{ g}$

29. $500 \text{ mg} = 0.0005 \text{ kg}$

$\approx \dfrac{0.0005 \text{ kg}}{1} \cdot \dfrac{2.2 \text{ lb.}}{1 \text{ kg}} \cdot \dfrac{16 \text{ oz.}}{1 \text{ lb.}}$

$\approx 0.0176 \text{ oz.}$

31. $55 \text{ kg} \approx \dfrac{55 \text{ kg}}{1} \cdot \dfrac{2.2 \text{ lb.}}{1 \text{ kg}} \approx 121 \text{ lb.}$

33. $F = \dfrac{9}{5}C + 32$

$F = \dfrac{9}{5} \cdot \dfrac{54\frac{4}{9}}{1} + 32$

$F = \dfrac{9}{5} \cdot \dfrac{490}{9} + 32$

$F = 98 + 32$

$F = 130°$

35. $F = \dfrac{9}{5}C + 32$

$F = \dfrac{9}{5} \cdot \dfrac{6}{1} + 32$

$F = \dfrac{54}{5} + \dfrac{160}{5}$

$F = \dfrac{214}{5}$

$F = 42.8°$

37. $F = \dfrac{9}{5}C + 32$

$F = \dfrac{9}{5} \cdot \dfrac{-273.15}{1} + 32$

$F = \dfrac{-2458.35}{5} + 32$

$F = -491.67 + 32$

$F = -459.67°$

39. $C = \dfrac{5}{9}(F - 32)$

$C = \dfrac{5}{9}(98.6 - 32)$

$C = \dfrac{5}{9} \cdot \dfrac{66.6}{1}$

$C = \dfrac{333}{9}$

$C = 37°$

41. $C = \dfrac{5}{9}(F - 32)$

$C = \dfrac{5}{9}(102 - 32)$

$C = \dfrac{5}{9} \cdot \dfrac{70}{1}$

$C = \dfrac{350}{9}$

$C = 38.\overline{8}°$

43. $C = \dfrac{5}{9}(F - 32)$

$C = \dfrac{5}{9}(-5 - 32)$

$C = \dfrac{5}{9} \cdot \dfrac{-37}{1}$

$C = \dfrac{-185}{9}$

$C = -20.\overline{5}°$

Review Exercises

1. $3200\overline{)896.00}$ with quotient 0.28

$\quad \underline{640\ 0}$

$\quad 256\ 00$

$\quad \underline{256\ 00}$

$\quad \quad \quad 0$

2. $(90 + 110 + 84) \div 2400 = 284 \div 2400$

$= 0.118\overline{3}$

3.

$-\dfrac{1}{3}x + \dfrac{2}{5}y - x - \dfrac{3}{4}y = -\dfrac{1}{3}x - \dfrac{3}{3}x + \dfrac{2 \cdot 4}{5 \cdot 4}y - \dfrac{3 \cdot 5}{4 \cdot 5}y$

$= -\dfrac{4}{3}x + \dfrac{8}{20}y - \dfrac{15}{20}y$

$= -\dfrac{4}{3}x - \dfrac{7}{20}y$

4. $2(t+1) = 5t$ Check: $2\left(\dfrac{2}{3}+1\right) = 5\left(\dfrac{2}{3}\right)$

$\;\; 2t + 2 = 5t$

$\;\; \dfrac{-2t }{0 + 2 = 3t} = \dfrac{-2t}{}$ $2\left(1\dfrac{2}{3}\right) = \dfrac{10}{3}$

$\;\; \dfrac{2}{3} = \dfrac{3t}{3}$ $\dfrac{2}{1}\cdot\dfrac{5}{3} = \dfrac{10}{3}$

$\;\; \dfrac{2}{3} = t$ $\dfrac{10}{3} = \dfrac{10}{3}$

5. $\dfrac{m}{20.4} = \dfrac{8}{15}$ Check: $\dfrac{10.88}{20.4} = \dfrac{8}{15}$

$\;\; 15\cdot m = 20.4\cdot\left(8\right)$ $15\cdot 10.88 = 20.4\cdot 8$

$\;\; 15m = 163.2$ $163.2 = 163.2$

$\;\; \dfrac{15m}{15} = \dfrac{163.2}{15}$

$\;\; m = 10.88$

Exercise Set 7.6

1. A front-end ratio is the ratio of the total monthly house payment to gross monthly income.

3. A PITI payment includes principal, interest, taxes, and insurance.

5. **Understand:** We are to decide whether the Wu family qualifies for the loan. Because they are applying for a conventional loan, we must consider their front-end ratio, back-end ratio, and credit score.

   Plan: 1. Calculate their front-end ratio. Their front-end ratio should not exceed 0.28.

   $$\text{front-end ratio} = \frac{\text{monthly PITI payment}}{\text{gross monthly income}}$$

   2. Calculate their back-end ratio. Their back-end ratio should not exceed 0.36.

   $$\text{back-end ratio} = \frac{\text{total monthly debt payments}}{\text{gross monthly income}}$$

   3. Consider their credit score.

   Execute: front-end ratio $= \dfrac{845}{3850} \approx 0.22$

   back-end ratio $= \dfrac{435+845}{3850} = \dfrac{1280}{3850} \approx 0.33$

   Credit score $= 700$

   Answer: The front-end ratio, back-end ratio, and credit score fall within the guidelines for a conventional loan. Considering only the two ratios and credit score, the Wu family qualifies.

7. **Understand:** We are to decide whether the Rivers family qualifies for the loan. Because they are applying for a VA loan, we must consider only their back-end ratio.

   Plan: Calculate their back-end ratio. Their back-end ratio should not exceed 0.41. We must find the total monthly debt payments.

   $$\text{back-end ratio} = \frac{\text{total monthly debt payments}}{\text{gross monthly income}}$$

   Execute: Total debt payments $= 30 + 45 + 279 + 93 + 756 = \$1203$

   back-end ratio $= \dfrac{1203}{2845} \approx 0.42$

   Answer: The back-end ratio does not fall within the guidelines for a VA loan. Considering only the back-end ratio, the Rivers family does not qualify.

9. **Understand:** We are to decide whether the Bishop family qualifies for the loan. Because they are applying for a FHA loan, we must consider their front-end ratio and back-end ratio.

   Plan: 1. Calculate their font-end ratio. Their front-end ratio should not exceed 0.29.

   $$\text{front-end ratio} = \frac{\text{monthly PITI payment}}{\text{gross monthly income}}$$

   2. Calculate their back-end ratio. Their back-end ratio should not exceed 0.41.

   $$\text{back-end ratio} = \frac{\text{total monthly debt payments}}{\text{gross monthly income}}$$

   Execute: Front-end ratio $= \dfrac{785}{2530} \approx 0.31$

   Total debt payments $= 40 + 249 + 65 + 785 = \$1139$

   back-end ratio $= \dfrac{1139}{2530} \approx 0.45$

   Answer: The front-end ratio and back-end ratio do not fall within the guidelines for a FHA loan. Considering only the two ratios, the Bishop family does not qualify.

11. **Understand:** We are to calculate the maximum monthly PITI payment that would meet the front-end ratio qualifications for a conventional loan. The general guideline is that the font-end ratio should not exceed 0.28. The front-end ratio is calculated as follows.

$$\text{front-end ratio} = \frac{\text{monthly PITI payment}}{\text{gross monthly income}}$$

Plan: Let p represent the monthly PITI payment. Replace each part of the front-end ratio formula with appropriate values and solve for p.

Execute: $0.28 = \dfrac{p}{4800}$

$$\frac{4800}{1} \cdot \frac{0.28}{1} = \frac{p}{4800} \cdot \frac{4800}{1}$$

$$\$1344 = p$$

Answer: The maximum monthly PITI payment that would meet the font-end ratio qualification for a conventional loan with a gross monthly income of $4800 is $1344.

13. **Understand:** We are to calculate the maximum monthly PITI payment that would meet the front-end ratio qualifications for a FHA loan. The general guideline is that the font-end ratio should not exceed 0.29. The front-end ratio is calculated as follows.

$$\text{front-end ratio} = \frac{\text{monthly PITI payment}}{\text{gross monthly income}}$$

Plan: Let p represent the monthly PITI payment. Replace each part of the front-end ratio formula with appropriate values and solve for p.

Execute: $0.29 = \dfrac{p}{1840}$

$$\frac{1840}{1} \cdot \frac{0.29}{1} = \frac{p}{1840} \cdot \frac{1840}{1}$$

$$\$533.6 = p$$

Answer: The maximum monthly PITI payment that would meet the font-end ratio qualification for a FHA loan with a gross monthly income of $1840 is $533.60.

15. **Understand:** We are to calculate the maximum debt that would meet the back-end ratio qualifications for a conventional loan. The general guideline is that the back-end ratio should not exceed 0.36. The back-end ratio is calculated as follows.

$$\text{back-end ratio} = \frac{\text{total monthly debt payments}}{\text{gross monthly income}}$$

Plan: Let d represent the total monthly debt. Replace each part of the back-end ratio formula with appropriate values and solve for d.

Execute: $0.36 = \dfrac{d}{3480}$

$$\frac{3480}{1} \cdot \frac{0.36}{1} = \frac{p}{3480} \cdot \frac{3480}{1}$$

$$\$1252.8 = p$$

Answer: The maximum monthly debt that would meet the back-end ratio qualification for a conventional loan with a gross monthly income of $3480 is $1252.80.

17. **Understand:** We are to calculate the maximum debt that would meet the back-end ratio qualifications for a VA loan. The general guideline is that the back-end ratio should not exceed 0.41. The back-end ratio is calculated as follows.

$$\text{back-end ratio} = \frac{\text{total monthly debt payments}}{\text{gross monthly income}}$$

Plan: Let d represent the total monthly debt. Replace each part of the back-end ratio formula with appropriate values and solve for d.

Execute: $0.41 = \dfrac{d}{1675}$

$$\frac{1675}{1} \cdot \frac{0.41}{1} = \frac{p}{1675} \cdot \frac{1675}{1}$$

$$\$686.75 = p$$

Answer: The maximum monthly debt that would meet the back-end ratio qualification for a VA loan with a gross monthly income of $1675 is $686.75.

19. **Understand:** We are to calculate the maximum debt that would meet the back-end ratio qualifications for a FHA loan. The general guideline is that the back-end ratio should not exceed 0.41. The back-end ratio is calculated as follows.

$$\text{back-end ratio} = \frac{\text{total monthly debt payments}}{\text{gross monthly income}}$$

Plan: Let d represent the total monthly debt. Replace each part of the back-end ratio formula with appropriate values and solve for d.

Execute: $0.41 = \dfrac{d}{2375}$

$$\frac{2375}{1} \cdot \frac{0.41}{1} = \frac{p}{2375} \cdot \frac{2375}{1}$$

$$\$973.75 = p$$

Answer: The maximum monthly debt that would meet the back-end ratio qualification for a conventional loan with a gross monthly income of $2375 is $973.75.

21. **Understand:** To find the time when Brian catches up to Terese, we must calculate the amount of time it will take him to catch up to her. We can then add that time to 3:30 P.M.

He passed the same exit 1 hour after she did. Let's begin filling in the chart:

Because we want to find Brian's time to catch up we let t represent the amount of time it takes him to catch up. At the time Brian catches up, Terese's travel time is 1 hour more than Brian's travel time from the same exit.

Because $d = rt$, we multiply the rate and time columns to get the distance column.

Categories	Rate	Time	Distance
Terese	60	$t + 1$	$60(t+1)$
Brian	70	t	$70t$

Plan: Because they will have gone the same distance past Exit 62 when Brian catches up, we set the expression of their individual distances equal and solve for t.

Execute:
$$60(t+1) = 70t$$
$$60t + 60 = 70t$$
$$\underline{-60t \qquad = -60t}$$
$$0 + 60 = 10t$$
$$\frac{60}{10} = \frac{10t}{10}$$
$$6 = t$$

Answer: It will take Brian 6 hours after passing Exit 62 to catch up to Terese. If it was 3:30 P.M. when he passed Exit 62, then it will be 9:30 P.M. when he catches up.

Check: If Brian travels for 6 hours after Exit 62, then Terese travels for $6 + 1 = 7$ hours after Exit 62. We can check by verifying that they indeed are the exact same distance after traveling for their respective amount of time.

Brian: $d = rt$ Terese: $d = rt$
$\quad d = (70)(6)$ $\quad d = (60)(7)$
$\quad d = 420$ mi. $\quad d = 420$ mi.

23. **Understand:** To find the time when Tyler catches up to Liz, we must calculate the amount of time it will take him to catch up to her. We can then add that time to 12:33 P.M.

He passed the same marker 15 minutes or 0.25 hours after she did. Let's begin filling in the chart:

Because we want to find Tyler's time to catch up we let t represent the amount of time it takes him to catch up. At the time Tyler catches up, Liz's travel time is 0.25 hours more than Tyler's travel time from the same marker.

Because $d = rt$, we multiply the rate and time columns to get the distance column.

Categories	Rate	Time	Distance
Liz	8	$t + 0.25$	$8(t + 0.25)$
Tyler	10	t	$10t$

Plan: Because they will have gone the same distance past a marker when Tyler catches up, we set the expression of their individual distances equal and solve for t.

Execute: $8(t + 0.25) = 10t$
$$8t + 2 = 10t$$
$$\underline{-8t \qquad = -8t}$$
$$0 + 2 = 2t$$
$$\frac{2}{2} = \frac{2t}{2}$$
$$1 = t$$

Answer: It will take Tyler 1 hour after passing a marker to catch up to Liz. If it was 12:33 P.M. when he passed a marker, then it will be 1:33 P.M. when he catches up.

Check: If Tyler travels for 1 hour after a marker, then Liz travels for $1 + 0.25 = 1.25$ hours after a market. We can check by verifying that they indeed are the exact same distance after traveling for their respective amount of time.

Tyler: $d = rt$ Liz: $d = rt$
$\quad d = (10)(1)$ $\quad d = (8)(1.25)$
$\quad d = 10$ mi. $\quad d = 10$ mi.

25. **Understand:** The order of 9.5 mg per kg means that the patient should receive 9.5 mg of the antibiotic for every 1 kg of the patient's mass.

Plan: Multiplying the patients mass by the dosage per unit mass should yield the total dose.

Execute: patient's mass × dosage per unit of mass = total dose

$$\frac{55 \text{ kg}}{1} \cdot \frac{9.5 \text{ mg}}{1 \text{ kg}} = 522.5 \text{ mg}$$

Answer: The patient should receive 522.5 mg of the antibiotic.

Check: Verify that if a 55 kg patient is receiving 522.5 mg of antibiotic, then he or she is receiving 9.5 mg per kg of mass.

$$\frac{522.5 \text{ mg}}{55 \text{ kg}} = 9.5 \text{ mg/kg}$$

27. **Understand:** The order of 0.04 mg/kg of clonazepam means that the patient should receive 0.04 mg/kg of clonazepam for every 1 kg of the patient's mass. We will need to find the mass of the patient in kg.

Plan: Multiplying the patients mass by the dosage per unit mass should yield the total does.

Execute: The mass of the patient is:

$$\frac{132 \text{ lb.}}{1} \cdot \frac{1 \text{ kg}}{2.2 \text{ lb.}} = 60 \text{ kg}$$

patient's mass × dosage per unit of mass = total dose

$$\frac{60 \text{ kg}}{1} \cdot \frac{0.04 \text{ mg}}{1 \text{ kg}} = 2.4 \text{ mg}$$

Answer: The patient should receive 2.4 mg of the clonazepam. Since each tablet is 0.5 mg, the patient should receive $2.4 \div 0.5 \approx 5$ tablets.

Check: Verify that if a 60 kg patient is receiving 2.4 mg of clonazepam, then he or she is receiving 0.04 mg per kg of mass.

$$\frac{2.4 \text{ mg}}{60 \text{ kg}} = 0.04 \text{ mg/kg}$$

29. **Understand:** The order is for 250 ml over 6 hr, With this particular IV, 15 drops of the solution is equal to 1 ml. We can write these rates as:

$$\frac{250 \text{ ml}}{6 \text{ hr.}} \quad \text{and} \quad \frac{15 \text{ drops}}{1 \text{ ml}}$$

We want to end up with a description of the number of drops per minute. This means that the milliliters and hours are undesired units. We need to convert hours to minutes and divide out milliliters completely.

Plan: Multiply the above rates to eliminate milliliters. Use the fact that 1 hr. is 60 min. to convert hours to minutes.

Execute:

$$\frac{250 \text{ ml}}{6 \text{ hr.}} \cdot \frac{15 \text{ drops}}{1 \text{ ml}} \cdot \frac{1 \text{ hr.}}{60 \text{ min.}} = \frac{3750 \text{ drops}}{360 \text{ min.}}$$
$$= 10.41\overline{7} \text{ drops/min.}$$

Answer: The patient should receive 10-11 drops per minute.

Check: Verify that a rate of 10 drops/min. with every 15 drops equal to 1 ml is a total of 250 ml after 6 hours.

$$\frac{6 \text{ hr.}}{1} \cdot \frac{60 \text{ min.}}{1 \text{ hr.}} \cdot \frac{10 \text{ drops}}{1 \text{ min.}} = 3600 \text{ drops}$$

If the patient receives 3600 drops in 6 hr. and if every 15 drops is 1 ml, we can calculate the total number of ml that the patient receives:

$$\frac{3600 \text{ drops}}{1} \cdot \frac{1 \text{ ml}}{15 \text{ drops}} = \frac{3600}{15} \text{ ml} = 240 \text{ ml}$$

Because we round the number of drops to whole amounts, this result does not match 250 ml exactly. However, it is reasonable.

31. **Understand:** The order is for 10 ml per hour. And a 500 ml solution contains 20,000 units. We can write these rates as:

$$\frac{10 \text{ ml}}{1 \text{ hr.}} \quad \text{and} \quad \frac{20,000 \text{ units}}{500 \text{ ml}}$$

We want to find how many units of heparin the patient receives each hour. This means that the milliliters are undesired units. We need divide out milliliters completely.

Plan: Multiply the above rates to eliminate milliliters.

Execute:

$$\frac{10 \text{ ml}}{1 \text{ hr.}} \cdot \frac{20,000 \text{ units}}{500 \text{ ml}} = \frac{200,000}{500} \text{ units/hr.}$$
$$= 400 \text{ units/hr.}$$

Answer: The patient receives 400 units of heparin each hour.

Check: We can use a proportion to verify our answer.

$$\frac{10 \text{ ml}}{x \text{ units}} = \frac{500 \text{ ml}}{20{,}000 \text{ units}}$$

$$20{,}000 \cdot 10 = 500 \cdot x$$

$$200{,}000 = 500x$$

$$\frac{200{,}000}{500} = \frac{500x}{500}$$

$$400 = x$$

Review Exercises

1. $\dfrac{0.14}{0.5} = \dfrac{\frac{14}{100}}{\frac{5}{10}}$

   $= \dfrac{14}{100} \div \dfrac{5}{10}$

   $= \dfrac{14}{\underset{10}{\cancel{100}}} \cdot \dfrac{\cancel{10}}{5}$

   $= \dfrac{14}{50}$

   $= \dfrac{7}{25}$

2. There are 4 Jacks and 4 Kings in a standard deck of 52 cards. $P = \dfrac{8}{52} = \dfrac{2}{13}$

3. $u = \dfrac{p}{q}$

   $u = \dfrac{\$2.29}{16 \text{ oz.}}$

   $u \approx \$0.14 \,/\, \text{oz.}$

4. $75 \text{ ft.} = \dfrac{75 \,\cancel{\text{ft.}}}{1} \cdot \dfrac{1 \text{ yd.}}{3 \,\cancel{\text{ft.}}} = \dfrac{75}{3} \text{ yd.} = 25 \text{ yd.}$

5. Because mg is 3 units to the right of g, we move the decimal point 3 places to the right in 8.5.

   $8.5 \text{ g} = 8500 \text{ mg}$

Chapter 7 Review Exercises

1. false   2. true   3. true

4. false   5. false   6. false

7. favorable; possible

8. denominator; numerator

9. undesired; desired   10. 2

11. $\dfrac{22 \div 2}{32 \div 2} = \dfrac{11}{16}$ 12. $\dfrac{2.5}{10.8} = \dfrac{2.5 \cdot 10}{10.8 \cdot 10} = \dfrac{25}{108}$

13. There are 4 Queens and 4 Kings in a standard deck of 52 cards. $P = \dfrac{8}{52} = \dfrac{2}{13}$

14. There are 2 favorable outcomes out of six possible outcomes. $P = \dfrac{2}{6} = \dfrac{1}{3}$

15. Students-to-faculty $= \dfrac{4500}{275} \approx 16.4$

16. Selling price-to-annual earnings ratio $= \dfrac{12\frac{5}{8}}{1.78}$

    $= \dfrac{12.625}{1.78}$

    ≈ 7.09

17. $r = \dfrac{d}{t}$

    $r = \dfrac{210.6 \text{ mi.}}{3 \text{ hr.}}$

    $r = 70.2 \text{ mi./hr.}$

18. $u = \dfrac{p}{q}$

    $u = \dfrac{\$2.52}{28 \text{ min.}}$

    $u = \dfrac{252 \,¢}{28 \text{ min.}}$

    $u = 9 \,¢/\text{min.}$

19. $u = \dfrac{p}{q}$

    $u = \dfrac{\$1.44}{32 \text{ oz.}}$

    $u = \dfrac{144 \,¢}{32 \text{ oz.}}$

    $u = 4.5 \,¢/\text{oz.}$

20. $u = \dfrac{p}{q}$

    $u = \dfrac{\$2.49}{64 \text{ oz.}}$

    $u = \dfrac{249 \,¢}{64 \text{ oz.}}$

    $u = 3.9 \,¢/\text{oz.}$

21. Let F represent the unit price of the 24-oz. cup of soda and T represent the unit price of the 32-oz. cup of soda.

    $F = \dfrac{\$0.55}{24 \text{ oz.}}$ $T = \dfrac{\$0.79}{32 \text{ oz.}}$

    $F = \dfrac{55 \,¢}{24 \text{ oz.}}$ $T = \dfrac{79 \,¢}{32 \text{ oz.}}$

    $F \approx 2.29 \,¢/\text{oz.}$ $T \approx 2.47 \,¢/\text{oz.}$

    Because 2.29 cents is the smaller unit price, the 24-oz. cup of soda is the better buy.

22. Buying 2 of the 12-oz. cans means that we are buying $2(12) = 24$ oz. of store brand frozen juice for $2.58. Buying 4 of the 8-oz. cans means that we are buying $4(8) = 32$ oz. of the name brand frozen juice for $3.90. Let S represent the unit price of the Store brand can of frozen juice and N represent the unit price of the name brand can of frozen juice.

$S = \dfrac{\$2.58}{24 \text{ oz.}}$ $N = \dfrac{\$3.90}{32 \text{ oz.}}$

$S = \dfrac{258 \text{ ¢}}{24 \text{ oz.}}$ $N = \dfrac{390 \text{ ¢}}{32 \text{ oz.}}$

$S = 10.75 \text{ ¢ / oz.}$ $N \approx 12.19 \text{ ¢ / oz.}$

Because 10.75¢ is the smaller unit price, the 2 cans of 12-oz. is a better buy.

23. $\dfrac{5}{9} \overset{?}{=} \dfrac{7}{12}$

$5 \cdot 12 \overset{?}{=} 9 \cdot 7$

$60 \neq 63$

No

24. $\dfrac{3}{7.2} \overset{?}{=} \dfrac{0.75}{1.8}$

$(3)(1.8) \overset{?}{=} (7.2)(0.75)$

$5.4 = 5.4$

Yes

25. $\dfrac{1.8}{-5} = \dfrac{k}{12.5}$

$12.5 \cdot (1.8) = -5 \cdot k$

$22.5 = -5k$

$\dfrac{22.5}{-5} = \dfrac{-5k}{-5}$

$-4.5 = k$

Check: $\dfrac{1.8}{-5} = \dfrac{-4.5}{12.5}$

$12.5 \cdot (1.8) = -5 \cdot (-4.5)$

$22.5 = 22.5$

26. $\dfrac{2}{m} = \dfrac{3\frac{1}{4}}{15}$

$15 \cdot 2 = m \cdot 3\frac{1}{4}$

$30 = \dfrac{13}{4} m$

$\dfrac{4}{13} \cdot \dfrac{30}{1} = \dfrac{13}{4} m \cdot \dfrac{4}{13}$

$\dfrac{120}{13} = 1m$

$9\dfrac{3}{13} = m$

Check: $\dfrac{2}{9\frac{3}{13}} = \dfrac{3\frac{1}{4}}{15}$

$15 \cdot 2 = 9\dfrac{3}{13} \cdot 3\dfrac{1}{4}$

$30 = \dfrac{\overset{30}{\cancel{120}}}{\cancel{13}} \cdot \dfrac{\cancel{13}}{\cancel{4}_1}$

$30 = 30$

27. 3.78 l covers about 25 m² translates to $\dfrac{3.78\,l}{25 \text{ m}^2}$

and 4.52 l covers about ? m² translates to $\dfrac{4.52\,l}{x \text{ m}^2}$

So, $\dfrac{3.78\,l}{25 \text{ m}^2} = \dfrac{4.52\,l}{x \text{ m}^2}$

$x \cdot (3.78) = 25 \cdot (4.52)$

$3.78x = 113$

$\dfrac{3.78x}{3.78} = \dfrac{113}{3.78}$

$x = 29.9 \text{ m}^2$

The 4.52 l will cover about 29.9 m².

28. ¼ in. represents 20 mi. translates to $\dfrac{\frac{1}{4} \text{ in.}}{20 \text{ mi.}}$ and

3 ½ in. represents ? mi. translates to $\dfrac{3\frac{1}{2} \text{ in.}}{x \text{ mi.}}$

So, $\dfrac{\frac{1}{4} \text{ in.}}{20 \text{ mi.}} = \dfrac{3\frac{1}{2} \text{ in.}}{x \text{ mi.}}$

$x \cdot \dfrac{1}{4} = 20 \cdot 3\dfrac{1}{2}$

$\dfrac{1}{4} x = \dfrac{\overset{10}{\cancel{20}}}{1} \cdot \dfrac{7}{\cancel{2}}$

$\dfrac{1}{4} x = 70$

$\dfrac{\cancel{4}}{1} \cdot \dfrac{1}{\cancel{4}} x = 70 \cdot 4$

$x = 280 \text{ mi.}$

3 ½ in. represents 280 mi.

29. $\dfrac{3.2}{4.8} = \dfrac{2.4}{d}$

$3.2 \cdot d = 4.8(2.4)$

$3.2d = 11.52$

$\dfrac{3.2d}{3.2} = \dfrac{11.52}{3.2}$

$d = 3.6 \text{ cm}$

30. a) $\dfrac{4\frac{1}{2}}{3\frac{3}{8}} = \dfrac{9}{a}$

$4\frac{1}{2} \cdot a = 3\frac{3}{8} \cdot 9$

$\dfrac{9}{2}a = \dfrac{27}{8} \cdot \dfrac{9}{1}$

$\dfrac{9}{2}a = \dfrac{243}{8}$

$\dfrac{2}{9} \cdot \dfrac{9}{2}a = \dfrac{2}{9} \cdot \dfrac{243}{8}$

$\dfrac{\cancel{2}}{\cancel{9}} \cdot \dfrac{\cancel{9}}{\cancel{2}}a = \dfrac{^1\cancel{2}}{_1\cancel{9}} \cdot \dfrac{\cancel{243}^{27}}{\cancel{8}_4}$

$a = \dfrac{27}{4} = 6\frac{3}{4}$ ft.

b)

$\dfrac{4\frac{1}{2}}{3\frac{3}{8}} = \dfrac{4\frac{1}{2}}{b}$

Since the numerators are equivalent, the denominators must also be equivalent. Therefore, $b = 3\frac{3}{8}$ ft.

c) $\dfrac{4\frac{1}{2}}{3\frac{3}{8}} = \dfrac{7}{c}$

$4\frac{1}{2} \cdot c = 3\frac{3}{8} \cdot 7$

$\dfrac{9}{2}c = \dfrac{27}{8} \cdot \dfrac{7}{1}$

$\dfrac{9}{2}c = \dfrac{189}{8}$

$\dfrac{2}{9} \cdot \dfrac{9}{2}c = \dfrac{2}{9} \cdot \dfrac{189}{8}$

$\dfrac{\cancel{2}}{\cancel{9}} \cdot \dfrac{\cancel{9}}{\cancel{2}}c = \dfrac{^1\cancel{2}}{_1\cancel{9}} \cdot \dfrac{\cancel{189}^{21}}{\cancel{8}_4}$

$c = \dfrac{21}{4} = 5\frac{1}{4}$ ft.

31. $12 \text{ yd.} = \dfrac{12 \cancel{yd.}}{1} \cdot \dfrac{3 \cancel{ft.}}{1 \cancel{yd.}} \cdot \dfrac{12 \text{ in.}}{1 \cancel{ft.}} = 432$ in.

32. $9152 \text{ yd.} = \dfrac{9152 \cancel{yd.}}{1} \cdot \dfrac{3 \cancel{ft.}}{1 \cancel{yd.}} \cdot \dfrac{1 \text{ mi.}}{5280 \cancel{ft.}}$

$= \dfrac{27,456}{5280}$ mi.

$= 5.2$ mi.

33. $40 \text{ oz.} = \dfrac{40 \cancel{oz.}}{1} \cdot \dfrac{1 \text{ lb.}}{16 \cancel{oz.}} = \dfrac{40}{16}$ lb. $= 2.5$ lb.

34. $4800 \text{ lb.} = \dfrac{4800 \cancel{lb.}}{1} \cdot \dfrac{1 \text{ T}}{2000 \cancel{lb.}}$

$= \dfrac{4800}{2000}$ T

$= 2.4$ T

35. $6 \text{ qt.} = \dfrac{6 \cancel{qt.}}{1} \cdot \dfrac{2 \cancel{pt.}}{1 \cancel{qt.}} \cdot \dfrac{2 \cancel{c.}}{1 \cancel{pt.}} \cdot \dfrac{8 \text{ oz.}}{1 \cancel{c.}} = 192$ oz.

36. $49.6 \text{ pt.} = \dfrac{49.6 \cancel{pt.}}{1} \cdot \dfrac{1 \cancel{qt.}}{2 \cancel{pt.}} \cdot \dfrac{1 \text{ gal.}}{4 \cancel{qt.}}$

$= \dfrac{49.6}{8}$ gal.

$= 6.2$ gal.

37. $2\frac{1}{4} = 2.25 \text{ hr.} = \dfrac{2.25 \cancel{hr.}}{1} \cdot \dfrac{60 \text{ min.}}{1 \cancel{hr.}} = 135$ min.

38. $4 \text{ yr.} = \dfrac{4 \cancel{yr.}}{1} \cdot \dfrac{365.25 \text{ d.}}{1 \cancel{yr.}} = 1461$ d.

39. $A = lw$ and $217 \text{ ft.}^2 = \dfrac{217 \cancel{ft.^2}}{1} \cdot \dfrac{1 \text{ yd.}^2}{9 \cancel{ft.^2}}$

$A = 15.5 \cdot (14)$

$A = 217 \text{ ft.}^2$ $= \dfrac{217}{9}$ yd.^{2}

$= 24.\overline{1}$ yd.^{2}

40. $116,160$ ft./sec.

$= \dfrac{116,160 \cancel{ft.}}{1 \cancel{sec.}} \cdot \dfrac{1 \text{ mi.}}{5280 \cancel{ft.}} \cdot \dfrac{60 \cancel{sec.}}{1 \cancel{min.}} \cdot \dfrac{60 \cancel{min.}}{1 \text{ hr.}}$

$= \dfrac{418,176,000}{5280}$ mi./hr.

$= 79,200$ mi./hr.

41. Because dm is 1 unit to the right of m, we move the decimal point 1 place the right in 5.

$5 \text{ m} = 50$ dm

42. Because m is 3 decimal places to the left of mm, we move the decimal point 3 places to the left in 375.

375 mm = 0.375 m

43. Because g is 3 units to the right of kg, we move the decimal point 3 places to the right in 0.26.

    0.26 kg = 260 g

44. 950,000 g = 950 kg

$$= \frac{950 \ \cancel{kg}}{1} \cdot \frac{1 \ t}{1000 \ \cancel{kg}}$$

$$= \frac{950}{1000} = 0.95 \ t$$

45. Because L is 1 unit to the left of dl, we move the decimal point 1 place to the left in 285.

    285 dl = 28.5 L

46. $1 \ cm^3 = 1 \ ml$: Because ml is 3 units to the right of l, we move the decimal point 3 places to the right in 0.075.

    0.075 l = 75 ml = 75 cc

47. $20 \ ft. \approx \frac{20 \ \cancel{ft.}}{1} \cdot \frac{1 \ m}{3.281 \ \cancel{ft.}} \approx \frac{20}{3.281} \ m \approx 6.10 \ m$

48. $12 \ mi. \approx \frac{12 \ \cancel{mi.}}{1} \cdot \frac{1.609 \ km}{1 \ \cancel{mi.}} \approx 19.31 \ km$

49. $145 \ lb. \approx \frac{145 \ \cancel{lb.}}{1} \cdot \frac{0.454 \ kg}{1 \ \cancel{lb.}} \approx 65.83 \ kg$

50. $3 \ l \approx \frac{3 \ \cancel{l}}{1} \cdot \frac{1.057 \ \cancel{qt.}}{1 \ \cancel{l}} \cdot \frac{1 \ gal.}{4 \ \cancel{qt.}}$

    $\approx \frac{3.171}{4} \ gal.$

    $\approx 0.79 \ gal.$

51. $C = \frac{5}{9}(F - 32)$ 52. $F = \frac{9}{5}C + 32$

    $C = \frac{5}{9}(90 - 32)$ $F = \frac{9}{5} \cdot \frac{-4}{1} + 32$

    $C = \frac{5}{9} \cdot \frac{58}{1}$ $F = \frac{-36}{5} + \frac{160}{5}$

    $C = \frac{290}{9}$ $F = \frac{124}{5}$

    $C = 32.\overline{2}°$ $F = 24.8°$

53. **Understand:** The order is for 450 ml over 6 hr. With this particular IV, 10 drops of the solution is equal to 1 ml. We can write these rates as:

    $\frac{450 \ ml}{6 \ hr.}$ and $\frac{10 \ drops}{1 \ ml}$

We want to end up with a description of the number of drops per minute. This means that the milliliters and hours are undesired units. We need to convert hours to minutes and divide out milliliters completely.

Plan: Multiply the above rates to eliminate milliliters. Use the fact that 1 hr. is 60 min. to convert hours to minutes.

Execute:

$$\frac{450 \ \cancel{ml}}{6 \ \cancel{hr.}} \cdot \frac{10 \ drops}{1 \ \cancel{ml}} \cdot \frac{1 \ \cancel{hr.}}{60 \ min.} = \frac{4500 \ drops}{360 \ min.}$$

$$= 12.5 \ drops/min.$$

Answer: The patient should receive 12-13 drops per minute.

Check: Verify that a rate of 13 drops/min. with every 10 drops equal to 1 ml is a total of 450 ml after 6 hours.

$$\frac{6 \ \cancel{hr.}}{1} \cdot \frac{60 \ \cancel{min.}}{1 \ \cancel{hr.}} \cdot \frac{13 \ drops}{1 \ \cancel{min.}} = 4680 \ drops$$

If the patient receives 4680 drops in 6 hr. and every 10 drops is 1 ml, we can calculate the total number of ml that the patient receives:

$$\frac{4680 \ \cancel{drops}}{1} \cdot \frac{1 \ ml}{10 \ \cancel{drops}} = \frac{4680}{10} \ ml = 468 \ ml$$

Because we round the number of drops to whole amounts, this result does not match 450 ml exactly. However, it is reasonable.

54. **Understand:** The order of 2.9 mg per kg means that the patient should receive 2.9 mg of the antibiotic for every 1 kg of the patient's mass.

Plan: Multiplying the patients mass by the dosage per unit mass should yield the total does.

Execute: patient's mass × dosage per unit of mass = total dose

$$\frac{32 \ \cancel{kg}}{1} \cdot \frac{2.9 \ mg}{1 \ \cancel{kg}} = 92.8 \ mg$$

Answer: The patient should receive 92.8 mg of the antibiotic.

Check: Verify that if a 32 kg patient is receiving 92.8 mg of antibiotic, then he or she is receiving 2.9 mg per kg of mass.

$$\frac{92.8 \ mg}{32 \ kg} = 2.9 \ mg/kg$$

55. front-end ratio $= \dfrac{\text{monthly PITI payment}}{\text{gross monthly income}}$

$= \dfrac{978}{3450}$

≈ 0.28

56. back-end ratio $= \dfrac{\text{total monthly debt payments}}{\text{gross monthly income}}$

$= \dfrac{960}{2420}$

≈ 0.40

57. **Understand:** We are to calculate the maximum monthly PITI payment that would meet the front-end ratio qualifications for a FHA loan. The general guideline is that the font-end ratio should not exceed 0.29. The front-end ratio is calculated as follows.

front-end ratio $= \dfrac{\text{monthly PITI payment}}{\text{gross monthly income}}$

Plan: Let p represent the monthly PITI payment. Replace each part of the front-end ratio formula with appropriate values and solve for p.

Execute: $0.29 = \dfrac{p}{1970}$

$\dfrac{1970}{1} \cdot \dfrac{0.29}{1} = \dfrac{p}{\cancel{1970}} \cdot \dfrac{\cancel{1970}}{1}$

$\$571.3 = p$

Answer: The maximum monthly PITI payment that would meet the font-end ratio qualification for a FHA loan with a gross monthly income of $\$1970$ is $\$571.30$.

58. **Understand:** We are to calculate the maximum debt that would meet the back-end ratio qualifications for a VA loan. The general guideline is that the back-end ratio should not exceed 0.41. The back-end ratio is calculated as follows.

back-end ratio $= \dfrac{\text{total monthly debt payments}}{\text{gross monthly income}}$

Plan: Let d represent the total monthly debt. Replace each part of the back-end ratio formula with appropriate values and solve for d.

Execute: $0.41 = \dfrac{d}{4200}$

$\dfrac{4200}{1} \cdot \dfrac{0.41}{1} = \dfrac{p}{\cancel{4200}} \cdot \dfrac{\cancel{4200}}{1}$

$\$1722 = p$

Answer: The maximum monthly debt that would meet the back-end ratio qualification for a VA loan with a gross monthly income of $\$4200$ is $\$1722$.

59. **Understand:** To find the time when Tanya catches up to Dale, we must calculate the amount of time it will take her to catch up to him.

She passed the same exit ½ hour after he did. Let's begin filling in the chart:

Because we want to find Tanya's time to catch up we let t represent the amount of time it takes her to catch up. At the time Tanya catches up, Dale's travel time is ½ hour more than Tanya's travel time from the same exit.

Because $d = rt$, we multiply the rate and time columns to get the distance column.

Categories	Rate	Time	Distance
Dale	60	$t + \frac{1}{2}$	$60(t + \frac{1}{2})$
Tanya	70	t	$70t$

Plan: Because they will have gone the same distance past Exit 50 when Tanya catches up, we set the expression of their individual distances equal and solve for t.

Execute: $60\left(t + \dfrac{1}{2}\right) = 70t$

$60t + 30 = 70t$

$\underline{-60t \qquad = -60t}$

$0 + 30 = 10t$

$\dfrac{30}{10} = \dfrac{10t}{10}$

$3 = t$

Answer: It will take Tanya 3 hours after passing Exit 50 to catch up to Dale.

Check: If Tanya travels for 3 hours after Exit 50, then Dale travels for $3 + ½ = 3\,½$ hours after Exit 50. We can check by verifying that they indeed are the exact same distance after traveling for their respective amount of time.

Tanya: $d = rt$ Dale: $d = rt$

$d = (70)(3)$

$d = 210$ mi. $d = (60)\left(3\dfrac{1}{2}\right)$

$d = \dfrac{\overset{30}{\cancel{60}}}{1} \cdot \dfrac{7}{\cancel{2}}$

$d = 210$ mi.

60. **Understand:** To find the time when Micah catches up to Vijay, we must calculate the amount of time it will take him to catch up to her. We can then add that time to 11:02 A.M.

He passed the same drink station 20 minutes or 1/3 hours after she did. Let's begin filling in the chart:

Because we want to find Micah's time to catch up we let t represent the amount of time it takes him to catch up. At the time Micah catches up, Vijay's travel time is 1/3 hours more than Micah's travel time from the same drink station.

Because $d = rt$, we multiply the rate and time columns to get the distance column.

Categories	Rate	Time	Distance
Vijay	3	$t + 1/3$	$3(t + 1/3)$
Micah	5	t	$5t$

Plan: Because they will have gone the same distance past the drink station when Micah catches up, we set the expression of their individual distances equal and solve for t.

Execute: $3\left(t + \dfrac{1}{3}\right) = 5t$

$$3t + 3 \cdot \dfrac{1}{3} = 5t$$
$$3t + 1 = 5t$$
$$\underline{-3t \qquad = -3t}$$
$$0 + 1 = 2t$$
$$\dfrac{1}{2} = \dfrac{2t}{2}$$
$$\dfrac{1}{2} = t$$

Answer: It will take Micah ½ hour or 30 minutes after passing the drink station to catch up to Vijay. If it was 11:02 A.M. when he passed the drink station, then it will be 11:32 A.M. when he catches up.

Check: If Micah travels for ½ hours after the drink station, then Vijay travels for $\dfrac{1}{2} + \dfrac{1}{3} = \dfrac{5}{6}$ hours or 50 min. after a drink station. We can check by verifying that they indeed are the exact same distance after traveling for their respective amount of time.

Micah: $d = rt$ Vijay: $d = rt$

$$d = 5\left(\dfrac{1}{2}\right) \qquad d = 3\left(\dfrac{5}{6}\right)$$
$$d = 2.5 \text{ mi.} \qquad d = 2.5 \text{ mi.}$$

Chapter 7 Practice Test

1. $\dfrac{9.5}{13} = \dfrac{9.5 \cdot 10}{13 \cdot 10} = \dfrac{95}{130} = \dfrac{95 \div 5}{130 \div 5} = \dfrac{19}{26}$

2. There are 22 favorable outcomes out of 4578 possible outcomes. $P = \dfrac{22}{4578} = \dfrac{11}{2289}$

3. There are 2 favorable outcomes out of ten possible outcomes. $P = \dfrac{2}{10} = \dfrac{1}{5}$

4. Students-to-faculty ratio $= \dfrac{3850}{125} \approx 30.8$

5. $r = \dfrac{d}{t}$

   $r = \dfrac{158.6 \text{ mi.}}{2.2\underline{5} \text{ hr.}}$

   $r = 70.48 \text{ mi./hr.}$

6. $u = \dfrac{p}{q}$

   $u = \dfrac{\$0.65}{15 \text{ oz.}}$

   $u = \dfrac{65\cancel{c}}{15 \text{ oz.}}$

   $u \approx 4.3\cancel{c}/\text{oz.}$

7. Let F represent the unit price of the 20-oz. can of pineapple and T represent the unit price of the 8-oz. can of pineapple.

   $F = \dfrac{\$1.19}{20 \text{ oz.}}$ $T = \dfrac{\$0.79}{8 \text{ oz.}}$

   $F = \dfrac{119\cancel{c}}{20 \text{ oz.}}$ $T = \dfrac{79\cancel{c}}{8 \text{ oz.}}$

   $F = 5.95\cancel{c}/\text{oz.}$ $T = 9.875\cancel{c}/\text{oz.}$

   Because 5.95 cents is the smaller unit price, the 20 oz. can of pineapple is the better buy.

8. $\dfrac{-6.5}{12} = \dfrac{n}{10.8}$

   $10.8 \cdot (-6.5) = 12 \cdot n$

   $-70.2 = 12n$

   $\dfrac{-70.2}{12} = \dfrac{12n}{12}$

   $-5.85 = n$

9. $\dfrac{5.6}{a} = \dfrac{8.5}{11.9}$

$8.5a = 5.6(11.9)$

$8.5a = 66.64$

$\dfrac{8.5a}{8.5} = \dfrac{66.64}{8.5}$

$a = 7.84 \text{ cm}$

10. \$49 to install 20 ft.^{2} translates to $\dfrac{\$49}{20 \text{ ft.}^2}$ and

? to install 210 ft.^{2} translates to $\dfrac{\$x}{210 \text{ ft.}^2}$

So, $\dfrac{\$49}{20 \text{ ft.}^2} = \dfrac{\$x}{210 \text{ ft.}^2}$

$210 \cdot 49 = 20 \cdot x$

$10,290 = 20x$

$\dfrac{10,290}{20} = \dfrac{20x}{20}$

$\$514.5 = x$

To install the same floor in a 210 sq. ft. room will cost \$514.50.

11. $\dfrac{1}{4}$ in. represents 50 mi. translates to $\dfrac{\frac{1}{4} \text{ in.}}{50 \text{ mi.}}$ and

$4\dfrac{1}{2}$ in. represents ? mi. translates to $\dfrac{4\frac{1}{2} \text{ in.}}{x \text{ mi.}}$

So, $\dfrac{\frac{1}{4} \text{ in.}}{50 \text{ mi.}} = \dfrac{4\frac{1}{2} \text{ in.}}{x \text{ mi.}}$

$x \cdot \dfrac{1}{4} = 50 \cdot 4\dfrac{1}{2}$

$\dfrac{1}{4}x = \dfrac{\overset{25}{\cancel{50}}}{1} \cdot \dfrac{9}{\cancel{2}}$

$\dfrac{1}{4}x = 225$

$\dfrac{4}{1} \cdot \dfrac{1}{4}x = 225 \cdot 4$

$x = 900 \text{ mi.}$

$4\dfrac{1}{2}$ in. represents 900 mi.

12. $14 \text{ ft.} = \dfrac{14 \cancel{\text{ ft.}}}{1} \cdot \dfrac{12 \text{ in.}}{1 \cancel{\text{ ft.}}} = 168 \text{ in.}$

13. $20 \text{ lb.} = \dfrac{20 \cancel{\text{ lb.}}}{1} \cdot \dfrac{16 \text{ oz.}}{1 \cancel{\text{ lb.}}} = 320 \text{ oz.}$

14. $36 \text{ pt.} = \dfrac{36 \cancel{\text{ pt.}}}{1} \cdot \dfrac{1 \cancel{\text{ qt.}}}{2 \cancel{\text{ pt.}}} \cdot \dfrac{1 \text{ gal.}}{4 \cancel{\text{ qt.}}} = \dfrac{36}{8} \text{ gal.} = 4.5 \text{ gal.}$

15. $150 \text{ min.} = \dfrac{150 \cancel{\text{ min.}}}{1} \cdot \dfrac{1 \text{ hr.}}{60 \cancel{\text{ min.}}}$

$= \dfrac{150}{60} \text{ hr.}$

$= 2.5 \text{ hr.}$

16. $A = lw$ and $250 \text{ ft.}^2 = \dfrac{250 \cancel{\text{ ft.}^2}}{1} \cdot \dfrac{1 \text{ yd.}^2}{9 \cancel{\text{ ft.}^2}}$

$A = 20 \cdot (12.5)$

$A = 250 \text{ ft.}^2$ $= \dfrac{250}{9} \text{ yd.}^2$

$= 27.\overline{7} \text{ yd.}^2$

17. a) Because cm is 2 units to the right of m, we move the decimal point 2 places the right in 0.058.   0.058 m = 5.8 cm

    b) Because km is 3 units to the left of m, we move the decimal point 3 places to the left in 420.   420 m = 0.42 km

18. a) Because kg is 3 units to the left of g, we move the decimal point 3 places to the left in 24.   24 g = 0.024 kg

    b) $0.091 \text{ t} = \dfrac{0.091 \cancel{\text{ t}}}{1} \cdot \dfrac{1000 \text{ kg}}{1 \cancel{\text{ t}}}$

$= 91 \text{ kg}$

$= 91,000 \text{ g}$

19. a) $280 \text{ cm}^2 = \dfrac{280 \cancel{\text{ cm}^2}}{1} \cdot \dfrac{1 \text{ m}^2}{10,000 \cancel{\text{ cm}^2}}$

$= \dfrac{280}{10,000} \text{ m}^2$

$= 0.028 \text{ m}^2$

    b) The dm are the first unit to the right of m. With area units, we move the decimal point twice the number of places, so that means the decimal point moves to the right 2 places. $0.8 \text{ m}^2 = 80 \text{ dm}^2$

20. a) Because l is 1 unit to the left of dl, we move the decimal point 1 place to the left in 6.5.   6.5 dl = 0.65 l

    b) $1 \text{ cm}^3 = 1 \text{ ml}$: Because ml is 3 units to the right of l, we move the decimal point 3 places to the right in 0.8.   0.8 l = 800 cc

21. $75 \text{ ft.} \approx \dfrac{75 \cancel{\text{ft.}}}{1} \cdot \dfrac{1 \text{ m}}{3.281 \cancel{\text{ft.}}}$

$\approx \dfrac{75}{3.281} \text{ m}$

$\approx 22.86 \text{ m}$

22. $C = \dfrac{5}{9}(F - 32)$

$C = \dfrac{5}{9}(30 - 32)$

$C = \dfrac{5}{9} \cdot \left(\dfrac{-2}{1}\right)$

$C = \dfrac{-10}{9}$

$C = -1.\overline{1}°$

23. We must first find the total monthly debt.
$128 + 78 + 295 + 324 + 714 = \$1539$

$\text{back-end ratio} = \dfrac{\text{monthly debt}}{\text{gross monthly income}}$

$= \dfrac{1539}{3420}$

≈ 0.45

24. **Understand:** The order is for 750 ml over 8 hr. With this particular IV, 10 drops of the solution is equal to 1 ml. We can write these rates as:

$\dfrac{750 \text{ ml}}{8 \text{ hr.}}$ and $\dfrac{10 \text{ drops}}{1 \text{ ml}}$

We want to end up with a description of the number of drops per minute. This means that the milliliters and hours are undesired units. We need to convert hours to minutes and divide out milliliters completely.

Plan: Multiply the above rates to eliminate milliliters. Use the fact that 1 hr. is 60 min. to convert hours to minutes.

Execute:

$\dfrac{750 \cancel{\text{ml}}}{8 \cancel{\text{hr.}}} \cdot \dfrac{10 \text{ drops}}{1 \cancel{\text{ml}}} \cdot \dfrac{1 \cancel{\text{hr.}}}{60 \text{ min.}} = \dfrac{7500 \text{ drops}}{480 \text{ min.}}$

$= 15.625 \text{ drops/min.}$

Answer: The patient should receive 15-16 drops per minute.

Check: Verify that a rate of 16 drops/min. with every 10 drops equal to 1 ml is a total of 750 ml after 8 hours.

$\dfrac{8 \cancel{\text{hr.}}}{1} \cdot \dfrac{60 \cancel{\text{min.}}}{1 \cancel{\text{hr.}}} \cdot \dfrac{16 \text{ drops}}{1 \cancel{\text{min.}}} = 7680 \text{ drops}$

If the patient receives 7680 drops in 8 hr. and every 10 drops is 1 ml, we can calculate the total number of ml that the patient receives:

$\dfrac{7680 \text{ drops}}{1} \cdot \dfrac{1 \text{ ml}}{10 \text{ drops}} = \dfrac{7680}{10} \text{ ml} = 768 \text{ ml}$

Because we round the number of drops to whole amounts, this result does not match 750 ml exactly. However, it is reasonable.

25. **Understand:** To find the time when Darryl catches up to Catrina, we must calculate the amount of time it will take him to catch up to her.

He passed the same mile marker 1 hour after she did. Let's begin filling in the chart:

Because we want to find Darryl's time to catch up we let t represent the amount of time it takes him to catch up. At the time Darryl catches up, Catrina's travel time is 1 hour more than Darryl's travel time from the same mile marker.

Because $d = rt$, we multiply the rate and time columns to get the distance column.

Categories	Rate	Time	Distance
Catrina	65	$t + 1$	$65(t + 1)$
Darryl	70	t	$70t$

Plan: Because they will have gone the same distance past mile marker 80 when Darryl catches up, we set the expression of their individual distances equal and solve for t.

Execute: $65(t + 1) = 70t$

$65t + 65 = 70t$

$\underline{-65t \qquad = -65t}$

$0 + 65 = 5t$

$\dfrac{65}{5} = \dfrac{5t}{5}$

$13 = t$

Answer: It will take Darryl 13 hours after passing mile marker 80 to catch up to Catrina.

Check: If Darryl travels for 13 hours after mile marker 80, then Catrina travels for $13 + 1 = 14$ hours after mile marker 80. We can check by verifying that they indeed are the exact same distance after traveling for their respective amount of time.

Darryl: $d = rt$ Catrina: $d = rt$

$d = (70)(13)$ $d = (65)(14)$

$d = 910$ mi. $d = 910$ mi.

Chapter 7 Cumulative Exercises

1. false   2. true   3. false

4. true   5. false   6. false

7. positive   8. reciprocal   9. LCM

10. cross

11. forty-nine thousand, eight hundred two and seventy-six hundredths

12. 9.14×10^8

13.

14. 41.33   15. -1 16. 5

17. $[10 - 3(7)] - [16 \div 2(7 + 3)]$
$= [10 - 21] - [16 \div 2 \cdot 10]$
$= -11 - [8 \cdot 10]$
$= -11 - 80$
$= -91$

18. $5\dfrac{2}{3} \div \left(-2\dfrac{1}{6}\right) = \dfrac{17}{3} \div \left(\dfrac{-13}{6}\right)$
$= \dfrac{17}{\cancel{3}} \cdot \left(\dfrac{\cancel{6}^2}{-13}\right)$
$= \dfrac{34}{-13}$
$= -2\dfrac{8}{13}$

19. $10\dfrac{1}{2} - 4\dfrac{3}{5} = 10\dfrac{1 \cdot 5}{2 \cdot 5} - 4\dfrac{3 \cdot 2}{5 \cdot 2}$
$= 10\dfrac{5}{10} - 4\dfrac{6}{10}$
$= 9\dfrac{15}{10} - 4\dfrac{6}{10}$
$= 5\dfrac{9}{10}$

20. $|12.5 - 18.65| \div 0.2 = |-6.15| \div 0.2$
$= 6.15 \div 0.2$
$= 30.75$

21. $\left(8\dfrac{2}{3}\right)(-3.6) = \dfrac{\overset{13}{\cancel{26}}}{\underset{1}{\cancel{3}}} \cdot \left(\dfrac{-\overset{12}{\cancel{36}}}{\underset{5}{\cancel{10}}}\right)$
$= \dfrac{-156}{5}$
$= -31.2$

22. $\left(\dfrac{3}{4}\right)^2 = \dfrac{9}{16}$ 23. $\sqrt{\dfrac{45}{5}} = \sqrt{9} = 3$

24. $(-0.3)^4 = (-0.3)(-0.3)(-0.3)(-0.3) = 0.0081$

25. $\dfrac{1}{2} \cdot (38) \cdot (0.4^2) = 0.5 \cdot (38) \cdot (0.16)$
$= 3.04$

26. $(12.4y^3 - 8.2y^2 + 9) - (4.1y^3 + y - 1.2)$
$= (12.4y^3 - 8.2y^2 + 9) + (-4.1y^3 - y + 1.2)$
$= 8.3y^3 - 8.2y^2 - y + 10.2$

27. $\left(\dfrac{2}{3}a - 1\right)\left(\dfrac{3}{4}a - 7\right)$
$= \dfrac{2}{3}a \cdot \dfrac{3}{4}a + \dfrac{2}{3}a \cdot (-7) + (-1) \cdot \dfrac{3}{4}a - 1 \cdot (-7)$
$= \dfrac{6}{12}a^2 - \dfrac{14}{3}a - \dfrac{3}{4}a - (-7)$
$= \dfrac{1}{2}a^2 - \dfrac{56}{12}a - \dfrac{9}{12}a + 7$
$= \dfrac{1}{2}a^2 - \dfrac{65}{12}a + 7$

28. $2400 = 2^5 \cdot 3 \cdot 5^2$

29. $40xy^2 = 2^3 \cdot 5 \cdot x \cdot y^2$
$60x^5 = 2^3 \cdot 3 \cdot 5 \cdot x^5$
$\text{GCF}(40xy^2, 60x^5) = 2^2 \cdot 5x$
$= 4 \cdot 5x$
$= 20x$

30. $\dfrac{18x^6}{3x^2} = \dfrac{18}{3}x^{6-2} = 6x^4$

31. $32m^4 + 24m^2 - 16m$
$= 8m\left(\dfrac{32m^4 + 24m^2 - 16m}{8m}\right)$
$= 8m\left(\dfrac{32m^4}{8m} + \dfrac{24m^2}{8m} - \dfrac{16m}{8m}\right)$
$= 8m(4m^3 + 3m - 2)$

32. $\dfrac{9x^2}{20y} \cdot \dfrac{5y}{12x^6}$

$= \dfrac{\cancel{3} \cdot 3 \cdot \cancel{x} \cdot x}{2 \cdot 2 \cdot \cancel{5} \cdot \cancel{y}} \cdot \dfrac{\cancel{5} \cdot \cancel{y}}{2 \cdot 2 \cdot \cancel{3} \cdot \cancel{x} \cdot \cancel{x} \cdot x \cdot x \cdot x \cdot x}$

$= \dfrac{3}{16x^4}$

33.
$$8k^2 = 2^3 \cdot k^2$$
$$10k = 2 \cdot 5 \cdot k$$
$$\text{LCM}(8k^2, 10k) = 2^3 \cdot 5 \cdot k^2 = 8 \cdot 5 \cdot k^2 = 40k^2$$

34. $\dfrac{2}{x} + \dfrac{5}{9} = \dfrac{2 \cdot 9}{x \cdot 9} + \dfrac{5 \cdot x}{9 \cdot x}$

$= \dfrac{18}{9x} + \dfrac{5x}{9x}$

$= \dfrac{18 + 5x}{9x}$

35.
$$2.5n - 16 = 30$$
$$10(2.5n - 16) = 10(30)$$
$$10 \cdot (2.5n) - 10 \cdot (16) = 10 \cdot (30)$$
$$25n - 160 = 300$$
$$\underline{+160 = +160}$$
$$25n + 0 = 460$$
$$\dfrac{25n}{25} = \dfrac{460}{25}$$
$$n = 18.4$$

Check: $2.5 \cdot (18.4) - 16 = 30$
$$46 - 16 = 30$$
$$30 = 30$$

36.
$$\dfrac{3}{4}x - 5 = \dfrac{1}{5}x + 3$$
$$20 \cdot \left(\dfrac{3}{4}x - 5\right) = 20 \cdot \left(\dfrac{1}{5}x + 3\right)$$
$$20 \cdot \dfrac{3}{4}x - 20 \cdot 5 = 20 \cdot \dfrac{1}{5}x + 20 \cdot 3$$
$$15x - 100 = 4x + 60$$
$$\underline{-4x \qquad = -4x}$$
$$11x - 100 = 0 + 60$$
$$\underline{+100 = \quad +100}$$
$$11x + 0 = 160$$
$$\dfrac{11x}{11} = \dfrac{160}{11}$$
$$x = 14\dfrac{6}{11}$$

Check: $\dfrac{3}{4} \cdot \left(14\dfrac{6}{11}\right) - 5 = \dfrac{1}{5} \cdot \left(14\dfrac{6}{11}\right) + 3$

$$\dfrac{3}{\cancel{4}} \cdot \dfrac{\cancel{160}^{40}}{11} - 5 = \dfrac{1}{\cancel{5}} \cdot \dfrac{\cancel{160}^{32}}{11} + 3$$

$$\dfrac{120}{11} - 5 = \dfrac{32}{11} + 3$$

$$\dfrac{120}{11} - \dfrac{55}{11} = \dfrac{32}{11} + \dfrac{33}{11}$$

$$\dfrac{65}{11} = \dfrac{65}{11}$$

37.
$$\dfrac{10.2}{b} = \dfrac{-12.5}{20}$$
$$20 \cdot (10.2) = b \cdot (-12.5)$$
$$204 = -12.5b$$
$$\dfrac{204}{-12.5} = \dfrac{-12.5b}{-12.5}$$
$$-16.32 = b$$

Check: $\dfrac{10.2}{-16.32} = \dfrac{-12.5}{20}$

$$20 \cdot (10.2) = -16.32 \cdot (-12.5)$$
$$204 = 204$$

38. $12.5 \text{ mi.} = \dfrac{12.5 \ \cancel{\text{mi.}}}{1} \cdot \dfrac{5280 \text{ ft.}}{1 \ \cancel{\text{mi.}}} = 66{,}000 \text{ ft.}$

39. Because g is 3 units to the right of kg, we move the decimal point 3 places to the right in 0.48.

    0.48 kg = 480 g

40. $F = \dfrac{9}{5}C + 32$

    $F = \dfrac{9}{5} \cdot \dfrac{60}{1} + 32$

    $F = \dfrac{540}{5} + 32$

    $F = 108 + 32$

    $F = 140°$

41. $V = lwh$

    $= (4)(2)(12)$

    $= 96m^3$

42.
$$N = R - C$$
$$45{,}698 = 96{,}408 - C$$
$$\underline{-96{,}408 \quad -96{,}408}$$
$$-50{,}710 = -C$$
$$\$50{,}710 = C$$

43. $A = bh$

    $A = (y+5)(y-9)$

    $A = y^2 - 9y + 5y - 45$

    $A = y^2 - 4y - 45$

44. To find the perimeter of a rectangle, we must use the formula $P = 2l + 2w$. We will let the length be l. The width is "3 less than the length" so we say $l - 3$. We also know that the perimeter is 50.

$$P = 2l + 2w$$
$$50 = 2l + 2(l - 3)$$
$$50 = 2l + 2l - 6$$
$$50 = 4l - 6$$
$$\underline{+6 = \qquad +6}$$
$$56 = 4l$$
$$\frac{56}{4} = \frac{4l}{4}$$
$$14 = w$$

The length is 14 m and the width is $14 - 3 = 11$m.

45. Area of a trapezoid $= = \dfrac{h(b_1 + b_2)}{2}$

$$= \dfrac{6\left(3\dfrac{1}{4} + 10\dfrac{1}{2}\right)}{2}$$

$$= \dfrac{6(3.25 + 10.5)}{2}$$

$$= \dfrac{6(13.75)}{2}$$

$$= \dfrac{82.5}{2}$$

$$= 41.25 \text{ or } 41\dfrac{1}{4} \text{ ft.}^2$$

46. Mean: $\dfrac{(4.6 + 3.1 + 6.2 + 8.4)}{4} = \dfrac{22.3}{4} = 5.575$

The mean is \$5.575 million

Median: 3.1, 4.6, 6.2, 8.4
The median is the mean of the middle two scores which are 4.6 and 6.2.
$$\dfrac{4.6 + 6.2}{2} = \dfrac{10.8}{2} = \$5.4 \text{ million}$$

47. There are 4 Jacks, 4 Queens, and 4 Kings in a standard deck of 52 cards. $P = \dfrac{12}{52} = \dfrac{3}{13}$

48. Let F represent the unit price of the 16-oz. can of beans and T represent the unit price of the 28-oz. can of beans.

$$F = \frac{\$0.49}{16 \text{ oz.}} \qquad T = \frac{\$1.19}{28 \text{ oz.}}$$
$$F = \frac{49\text{¢}}{16 \text{ oz.}} \qquad T = \frac{119\text{¢}}{28 \text{ oz.}}$$
$$F \approx 3.06\text{¢}/\text{oz.} \qquad T = 4.25\text{¢}/\text{oz.}$$

Because 3.06 cents is the smaller unit price, the 16- oz. can of beans is the better buy.

49.
$$\frac{6\,1}{7 \text{ days}} = \frac{x\,1}{365.25 \text{ days}}$$
$$365.25 \cdot \left(6\right) = 7 \cdot x$$
$$2191.5 = 7x$$
$$\frac{2191.5}{7} = \frac{7x}{7}$$
$$313 \approx x, \text{ so } x \approx 313 \text{ liters}$$

50.
$$\frac{7}{100} = \frac{6}{h}$$
$$7 \cdot h = 100 \cdot 6$$
$$7h = 600$$
$$\frac{7h}{7} = \frac{600}{7}$$
$$h \approx 85.7 \text{ m}$$

Chapter 8 Percents

Exercise Set 8.1

1. A percent is a ratio representing some part out of 100.

3. Write the percent as a ratio with 100 in the denominator, then divide. If the percent contains a mixed number or fraction, write the mixed number or fraction as a decimal number before dividing by 100.

5. $20\% = \dfrac{20}{100} = \dfrac{1}{5}$

7. $15\% = \dfrac{15}{100} = \dfrac{3}{20}$

9. $12.4\% = \dfrac{12.4}{100}$
$= \dfrac{12.4(10)}{100(10)}$
$= \dfrac{124}{1000}$
$= \dfrac{31}{250}$

11. $3.75\% = \dfrac{3.75}{100}$
$= \dfrac{3.75(100)}{100(100)}$
$= \dfrac{375}{10,000}$
$= \dfrac{3}{80}$

13. $45\dfrac{1}{2}\% = \dfrac{45\frac{1}{2}}{100}$
$= 45\dfrac{1}{2} \div 100$
$= \dfrac{91}{2} \cdot \dfrac{1}{100}$
$= \dfrac{91}{200}$

15. $33\dfrac{1}{3}\% = \dfrac{33\frac{1}{3}}{100}$
$= 33\dfrac{1}{3} \div 100$
$= \dfrac{\cancel{100}}{3} \cdot \dfrac{1}{\cancel{100}}$
$= \dfrac{1}{3}$

17. $75\% = \dfrac{75}{100} = 0.75$

19. $135\% = \dfrac{135}{100} = 1.35$

21. $12.9\% = \dfrac{12.9}{100} = 0.129$

23. $1.65\% = \dfrac{1.65}{100} = 0.0165$

25. $53\dfrac{2}{5}\% = 53.4\%$
$= \dfrac{53.4}{100}$
$= 0.534$

27. $16\dfrac{1}{6}\% = 16.1\overline{6}\%$
$= \dfrac{16.1\overline{6}}{100}$
$= 0.161\overline{6}$

29. $\dfrac{1}{4} = \dfrac{1}{\cancel{4}} \cdot \dfrac{\cancel{100}^{25}}{1}\%$
$= 25\%$

31. $\dfrac{2}{5} = \dfrac{2}{\cancel{5}} \cdot \dfrac{\cancel{100}^{20}}{1}\%$
$= 40\%$

33. $\dfrac{3}{8} = \dfrac{3}{\cancel{8}_2} \cdot \dfrac{\cancel{100}^{25}}{1}\%$
$= \dfrac{75}{2}\%$
$= 37\dfrac{1}{2}\%$ or 37.5%

35. $\dfrac{1}{6} = \dfrac{1}{\cancel{6}_3} \cdot \dfrac{\cancel{100}^{50}}{1}\%$
$= \dfrac{50}{3}\%$
$= 16\dfrac{2}{3}\%$ or $16.\overline{6}\%$

37. $\dfrac{1}{3} = \dfrac{1}{3} \cdot \dfrac{100}{1}\%$
$= \dfrac{100}{3}\%$
$= 33\dfrac{1}{3}\%$ or $33.\overline{3}\%$

39. $\dfrac{4}{9} = \dfrac{4}{9} \cdot \dfrac{100}{1}\%$
$= \dfrac{400}{9}\%$
$= 44\dfrac{4}{9}\%$ or $44.\overline{4}\%$

41. $0.96 = (0.96)(100\%)$
$= 96\%$

43. $0.8 = (0.8)(100\%)$
$= 80\%$

45. $0.09 = (0.09)(100\%)$
$= 9\%$

47. $1.2 = (1.2)(100\%)$
$= 120\%$

49. $0.028 = (0.028)(100\%) = 2.8\%$

51. $0.004 = (0.004)(100\%) = 0.4\%$

53. $0.\overline{6} = (0.66\overline{6})(100\%) = 66.\overline{6}\%$

55. $1.\overline{63} = (1.\overline{63})(100\%) = 163.\overline{63}\%$

Review Exercises

1. $0.6(24) = 14.4$

2. $0.45(3600) = 1620$

3. $0.05(76.8) = 3.84$

4. $0.15x = 200$
$100(0.15x) = 100(200)$
$15x = 20,000$
$\dfrac{15x}{15} = \dfrac{20,000}{15}$
$x = 1333.\overline{3}$

Check: $0.15\left(1333.\overline{3}\right)=200$

$$\frac{15}{100}\cdot 1333\frac{1}{3}=200$$

$$\frac{\overset{5}{\cancel{15}}}{\underset{1}{\cancel{100}}}\cdot\frac{\overset{40}{\cancel{4000}}}{\underset{1}{\cancel{3}}}=200$$

$$200=200$$

5. $\qquad 0.75y=28.5$

$\qquad 100\left(0.75y\right)=100\left(28.5\right)$

$\qquad\qquad 75y=2850$

$$\frac{75y}{75}=\frac{2850}{75}$$

$\qquad\qquad\quad y=38$

Check: $0.75\left(38\right)=28.5$

$\qquad\qquad\quad 28.5=28.5$

6. $\qquad \dfrac{3}{4}\cdot n=24$

$$\frac{\cancel{4}}{\cancel{3}}\cdot\frac{\cancel{3}}{\cancel{4}}n=\frac{4}{_1\cancel{3}}\cdot\frac{\overset{8}{\cancel{24}}}{1}$$

$\qquad\qquad\quad n=32$

Exercise Set 8.2

1. whole amount; part of the whole

3. multiplication

5. $\quad 40\%\cdot 350=c$

$\quad\left(0.4\right)\left(350\right)=c$

$\qquad\qquad 140=c$

7. $c=15\%\cdot 78$

$c=\left(0.15\right)\left(78\right)$

$c=11.7$

9. $\quad 150\%\cdot 60=c$

$\quad\left(1.5\right)\left(60\right)=c$

$\qquad\qquad 90=c$

11. $c=16\%\cdot 30\dfrac{1}{2}$

$c=\left(0.16\right)\left(30.5\right)$

$c=4.88$

13. $\quad 110\%\cdot 46.5=c$

$\quad\left(1.10\right)\left(46.5\right)=c$

$\qquad\quad 51.15=c$

15. $c=5\dfrac{1}{2}\%\cdot 280$

$c=\left(0.055\right)\left(280\right)$

$c=15.4$

17. $\qquad 66\dfrac{2}{3}\%\cdot 56.8=c$

$\quad\left(\dfrac{200}{3}\cdot\dfrac{1}{100}\right)\left(\dfrac{56.8}{1}\right)=c$

$\qquad\qquad\quad \dfrac{11360}{300}=c$

$\qquad\qquad\quad\ 37.86=c$

19. $c=8.5\%\cdot 54.82$

$c=\left(0.085\right)\left(54.82\right)$

$c=4.6597$

21. $12.8\%\cdot 36,000=c$

$\left(0.128\right)\left(36,000\right)=c$

$\qquad\qquad\quad 4608=c$

23. $\quad 35\%\cdot w=77$

$\qquad 0.35w=77$

$$\frac{0.35w}{0.35}=\frac{77}{0.35}$$

$\qquad\quad w=220$

25. $\quad 4800=40\%\cdot w$

$\quad 4800=0.4w$

$$\frac{4800}{0.4}=\frac{0.4w}{0.4}$$

$12,000=w$

27. $\qquad 605=2.5\%\cdot w$

$\qquad 605=0.025w$

$$\frac{605}{0.025}=\frac{0.025w}{0.025}$$

$24,200=w$

29. $105\%\cdot w=49.14$

$\quad 1.05w=49.14$

$$\frac{1.05w}{1.05}=\frac{49.14}{1.05}$$

$\qquad w=46.8$

31. $\qquad 47.25=10\dfrac{1}{2}\%\cdot w$

$\qquad 47.25=0.105w$

$$\frac{47.25}{0.105}=\frac{0.105w}{0.105}$$

$\qquad\ 450=w$

33. $p\cdot 68=17$

$68p=17$

$$\frac{68p}{68}=\frac{17}{68}$$

$p=0.25$

$p=\left(0.25\right)\left(100\%\right)$

$p=25\%$

35. $\qquad 2.142=p\cdot 35.7$

$\qquad 2.142=35.7p$

$$\frac{2.142}{35.7}=\frac{35.7p}{35.7}$$

$\qquad\ 0.06=p$

$\left(0.06\right)\left(100\%\right)=p$

$\qquad\qquad 6\%=p$

37. $p\cdot 24\dfrac{1}{2}=7\dfrac{7}{20}$

$\quad 24.5p=7.35$

$$\frac{24.5p}{24.5}=\frac{7.35}{24.5}$$

$\qquad p=0.3$

$\qquad p=\left(0.3\right)\left(100\%\right)$

$\qquad p=30\%$

39. $p=\dfrac{21}{60}$

$p=0.35$

$p=\left(0.35\right)\left(100\%\right)$

$p=35\%$

41. $\dfrac{25}{35}=p$

$0.714\approx p$

$\left(0.714\right)\left(100\%\right)\approx p$

$71.4\%\approx p$

43. $\qquad\dfrac{43}{65}=p$

$\qquad 0.6615\approx p$

$\left(0.6615\right)\left(100\%\right)\approx p$

$\qquad 66.15\%\approx p$

Review Exercises

1. $8\frac{1}{4}\cdot 10\frac{1}{2}=\frac{33}{4}\cdot\frac{21}{2}$

$\qquad =\frac{693}{8}$

$\qquad =86\frac{5}{8}$

2. $5\frac{3}{4}\cdot 100=\frac{23}{\cancel{4}}\cdot\frac{\cancel{100}^{25}}{1}$

$\qquad =575$

3. $12.8(100)=1280$

4. $\dfrac{x}{14}=\dfrac{3}{4}$

$4\cdot x=14\cdot 3$

$4x=42$

$\dfrac{4x}{4}=\dfrac{42}{4}$

$x=10.5$

Check: $\dfrac{10.5}{14}=\dfrac{3}{4}$

$4\cdot(10.5)=14\cdot 3$

$42=42$

5. $\dfrac{n}{100}=\dfrac{5}{8}$

$8\cdot n=100\cdot 5$

$8n=500$

$\dfrac{8n}{8}=\dfrac{500}{8}$

$n=62.5$

Check: $\dfrac{62.5}{100}=\dfrac{5}{8}$

$8\cdot 62.5=100\cdot 5$

$500=500$

6. $\dfrac{3}{20}=\dfrac{x}{2400}$

$2400\cdot 3=20\cdot x$

$7200=20x$

$\dfrac{7200}{20}=\dfrac{20x}{20}$

$360=x$

360 units will have the defect.

Exercise Set 8.3

1. Use the form $\text{percent}=\dfrac{\text{part}}{\text{whole}}$ where the percent is expressed as a ratio with a denominator of 100.

3. $\dfrac{30}{100}=\dfrac{c}{560}$

$560\cdot 30=100\cdot c$

$16,800=100c$

$\dfrac{16,800}{100}=\dfrac{100c}{100}$

$168=c$

5. $\dfrac{12}{100}=\dfrac{c}{85}$

$85\cdot 12=100\cdot c$

$1020=100c$

$\dfrac{1020}{100}=\dfrac{100c}{100}$

$10.2=c$

7. $\dfrac{125}{100}=\dfrac{c}{90}$

$90\cdot 125=100\cdot c$

$11,250=100c$

$\dfrac{11,250}{100}=\dfrac{100c}{100}$

$112.5=c$

9. $\dfrac{15}{100}=\dfrac{c}{9\frac{1}{2}}$

$9.5\cdot(15)=100\cdot c$

$142.5=100c$

$\dfrac{142.5}{100}=\dfrac{100c}{100}$

$1.425=c$

11. $\dfrac{102}{100}=\dfrac{c}{88.5}$

$88.5\cdot(102)=100\cdot c$

$9027=100c$

$\dfrac{9027}{100}=\dfrac{100c}{100}$

$90.27=c$

13. $\dfrac{12\frac{1}{2}}{100}=\dfrac{c}{440}$

$440\cdot(12.5)=100\cdot c$

$5500=100c$

$\dfrac{5500}{100}=\dfrac{100c}{100}$

$55=c$

15. $\dfrac{33\frac{1}{3}}{100}=\dfrac{c}{87.6}$

$\dfrac{87.6}{1}\cdot\dfrac{100}{3}=100\cdot c$

$\dfrac{8760}{3}=100c$

$2920=100c$

$\dfrac{2920}{100}=\dfrac{100c}{100}$

$29.2=c$

17. $\dfrac{6.5}{100}=\dfrac{c}{22,800}$

$22,800\cdot 6.5=100\cdot c$

$148,200=100c$

$\dfrac{148,200}{100}=\dfrac{100c}{100}$

$1482=c$

19. $\dfrac{16.9}{100}=\dfrac{c}{2450}$

$2450\cdot(16.9)=100\cdot c$

$41,405=100c$

$\dfrac{41,405}{100}=\dfrac{100c}{100}$

$414.05=c$

21. $\dfrac{65}{100}=\dfrac{52}{w}$

$w\cdot 65=100\cdot 52$

$65w=5200$

$\dfrac{65w}{65}=\dfrac{5200}{65}$

$w=80$

23. $\dfrac{80}{100}=\dfrac{4480}{w}$

$w\cdot 80=100\cdot 4480$

$80w=448,000$

$\dfrac{80w}{80}=\dfrac{448,000}{80}$

$w=5600$

25. $\dfrac{4.5}{100}=\dfrac{1143}{w}$

$w\cdot 4.5=100\cdot 1143$

$4.5w=114,300$

$\dfrac{4.5w}{4.5}=\dfrac{114,300}{4.5}$

$w=25,400$

27. $\dfrac{120}{100} = \dfrac{65.52}{w}$

$w \cdot 120 = 100 \cdot 65.52$

$120w = 6552$

$\dfrac{120w}{120} = \dfrac{6552}{120}$

$w = 54.6$

29. $\dfrac{5\frac{1}{2}}{100} = \dfrac{13.53}{w}$

$w \cdot 5.5 = 100 \cdot 13.53$

$5.5w = 1353$

$\dfrac{5.5w}{5.5} = \dfrac{1353}{5.5}$

$w = 246$

31. $\dfrac{P}{100} = \dfrac{24}{80}$

$80 \cdot P = 100 \cdot 24$

$80P = 2400$

$\dfrac{80P}{80} = \dfrac{2400}{80}$

$P = 30$

Answer: 30%

33. $\dfrac{P}{100} = \dfrac{2.597}{74.2}$

$74.2 \cdot P = 100 \cdot 2.597$

$74.2P = 259.7$

$\dfrac{74.2P}{74.2} = \dfrac{259.7}{74.2}$

$P = 3.5$

Answer: 3.5%

35. $\dfrac{P}{100} = \dfrac{18\frac{1}{5}}{30\frac{1}{3}}$

$30\frac{1}{3} \cdot P = 100 \cdot 18\frac{1}{5}$

$\dfrac{91}{3}P = \dfrac{100}{1} \cdot \dfrac{91}{5}$

$\dfrac{91}{3}P = \dfrac{9100}{5}$

$\dfrac{3}{91} \cdot \dfrac{91}{3}P = \dfrac{\overset{100}{\cancel{9100}}}{5} \cdot \dfrac{3}{\cancel{91}}$

$1P = \dfrac{300}{5}$

$P = 60$ Answer: 60%

37. $\dfrac{P}{100} = \dfrac{76}{80}$

$80 \cdot P = 100 \cdot 76$

$80P = 7600$

$\dfrac{80P}{80} = \dfrac{7600}{80}$

$P = 95$

Answer: 95%

39. $\dfrac{P}{100} = \dfrac{32}{45}$

$45 \cdot P = 100 \cdot 32$

$45P = 3200$

$\dfrac{45P}{45} = \dfrac{3200}{45}$

$P = 71.\overline{1}$

Answer: $71.\overline{1}\%$

41. $\dfrac{P}{100} = \dfrac{17}{19}$

$19 \cdot P = 100 \cdot 17$

$19P = 1700$

$\dfrac{19P}{19} = \dfrac{1700}{19}$

$P \approx 89.47$ Answer: 89.47%

43. $\dfrac{P}{100} = \dfrac{80}{75}$

$75 \cdot P = 100 \cdot 80$

$75P = 8000$

$\dfrac{75P}{75} = \dfrac{8000}{75}$

$P = 106.\overline{6}$ Answer: $106.\overline{6}\%$

Review Exercises

1. $\dfrac{\overset{1}{\cancel{5}}}{\underset{4}{\cancel{16}}} \cdot \dfrac{\overset{5}{\cancel{20}}}{\underset{3}{\cancel{27}}} = \dfrac{5}{12}$

2. $0.48(560) = 268.8$ **3.** $0.05\big(24.8\big) = 1.24$

4. $\dfrac{5}{8} \cdot 400 = \dfrac{5}{8} \cdot \dfrac{400}{1} = \dfrac{2000}{8} = 250$ disagreed.

5. $\dfrac{3}{\cancel{4}} \cdot \dfrac{\cancel{4}}{5} = \dfrac{3}{5}$ were females that agreed.

Exercise Set 8.4

1. 1. Determine whether the percent, the whole, or the part is unknown.
2. If needed, write the problem as a simple percent sentence.
3. Translate to an equation (word for word or proportion).
4. Solve for the unknown.

3. unknown; 4000; 2400

5. Simple percent sentence:
60% of 800 is what number?

Word-for-Word Translation Method:
$60\% \cdot 800 = c$

$0.6\big(800\big) = c$

$480 = c$

Proportion Method: $\dfrac{60}{100} = \dfrac{c}{800}$

$800 \cdot 60 = 100 \cdot c$

$48{,}000 = 100c$

$\dfrac{48{,}000}{100} = \dfrac{100c}{100}$

$480 = c$

480 ml of the solution is HCl.

7. a) Simple percent sentence:
90% of 30 is what number?

Word-for-Word Translation Method:
$$90\% \cdot 30 = c$$
$$0.9(30) = c$$
$$27 = c$$

Proportion Method:
$$\frac{90}{100} = \frac{c}{30}$$
$$30 \cdot 90 = 100 \cdot c$$
$$2700 = 100c$$
$$\frac{2700}{100} = \frac{100c}{100}$$
$$27 = c$$

Sabrina answered 27 questions correctly.

b) If she answered 90% correctly then she answered 100% – 90% = 10% incorrectly.

c) If she answered 27 questions correctly, then she answered 30 – 27 = 3 incorrectly.

9. Simple percent sentence:
5% of 113 million is what number?

Word-for-Word Translation Method:
$$5\% \cdot 113 = c$$
$$0.05(113) = c$$
$$5.65 = c$$

Proportion Method:
$$\frac{5}{100} = \frac{c}{113}$$
$$5 \cdot 113 = 100 \cdot c$$
$$565 = 100c$$
$$\frac{565}{100} = \frac{100c}{100}$$
$$5.65 = c$$

There were 5.65 million visitors to emergency rooms for fever.

11. Simple percent sentence:
10% of 3106 is what number?

Word-for-Word Translation Method:
$$10\% \cdot 3106 = c$$
$$0.1(3106) = c$$
$$310.6 = c$$

Proportion Method:
$$\frac{10}{100} = \frac{c}{3106}$$
$$3106 \cdot 10 = 100 \cdot c$$
$$31,060 = 100c$$
$$\frac{31,060}{100} = \frac{100c}{100}$$
$$310.6 = c$$

Wright's commission is $310.60.

13. Simple percent sentence:
8% of 4249 is what number?

Word-for-Word Translation Method:
$$8\% \cdot 4249 = c$$
$$0.08(4249) = c$$
$$339.92 = c$$

Proportion Method:
$$\frac{80}{100} = \frac{c}{4249}$$
$$4249 \cdot 80 = 100 \cdot c$$
$$339,920 = 100c$$
$$\frac{339,920}{100} = \frac{100c}{100}$$
$$339.92 = c$$

Liz earned $339.92 in commission. She also earned $6 per hour for 60 hours. This is $6 \times 60 =$ $360. Her total gross pay is 360 + 339.92 = $699.92.

15. profit = gross – cost: $P = \$78,950 - \$62,100$
$$P = \$16,850$$

Simple Percent Sentence:
25% of 16,850 is what number?

Word-for-Word Translation Method:
$$25\% \cdot 16,850 = c$$
$$0.25(16,850) = c$$
$$4212.5 = c$$

Proportion Method:
$$\frac{25}{100} = \frac{c}{16,850}$$
$$16,850 \cdot 25 = 100 \cdot c$$
$$421,250 = 100c$$
$$\frac{421,250}{100} = \frac{100c}{100}$$
$$4212.5 = c$$

Cush's commission is $4212.50.

17.

Reason for Choosing the Flight	Percent	Equation	Number of Passengers (in millions)
Cost	46%	$46\% \cdot 155.1 = c$ or $\frac{46}{100} = \frac{c}{155.1}$	≈ 71.3

Safety Record	32%	$32\% \cdot 155.1 = c$ or $\dfrac{32}{100} = \dfrac{c}{155.1}$	≈ 49.6
Time of Arrival or Departure	14%	$14\% \cdot 155.1 = c$ or $\dfrac{14}{100} = \dfrac{c}{155.1}$	≈ 21.7
Frequent Flyer Mileage	4%	$4\% \cdot 155.1 = c$ or $\dfrac{4}{100} = \dfrac{c}{155.1}$	≈ 6.2

Cost: $46\% \cdot 155.1 = c$
$0.46(155.1) = c$
$71.346 = c$
$71.3 \approx c$

$\dfrac{46}{100} = \dfrac{c}{155.1}$
$155.1 \cdot 46 = 100 \cdot c$
$7134.6 = 100c$
$\dfrac{7134.6}{100} = \dfrac{100c}{100}$
$71.346 = c$
$71.3 \approx c$

Safety Record: $32\% \cdot 155.1 = c$
$0.32(155.1) = c$
$49.632 = c$
$49.6 \approx c$

$\dfrac{32}{100} = \dfrac{c}{155.1}$
$155.1 \cdot 32 = 100 \cdot c$
$4963.2 = 100c$
$\dfrac{4963.2}{100} = \dfrac{100c}{100}$
$49.632 = c$
$49.6 \approx c$

Time of Arrival or Departure: $14\% \cdot 155.1 = c$
$0.14(155.1) = c$
$21.714 = c$
$21.7 \approx c$

$\dfrac{14}{100} = \dfrac{c}{155.1}$
$155.1 \cdot 14 = 100 \cdot c$
$2171.4 = 100c$
$\dfrac{2171.4}{100} = \dfrac{100c}{100}$
$21.714 = c$
$21.7 \approx c$

Frequent Flyer Mileage: $4\% \cdot 155.1 = c$
$0.04(155.1) = c$
$6.204 = c$
$6.2 \approx c$

$\dfrac{4}{100} = \dfrac{c}{155.1}$
$155.1 \cdot 4 = 100 \cdot c$
$620.4 = 100c$
$\dfrac{620.4}{100} = \dfrac{100c}{100}$
$6.204 = c$
$6.2 \approx c$

19. Simple percent sentence:
750 is 6% of what number?

Word-for-Word Translation Method:
$750 = 6\% \cdot w$
$750 = 0.06w$
$\dfrac{750}{0.06} = \dfrac{0.06w}{0.06}$
$12,500 = w$

Proportion Method: $\dfrac{6}{100} = \dfrac{750}{w}$
$w \cdot 6 = 100 \cdot 750$
$6w = 75,000$
$\dfrac{6w}{6} = \dfrac{75,000}{6}$
$w = 12,500$

The seller's net is \$12,500.

21. Simple percent sentence:
38 is 42% of what number?

Word-for-Word Translation Method:
$38 = 42\% \cdot w$
$38 = 0.42w$
$\dfrac{38}{0.42} = \dfrac{0.42w}{0.42}$
$91 \approx w$

Proportion Method: $\dfrac{42}{100} = \dfrac{38}{w}$
$w \cdot 42 = 100 \cdot 38$
$42w = 3800$
$\dfrac{42w}{42} = \dfrac{3800}{42}$
$w \approx 91$

There were about 91 shot attempts in the game.

23. Simple percent sentence:
245,000 is 49% of what number?

Word-for-Word Translation Method:
$$245,000 = 49\% \cdot w$$
$$245,000 = 0.49w$$
$$\frac{245,000}{0.49} = \frac{0.49w}{0.49}$$
$$500,000 = w$$

Proportion Method:
$$\frac{49}{100} = \frac{245,000}{w}$$
$$w \cdot 49 = 100 \cdot 245,000$$
$$49w = 24,500,000$$
$$\frac{49w}{49} = \frac{24,500,000}{49}$$
$$w = 500,000$$

The advertising budget is $500,000.

25. Simple percent sentence:
What percent is 48 out of 54?

Word-for-Word Translation Method:
$$p = \frac{48}{54}$$
$$p \approx 0.\overline{8}$$
$$p \approx 0.\overline{8}(100\%)$$
$$p \approx 88.\overline{8}\%$$

Proportion Method:
$$\frac{P}{100} = \frac{48}{54}$$
$$54 \cdot P = 100 \cdot 48$$
$$54P = 4800$$
$$\frac{54P}{54} = \frac{4800}{54}$$
$$P \approx 88.\overline{8}$$

Hunter answered about $88.\overline{8}\%$ correctly.

27. Simple percent sentence:
What percent is 36 out of 110?

Word-for-Word Translation Method:
$$p = \frac{36}{110}$$
$$p \approx 0.32\overline{72}$$
$$p \approx 0.32\overline{72}(100\%)$$
$$p \approx 32.\overline{72}\%$$

Proportion Method:
$$\frac{P}{100} = \frac{36}{110}$$
$$110 \cdot P = 100 \cdot 36$$
$$110P = 3600$$
$$\frac{110P}{110} = \frac{3600}{110}$$
$$P \approx 32.\overline{72}$$

The baseball player gets a hit about $32.\overline{72}\%$ of the times he is at bat.

29. Simple percent sentence:
What percent is 3 billion out of 6.3 billion?

Word-for-Word Translation Method:
$$p = \frac{3}{6.3}$$
$$p \approx 0.476$$
$$p \approx 0.476(100\%)$$
$$p \approx 47.6\%$$

Proportion Method:
$$\frac{P}{100} = \frac{3}{6.3}$$
$$6.3 \cdot P = 100 \cdot 3$$
$$6.3P = 300$$
$$\frac{6.3P}{6.3} = \frac{300}{6.3}$$
$$P \approx 47.6$$

About 47.6% of the world's population live on less than $2 per day.

31. We must find the number of hours left in the day.
$24 - 8 - 7 - 1 - 1 = 7$ hours left.

Simple percent sentence:
What percent is 7 out of 24?

Word-for-Word Translation Method:
$$p = \frac{7}{24}$$
$$p \approx 0.291\overline{6}$$
$$p \approx 0.291\overline{6}(100\%)$$
$$p \approx 29.1\overline{6}\%$$

Proportion Method:
$$\frac{P}{100} = \frac{7}{24}$$
$$24 \cdot P = 100 \cdot 7$$
$$24P = 700$$
$$\frac{24P}{24} = \frac{700}{24}$$
$$P \approx 29.1\overline{6}$$

There is $29.1\overline{6}\%$ of the day left.

33. We must find the number of minutes in a week. Since there are 24 hours in a day, 60 minutes in an hour and 7 days in a week, we must multiply.

$24 \cdot 60 \cdot 7 = 10,080$ minutes in a week

Simple percent sentence:

What percent is 1680 out of 10,080?

Word-for-Word Translation Method:

$$p = \frac{1680}{10{,}080}$$

$$p \approx 0.1\overline{6}$$

$$p \approx 0.1\overline{6}(100\%)$$

$$p \approx 16.\overline{6}\%$$

Proportion Method: $\dfrac{P}{100} = \dfrac{1680}{10{,}080}$

$$10{,}080 \cdot P = 100 \cdot 1680$$

$$10{,}080 P = 168{,}000$$

$$\frac{10{,}080 P}{10{,}080} = \frac{168{,}000}{10{,}080}$$

$$P \approx 16.\overline{6}$$

About $16.\overline{6}$ % of the week is spent watching television.

35. The volume of the sun:

$$V = \frac{4}{3}\pi r^3$$

$$V = \frac{4}{3}(3.14)\left(695{,}990^3\right)$$

$$V = 1.41 \times 10^{18}$$

The volume of the core:

$$V = \frac{4}{3}\pi r^3$$

$$V = \frac{4}{3}(3.14)\left(170{,}000^3\right)$$

$$V = 2.06 \times 10^{16}$$

Simple percent sentence:
What percent of 1.41×10^{18} is 2.06×10^{16} ?

Word-for-Word Translation Method:

$$p = \frac{\left(2.06 \times 10^{16}\right)}{\left(1.41 \times 10^{18}\right)}$$

$$p \approx 0.0146$$

$$p \approx 0.0146(100\%)$$

$$p \approx 1.46\%$$

Proportion Method:

$$\frac{x}{100} = \frac{2.03 \times 10^{16}}{1.41 \times 10^{18}}$$

$$\left(1.41 \times 10^{18}\right)x = 100\left(2.03 \times 10^{16}\right)$$

$$x = \frac{100\left(2.03 \times 10^{16}\right)}{\left(1.41 \times 10^{18}\right)}$$

$$x \approx 1.46$$

The core makes up about 1.46% of the Sun's total volume.

37. a)

	Agreed	Disagreed	No Opinion
Percent of Respondents	20%	65%	15%

What percent of respondents is each "response"?

The total number of respondents is 180 + 585 + 135 = 900.

Agreed: $p = \dfrac{180}{900}$

$$p = 0.2$$

$$p = 0.2(100\%)$$

$$p = 20\%$$

Disagreed: $p = \dfrac{585}{900}$

$$p = 0.65$$

$$p = 0.65(100\%)$$

$$p = 65\%$$

No Opinion: $p = \dfrac{135}{900}$

$$p = 0.15$$

$$p = 0.15(100\%)$$

$$p = 15\%$$

b)

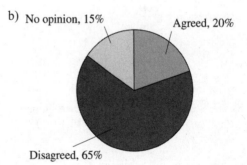

No opinion, 15% Agreed, 20%

Disagreed, 65%

c) 20% + 65% = 85%

d) 65% + 15% = 80%

39. The Smith family's total income is 3480.25 + 1879.75 = $5360.

a) We must find the percent of their income for each individual expense. We will use the direct translation method although the proportion method could also be used. The general basic percent sentence is:
What percent of 5360 is each "expense"?

Mortgage: $p = \dfrac{1580}{5360}$

$p \approx 0.295$

$p \approx 0.295(100\%)$

$p \approx 29.5\%$

Car loan 1: $p = \dfrac{420}{5360}$

$p \approx 0.078$

$p \approx 0.078(100\%)$

$p \approx 7.8\%$

Car loan 2: $p = \dfrac{345}{5360}$

$p \approx 0.064$

$p \approx 0.064(100\%)$

$p \approx 6.4\%$

Credit Card Payments: $p = \dfrac{280}{5360}$

$p \approx 0.052$

$p \approx 0.052(100\%)$

$p \approx 5.2\%$

Utilities: $p = \dfrac{540}{5360}$

$p \approx 0.101$

$p \approx 0.101(100\%)$

$p \approx 10.1\%$

Groceries: $p = \dfrac{600}{5360}$

$p \approx 0.112$

$p \approx 0.112(100\%)$

$p \approx 11.2\%$

b) The Smith family's total expenses in percent
   are $29.5 + 7.8 + 6.4 + 5.2 + 10.1 + 11.2$
   $\approx 70.2\%$.

c) After paying all expenses $100 - 70.2 \approx 29.8\%$
   is left.

d)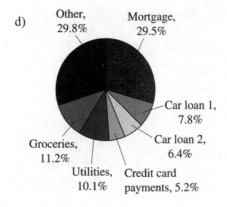

Review Exercises:

1. $5.1x + 9.8 + x - 12.4 = 5.1x + x + 9.8 - 12.4$
    $$= 6.1x - 2.6$$

2. $14.7 = x + 0.05x$

    $14.7 = 1.05x$

    $\dfrac{14.7}{1.05} = \dfrac{1.05x}{1.05}$

    $14 = x$

    Check: $14.7 = 14 + 0.05(14)$

    $14.7 = 14 + 0.7$

    $14.7 = 14.7$

3. $24.48 = y - 0.2y$

    $24.48 = 0.8y$

    $\dfrac{24.48}{0.8} = \dfrac{0.8y}{0.8}$

    $30.6 = y$

    Check: $24.48 = 30.6 - 0.2(30.6)$

    $24.48 = 30.6 - 6.12$

    $24.48 = 24.48$

4. $\dfrac{m}{2.4} = \dfrac{5.6}{8.4}$

    $8.4 \cdot m = 2.4 \cdot (5.6)$

    $8.4m = 13.44$

    $\dfrac{8.4m}{8.4} = \dfrac{13.44}{8.4}$

    $m = 1.6$

    Check: $\dfrac{1.6}{2.4} = \dfrac{5.6}{8.4}$

    $8.4(1.6) = 2.4(5.6)$

    $13.44 = 13.44$

5. $28,350 - 25,800 = \$2550$ is the amount of her
    raise.

Exercise Set 8.5

1. percent; initial amount; amount of increase or decrease

3. 25%; unknown; $80

5. **Understand:** We must calculate the sales tax and total amount of the purchase. We are given the tax rate and initial amount.

   Plan: Write a simple percent sentence, translate to an equation, and solve for the sales tax. Then to calculate the final amount, add the sales tax to the initial amount.

   Execute: The sales tax is 5% of the initial price. What is 5% of 185.95?

   $$c = 5\% \cdot 185.95$$
   $$c = 0.05(185.95)$$
   $$c \approx 9.30$$

   Answer: The sales tax is $9.30. Because the sales tax increases the initial price, we add the tax to the initial price to get the total amount.

   total $= 185.95 + 9.30 = \$195.25$

   Check: We can reverse the process as a check. We verify the tax by subtracting the initial amount from the total amount.

   $$195.25 - 185.95 = 9.30$$

   Now we can set up a proportion to verify that the tax rate is in fact 5%.

   $$\frac{P}{100} = \frac{9.3}{185.95}$$
   $$185.95 \cdot P = 100 \cdot 9.3$$
   $$185.95P = 930$$
   $$\frac{185.95P}{185.95} = \frac{930}{185.95}$$
   $$P = 5$$

7. **Understand:** We must calculate the sales tax and total amount of the purchase. We are given the tax rate and initial amount.

   Plan: Write a simple percent sentence, translate to an equation, and solve for the sales tax. Then to calculate the final amount, add the sales tax to the initial amount.

   Execute: The sales tax is 6% of the initial price. What is 6% of 26,450?

   $$c = 6\% \cdot 26,450$$
   $$c = 0.06(26,450)$$
   $$c = \$1587$$

   Answer: The sales tax is $1587. Because the sales tax increases the initial price, we add the tax to the initial price to get the total amount.

   total $= 26,450 + 1587 = \$28,037$

   Check: We can reverse the process as a check. We verify the tax by subtracting the initial amount from the total amount.

   $$28,037 - 26,450 = 1587$$

   Now we can set up a proportion to verify that the tax rate is in fact 6%.

   $$\frac{P}{100} = \frac{1587}{26,450}$$
   $$26,450 \cdot P = 100 \cdot 1587$$
   $$26,450P = 158,700$$
   $$\frac{26,450P}{26,450} = \frac{158,700}{26,450}$$
   $$P = 6$$

9. **Understand:** We must calculate the tip and total amount of the bill including tip. We are given the rate and bill.

   Plan: Write a simple percent sentence, translate to an equation, and solve for the tip. Once the tip is calculated, it must be added to the bill to get the total amount paid.

   Execute: The tip is 20% of the bill. What is 20% of 48.75?

   $$c = 20\% \cdot 48.75$$
   $$c = 0.2(48.75)$$
   $$c = \$9.75$$

   Answer: The tip is $9.75. The total amount paid is $48.75 + 9.75 = \$58.50$.

   Check: We can reverse the process as a check. We verify the tip by subtracting the bill from the total amount.

   $$58.50 - 48.75 = 9.75$$

   Now we can set up a proportion to verify that the rate is in fact 20%.

   $$\frac{P}{100} = \frac{9.75}{48.75}$$
   $$48.75 \cdot P = 100 \cdot 9.75$$
   $$48.75P = 975$$
   $$\frac{48.75P}{48.75} = \frac{975}{48.75}$$
   $$P = 20$$

11. **Understand:** We must calculate Michelle's new salary. We are given her initial salary and the percent of the raise. Because the 3.5% is added to 100% of her current salary, her new salary will be 103.5% of her current salary.

Plan: Write a simple percent sentence, translate to an equation, and solve for the new salary. Because we are calculating a final amount, we will use the variable a.

Execute: The new salary is 103.5% of the current salary. What is 103.5% of 29,600?

$$a = 103.5\% \cdot 29,600$$
$$a = 1.035(29,600)$$
$$a = \$30,636$$

Answer: Michelle's new salary will be $30,636.

Check: Note that if we need to know the amount of the raise, we can subtract the initial salary form the new salary.

$$\text{raise amount} = 30,636 - 29,600 = \$1036$$

We can reverse the process and verify that the raise of $1036 is in fact 3.5% of her initial salary. Now we can set up a proportion to verify that the raise is indeed 3.5%

$$\frac{P}{100} = \frac{1036}{29,600}$$
$$29,600 \cdot P = 100 \cdot 1036$$
$$29,600P = 103,600$$
$$\frac{29,600P}{29,600} = \frac{103,600}{29,600}$$
$$P = 3.5$$

13. **Understand:** We must calculate Jack's former salary. We are given the percent of increase and the amount of the raise.

Plan: Write a basic percent sentence, translate to an equation, and solve for the new salary. We will use the variable f to represent the former salary.

Execute: 4% of the former salary is the raise amount. 4% of what amount is 1008?

$$4\% \cdot f = 1008$$
$$0.04f = 1008$$
$$\frac{0.04f}{0.04} = \frac{1008}{0.04}$$
$$f = \$25,200$$

Answer: Jack's former salary was $25,200.

Check: We can verify that a 4% raise on an initial salary of $25,200 equals a raise of $1008.

4% of 25,200 is what raise? $4\% \cdot 25,200 = r$
$$0.04(25,200) = r$$
$$1008 = r$$

15. **Understand:** We must calculate Huang's former salary. We are given the new salary after the 4% raise. The new salary is the former salary added to the amount of the raise.

Plan: Write an equation, then solve.

Execute: Recall that the new salary is the former salary combined with the amount of the raise.

new salary = former salary + amount of raise

Recall that the amount of the raise is calculated by multiplying the former salary by 4%. We can incorporate this into the above equation.

new salary = former salary + 4% × former salary

If we let f represent the former salary, we can translate the above equation to:

$$30,420 = f + 0.04f$$

Now solve for f: $30,420 = f + 0.04f$
$$30,420 = 1.04f$$
$$\frac{30,420}{1.04} = \frac{1.04f}{1.04}$$
$$\$29,250 = f$$

Answer: Huang's former salary was $29,250.

Check: Because the raise increases his salary, we can reverse the process as a check. We verify the raise by subtracting the old salary from the new salary.

$$30,420 - 29,250 = 1170$$

Now we can set up a proportion to verify that the increase is in fact 4%.

$$\frac{P}{100} = \frac{1170}{29,250}$$
$$29,250 \cdot P = 100 \cdot 1170$$
$$29,250P = 117,000$$
$$\frac{29,250P}{29,250} = \frac{117,000}{29,250}$$
$$P = 4$$

17. **Understand:** We must calculate the initial price of the shoes. We are given the total after the 5% sale tax. The total cost is the initial cost added to the sales tax.

Plan: Write an equation, then solve.

Execute: Recall that the total cost is the initial price added to the amount of the sales tax.

total cost = initial cost + sales tax

Recall that the amount of the sales tax is calculated by multiplying the initial cost by 5%. We can incorporate this into the above equation.

total cost = initial cost + 5% × initial cost

If we let f represent the initial cost, we can translate the above equation to:

$$45.10 = f + 0.05f$$

Now solve for f: $\quad 45.10 = f + 0.05f$
$$45.10 = 1.05f$$
$$\frac{45.10}{1.05} = \frac{1.05f}{1.05}$$
$$\$42.95 \approx f$$

Answer: Rachel's initial cost was $42.95.

Check: Because the tax increases her bill, we can reverse the process as a check. We verify the tax by subtracting the original price from the final price.

$$45.10 - 42.95 = \$2.15$$

Now we can set up a proportion to verify that the tax is in fact 5%.

$$\frac{P}{100} = \frac{2.15}{42.95}$$
$$42.95 \cdot P = 100 \cdot 2.15$$
$$42.95P = 215$$
$$\frac{42.95P}{42.95} = \frac{215}{42.95}$$
$$P = 5$$

19. **Understand:** We must calculate the discount amount and final price. We are given the discount rate and the initial amount.

Plan: Write a simple percent sentence, translate to an equation, and solve for the discount amount. Then to calculate the final price, subtract the discount amount from the initial price.

Execute: The discount amount is 30% of the initial price. What amount is 30% of 65.99?

$$c = 30\% \cdot 65.99$$
$$c = 0.3(65.99)$$
$$c \approx \$19.80$$

Answer: The discount amount is $19.80. Because this discount is an amount of decrease, we subtract it from the initial price to get the final price.

final price = $65.99 - 19.80 = \$46.19$

Check: We can reverse the process to verify that the discount rate is correct. We verify the discount amount by subtracting the final price from the initial price.

$$65.99 - 46.19 = 19.80$$

Now we can set up a proportion to verify that the discount rate is 30%. $\quad \dfrac{P}{100} = \dfrac{19.80}{65.99}$
$$65.99 \cdot P = 100 \cdot 19.80$$
$$65.99P = 1980$$
$$\frac{65.99P}{65.99} = \frac{1980}{65.99}$$
$$P = 30$$

21. **Understand:** We must calculate the discount amount and final price. We are given the discount rate and the initial amount.

Plan: Write a simple percent sentence, translate to an equation, and solve for the discount amount. Then to calculate the final price, subtract the discount amount from the initial price.

Execute: The discount amount is 35% of the initial price. What amount is 35% of 699.99?

$$c = 35\% \cdot 699.99$$
$$c = 0.35(699.99)$$
$$c \approx \$245.00$$

Answer: The discount amount is $245. Because this discount is an amount of decrease, we subtract it from the initial price to get the final price.

final price = $699.99 - 245.00 = \$454.99$

Check: We can reverse the process to verify that the discount rate is correct. We verify the discount amount by subtracting the final price from the initial price.

$$699.99 - 454.99 = 245$$

Now we can set up a proportion to verify that the discount rate is 35%.

$$\frac{P}{100} = \frac{245}{699.99}$$
$$699.99 \cdot P = 100 \cdot 245$$
$$699.99P = 24,500$$
$$\frac{699.99P}{699.99} = \frac{24,500}{699.99}$$
$$P = 35$$

23. **Understand:** If 27% of her gross monthly salary is deducted, then her net pay will be 73% of her gross salary (100% − 27% = 73%).

Plan: Write a simple percent sentence. Translate to an equation or proportion, then solve. We will use the variable n to represent net pay.

Execute: Her net pay is 73% of her gross monthly salary.   What amount is 73% of 2708.33?

$$n = 73\% \cdot 2708.33$$
$$n = 0.73(2708.33)$$
$$n \approx \$1977.08$$

Answer: Her net pay will be $1977.08.

Check: We can reverse the process to verify that the deduction percent is correct. The deduction percent is equal to the ratio of the deduction amount to the gross pay. We first need the deduction amount by subtracting net pay from gross pay.

$$2708.33 - 1977.08 = 731.25$$

Now we can set up a proportion to verify that the deduction is 27%.

$$\frac{P}{100} = \frac{731.25}{2708.33}$$
$$2708.33 \cdot P = 100 \cdot 731.25$$
$$2708.33P = 73,125$$
$$\frac{2708.33P}{2708.33} = \frac{73,125}{2708.33}$$
$$P = 27$$

25. **Understand:** We must calculate the amount of the gross monthly salary contributed to Medicare.

Plan: Write a simple percent sentence. Translate to an equation or proportion. Then solve. We will use the variable m to represent Medicare.

Execute: The contribution is 1.45% of the gross monthly salary. What amount is 1.45% of 1546.67?

$$n = 1.45\% \cdot 1546.67$$
$$n = 0.0145(1546.67)$$
$$n \approx \$22.43$$

Answer: The contribution will be $22.43.

Check: We can reverse the process to verify that the deduction percent is correct. The deduction percent is equal to the ratio of the deduction amount to the gross pay.

Now we can set up a proportion to verify that the deduction is 1.45%.

$$\frac{P}{100} = \frac{22.43}{1546.67}$$
$$1546.67 \cdot P = 100 \cdot 22.43$$
$$1546.67P = 2243$$
$$\frac{1546.67P}{1546.67} = \frac{2243}{1546.67}$$
$$P = 1.45$$

27. **Understand:** We must calculate Gordon's gross monthly salary. We are given the percent of decrease and the total deductions.

Plan: Write a simple percent sentence, translate to an equation, and solve for the salary. We will use the variable g to represent his gross monthly salary.

Execute: 26% of his gross monthly salary is total deductions. 26% of what salary is 674.96?

$$26\% \cdot g = 674.96$$
$$0.26g = 674.96$$
$$\frac{0.26g}{0.26} = \frac{674.96}{0.26}$$
$$g = \$2596$$

Answer: Gordon's gross monthly salary is $2596.

Check: We can verify that 26% of $2596 equals the total deductions of $674.96. 26% of $2596 is what amount of decrease?

$$26\% \cdot 2596 = d$$
$$0.26 \cdot 2596 = d$$
$$674.96 = d$$

29. **Understand:** We must calculate cost of the table. We are given the percent of decrease and the discount amount

Plan: Write a simple percent sentence, translate to an equation, and solve for the cost of the table. We will use the variable c to represent the cost of the table.

Execute: 30% of the cost of the table is the discount amount. 30% of what amount is 44.09?

$$30\% \cdot c = 44.09$$
$$0.3c = 44.09$$
$$\frac{0.3c}{0.3} = \frac{44.09}{0.3}$$
$$c \approx \$146.97$$

Answer: The cost of the table is $146.97.

Check: We can verify that 30% of $146.97 equals the total deductions of $44.09. 30% of $146.97 is what amount?

$$30\% \cdot 146.97 = d$$
$$0.3 \cdot 146.97 = d$$
$$44.09 = d$$

31. **Understand:** We must calculate Luther's gross monthly salary. We are given the percent of decrease and the net pay.

Plan: Write an equation, then solve.

Execute: Recall that net pay is gross pay minus the total deductions.

net pay = gross monthly salary – total deductions

Also recall that the total of the deductions is 28% of the gross salary. We can incorporate this into the above equation.

net pay = gross monthly salary – 28% × gross pay

If we let g represent the gross monthly salary, we can translate the above equation to:

$$1850.40 = g - 0.28g$$

Now solve for g.
$$1850.40 = g - 0.28g$$
$$1850.40 = 0.72g$$
$$\frac{1850.40}{0.72} = \frac{0.72g}{0.72}$$
$$\$2570 = g$$

Answer: Luther's gross monthly salary is $2570.

Check: We can reverse the process to verify that the deduction rate is correct. We verify the deduction rate amount by subtracting the net pay from the gross amount.

$$2570 - 1850.40 = \$719.60$$

Now we can set up a proportion to verify that the deduction rate is 28%.
$$\frac{P}{100} = \frac{719.6}{2570}$$
$$2570 \cdot P = 100 \cdot 719.6$$
$$2570P = 71{,}960$$
$$\frac{2570P}{2570} = \frac{71{,}960}{2570}$$
$$P = 28$$

33. **Understand:** We must calculate the initial price of the curio stand. We are given the percent of decrease and the price after discount.

Plan: Write an equation, then solve.

Execute: Recall that price after discount is initial price minus the discount.

price after discount = initial price – discount

Also recall that the discount is 35% of the initial price. We can incorporate this into the above equation.

price after discount = initial price – (35% × initial price)

If we let p represent the initial price, we can translate the above equation to:

$$246.68 = p - 0.35p$$

Now solve for p.
$$246.68 = p - 0.35p$$
$$246.68 = 0.65p$$
$$\frac{246.68}{0.65} = \frac{0.65p}{0.65}$$
$$\$379.51 = p$$

Answer: The price before discount for the curio stand is $379.51.

Check: We can reverse the process to verify that the deduction rate is correct. We verify the deduction rate amount by subtracting the final price from the initial price.

$$379.51 - 246.68 = \$132.83$$

Now we can set up a proportion to verify that the discount rate is 35%.
$$\frac{P}{100} = \frac{132.83}{379.51}$$
$$379.51 \cdot P = 100 \cdot 132.83$$
$$379.51P = 13{,}283$$
$$\frac{379.51P}{379.51} = \frac{13{,}283}{379.51}$$
$$P = 35$$

35. **Understand:** We must calculate the sales tax rate, which is a percent of increase. We are given the initial amount (initial price) and the increase amount (sales tax).

Plan: Write a basic percent sentence. Translate to an equation or proportion. Then solve.

Execute: The percent of an initial amount is the amount of increase or decrease. What percent of 54.95 is 3.85?

$$\frac{p}{100} = \frac{3.85}{54.95}$$
$$54.95 \cdot p = 100 \cdot 3.85$$
$$54.95p = 385$$
$$\frac{54.95p}{54.95} = \frac{385}{54.95}$$
$$p \approx 7$$

Answer: The tax rate is 7%.

Check: We can verify that a tax rate of 7% applied to an initial price of $54.95 will equal a sales tax of $3.85. Sales tax is 7% of 54.95.

$$\text{sales tax} = 0.07(54.95) \approx 3.85$$

37. **Understand:** We must calculate the percent of increase. We are given the initial amount (former salary) and the increase amount (raise).

Plan: Write a simple percent sentence. Translate to an equation or proportion. Then solve.

Execute: The percent of an initial amount is the amount of increase or decrease. What percent of 24,260 is 849.10?

$$\frac{P}{100} = \frac{849.10}{24,260}$$
$$24,260 \cdot P = 100 \cdot 849.10$$
$$24,260P = 84,910$$
$$\frac{24,260P}{24,260} = \frac{84,910}{24,260}$$
$$p = 3.5$$

Answer: The percent of increase is 3.5%.

Check: We can verify that a percent of increase of 3.5% applied to a salary of $24,260 will equal a raise of $849.10.

$$\text{amount of increase} = 0.035(24,260) = 849.10$$

39. **Understand:** We must calculate the percent of decrease. We are given the initial price and the decrease price.

Plan: Write a simple percent sentence. Translate to an equation or proportion. Then solve.

Execute: The percent of an initial amount is the amount of decrease. What percent of 18,900 is 1512?

$$\frac{P}{100} = \frac{1512}{18,900}$$
$$18,900 \cdot P = 100 \cdot 1512$$
$$18,900P = 151,200$$
$$\frac{18,900P}{18,900} = \frac{151,200}{18,900}$$
$$P = 8$$

Answer: The percent of decrease is 8%.

Check: We can verify that a percent of decrease of 8% applied to an initial price of $18,900 will equal a discount of $1512. Amount of decrease is 8% of 18,900.

$$\text{amount of decrease} = 0.08(18,900) = 1512$$

41. **Understand:** We must calculate the percent of the increase. We are given Lena's initial hourly wage and her new hourly wage.

Plan: Because the percent of the increase is equal to the ratio of the amount of increase to the initial amount, we must first calculate the amount of increase. The amount can be calculated by subtracting the initial hourly wage from the new hourly wage. We then can write the proportion and solve.

Execute:
amount of increase
= new hourly wage − initial hourly wage
= 10.00 − 8.50
= $1.50

What percent of 8.50 is 1.5?
$$\frac{p}{100} = \frac{1.50}{8.50}$$
$$8.50 \cdot p = 100 \cdot 1.50$$
$$8.50p = 150$$
$$\frac{8.50p}{8.50} = \frac{150}{8.50}$$
$$p \approx 17.6$$

Answer: Lena received a 17.6% raise.

Check: We can verify that a 17.6% raise with a $8.50 initial hourly wage equals a raise of $1.50. The raise amount is 17.6% of 8.50.

$$R = 17.6\% \cdot 8.50$$
$$R = 0.176(8.50)$$
$$R \approx 1.50$$
$$8.5 + 1.50 = 10.00$$

43. **Understand:** We must calculate the percent of the increase. We are given a social workers initial case load and her new case load.

Plan: Because the percent of the increase is equal to the ratio of the amount of increase to the initial amount, we must first calculate the amount of increase. The amount can be calculated by subtracting the initial case load from the new case load. We then can write the proportion and solve.

Execute:

amount of increase

= initial case load − new case loaad

= 1540 − 1259

= 281

What percent of 1259 is 281?

$$\frac{P}{100} = \frac{281}{1259}$$
$$1259 \cdot P = 100 \cdot 281$$
$$1259P = 28,100$$
$$\frac{1259P}{1259} = \frac{28,100}{1259}$$
$$P \approx 22.3$$

Answer: The case load increased by 22.3%.

Check: We can verify that a 22.3% increase with a 1259 initial case load equals an increase of 281. The increase amount is 22.3% of 1259.

$$R = 22.3\% \cdot 1259$$
$$R = 0.223(1259)$$
$$R \approx 281$$
$$1259 + 281 = 1540$$

45. **Understand:** We must calculate the percent of the decrease. We are given DVD player's initial price and its new price.

Plan: Because the percent of the decrease is equal to the ratio of the amount of decrease to the initial amount, we must first calculate the amount of decrease. The amount can be calculated by subtracting the discounted price from the initial price. We then can write the proportion and solve.

Execute: amount of decrease

= initial price − discounted price

= 175.90 − 149.52

= $26.38

What percent of 175.90 is 26.38?

$$\frac{P}{100} = \frac{26.38}{175.90}$$
$$175.90 \cdot P = 100 \cdot 26.38$$
$$175.90P = 2638$$
$$\frac{175.90P}{175.90} = \frac{2638}{175.90}$$
$$P \approx 15$$

Answer: Van received a 15% discount.

Check: We can verify that a 15% discount with a $175.90 initial price equals a discount of 26.38. The discount amount is 15% of 175.90.

$$R = 15\% \cdot 175.90$$
$$R = 0.15(175.90)$$
$$R \approx 26.38$$

47. **Understand:** We must calculate the percent of the decrease. We are given The Jones family's initial bill total and their new bill total.

Plan: Because the percent of the decrease is equal to the ratio of the amount of decrease to the initial amount, we must first calculate the amount of decrease. The amount can be calculated by subtracting the new bill total from the initial bill total. We then can write the proportion and solve.

Execute: amount of decrease

= initial bill total − new bill total

= 1536.98 − 1020.57

= $516.41

What percent of 1536.98 is 516.41?

$$\frac{P}{100} = \frac{516.41}{1536.98}$$
$$1536.98 \cdot P = 100 \cdot 516.41$$
$$1536.98P = 51,641$$
$$\frac{1536.98P}{1536.98} = \frac{51,641}{1536.98}$$
$$P \approx 33.6$$

Answer: The Jones family decreased the total bills by 33.6%.

Check: We can verify that a 33.6% decrease with a $1536.98 initial bill total equals a decrease of $516.43. The decrease amount is 33.6% of 1536.98.

$$R = 33.6\% \cdot 1536.98$$
$$R = 0.336(1536.98)$$
$$R \approx 516.43$$

49. **Understand:** We must calculate the percent of the decrease. We are given the number of homes sold in January and February.

Plan: Because the percent of the decrease is equal to the ratio of the amount of decrease to the initial amount, we must first calculate the amount of decrease. The amount can be calculated by subtracting the number of homes sold in February from the number of homes sold in January. We then can write the proportion and solve.

Execute:
amount of decrease
= homes sold in Jan. – homes sold in Feb.
$= 1173 - 1038$
$= 135$

What percent of 1173 is 135?

$$\frac{P}{100} = \frac{135}{1173}$$
$$1173 \cdot P = 100 \cdot 135$$
$$1173P = 13,500$$
$$\frac{1173P}{1173} = \frac{13,500}{1173}$$
$$P \approx 11.5$$

Answer: The number of homes sold in February decreased by about 11.5%.

Check: We can verify that an 11.5% decrease with a 1173 initial amount equals a decrease of 135. The amount of decrease is 11.5% of 1173.

$$R = 11.5\% \cdot 1173$$
$$R = 0.115(1173)$$
$$R \approx 135$$

51. **Understand:** We must calculate the percent of the decrease. We are given the number of homes sold in April and July.

    Plan: Because the percent of the decrease is equal to the ratio of the amount of decrease to the initial amount, we must first calculate the amount of decrease. The amount can be calculated by subtracting the number of homes sold in July from the number of homes sold in April. We then can write the proportion and solve.

    Execute:
    amount of decrease
    = homes sold in Apr. – homes sold in July
    $= 1121 - 979$
    $= 142$

    What percent of 1121 is 142?

    $$\frac{P}{100} = \frac{142}{1121}$$
    $$1121 \cdot P = 100 \cdot 142$$
    $$1121P = 14,200$$
    $$\frac{1121P}{1121} = \frac{14,200}{1121}$$
    $$P \approx 12.7$$

    Answer: The number of homes sold in July decreased by about 12.7%.

Check: We can verify that a 12.7% decrease with a 1121 initial amount equals a decrease of 142. The amount of decrease is 12.7% of 1121.

$$R = 12.7\% \cdot 1121$$
$$R = 0.127(1121)$$
$$R \approx 142$$

53. **Understand:** We must calculate the percent of increase. We are given the number of homes sold in February and March.

    Plan: Because the percent of the increase is equal to the ratio of the amount of increase to the initial amount, we must first calculate the amount of increase. The amount can be calculated by subtracting the number of homes sold in February from the number of homes sold in March. We then can write the proportion and solve.

    Execute:
    amount of increase
    = homes sold in Mar. – homes sold in Feb.
    $= 1121 - 1038$
    $= 83$
    What percent of 1038 is 83?

    $$\frac{P}{100} = \frac{83}{1038}$$
    $$1038 \cdot P = 100 \cdot 83$$
    $$1038P = 8300$$
    $$\frac{1038P}{1038} = \frac{8300}{1038}$$
    $$P \approx 8$$

    Answer: The number of homes sold in March increased by about 8%.

    Check: We can verify that an 8% increase with a 1038 initial amount equals an increase of 83. The amount of increase is 8% of 1038.

    $$R = 8\% \cdot 1038$$
    $$R = 0.08(1038)$$
    $$R \approx 83$$
    $$1038 + 83 = 1121$$

55. **Understand:** We must calculate the percent of increase. We are given the number of homes sold in July and August.

    Plan: Because the percent of the increase is equal to the ratio of the amount of increase to the initial amount, we must first calculate the amount of increase. The amount can be calculated by subtracting the number of homes sold in July from the number of homes sold in September. We then can write the proportion and solve.

Execute:

amount of increase

= homes sold in Sept. − homes sold in July

= 1022 − 979

= 43

What percent of 979 is 43?

$$\frac{P}{100} = \frac{43}{979}$$

$$979 \cdot P = 100 \cdot 43$$

$$979P = 4300$$

$$\frac{979P}{979} = \frac{4300}{979}$$

$$P \approx 4.4$$

Answer: The number of homes sold in September increased by about 4.4%.

Check: We can verify that a 4.4% increase with a 979 initial amount equals an increase of 43. The amount of increase is 4.4% of 979.

$$R = 4.4\% \cdot 979$$

$$R = 0.044(979)$$

$$R \approx 43$$

$$979 + 43 = 1022$$

57. **Understand:** We must calculate the percent of the decrease. We are given the price of gasoline in January 1997 and January 1998.

Plan: Because the percent of the decrease is equal to the ratio of the amount of decrease to the initial amount, we must first calculate the amount of decrease. The amount can be calculated by subtracting the price of gasoline in 1998 from the price in 1997. We then can write the proportion and solve.

Execute:

amount of decrease

= price gas Jan. 1997 − price gas Jan. 1998

= 122 − 108.9

= 13.1

What percent of 122 is 13.1?

$$\frac{P}{100} = \frac{13.1}{122}$$

$$122 \cdot P = 100 \cdot 13.1$$

$$122P = 1310$$

$$\frac{122P}{122} = \frac{1310}{122}$$

$$P \approx 10.7$$

Answer: The price of a gallon of gasoline decreased by about 10.7% from January 1997 to January 1998.

Check: We can verify that a 10.7% decrease with a 122 initial amount equals a decrease of 13.1. The amount of decrease is 10.7% of 122.

$$R = 10.7\% \cdot 122$$

$$R = 0.107(122)$$

$$R \approx 13.1$$

59. **Understand:** We must calculate the percent of the decrease. We are given the price of gasoline in January 2001 and January 2002.

Plan: Because the percent of the decrease is equal to the ratio of the amount of decrease to the initial amount, we must first calculate the amount of decrease. The amount can be calculated by subtracting the price of gasoline in 2002 from the price in 2001. We then can write the proportion and solve.

Execute:

amount of decrease

= price gas Jan. 2001 − price gas Jan. 2002

= 137.7 − 110.9

= 26.8

What percent of 137.7 is 26.8?

$$\frac{P}{100} = \frac{26.8}{137.7}$$

$$137.7 \cdot P = 100 \cdot 26.8$$

$$137.7P = 2680$$

$$\frac{137.7P}{137.7} = \frac{2680}{137.7}$$

$$P \approx 19.5$$

Answer: The price of a gallon of gasoline decreased by about 19.5% from January 2001 to January 2002.

Check: We can verify that a 19.5% decrease with a 137.7 initial amount equals a decrease of 26.8. The amount of decrease is 19.5% of 137.7.

$$R = 19.5\% \cdot 137.7$$

$$R = 0.195(137.7)$$

$$R \approx 26.8$$

61. **Understand:** We must calculate the percent of increase. We are given the price of gasoline in January 1999 and January 2000.

Plan: Because the percent of the increase is equal to the ratio of the amount of increase to the initial amount, we must first calculate the amount of increase. The amount can be calculated by subtracting the price of gasoline in 1999 from the price in 2000. We then can write the proportion and solve.

Execute:

amount of increase

= price gas Jan. 2000 – price gas Jan. 1999

$= 126 - 91.3$

$= 34.7$

What percent of 91.3 is 34.7?

$$\frac{P}{100} = \frac{34.7}{91.3}$$

$$91.3 \cdot P = 100 \cdot 34.7$$

$$91.3P = 3470$$

$$\frac{91.3P}{91.3} = \frac{3470}{91.3}$$

$$P \approx 38.0$$

Answer: The price of a gallon of gasoline increased by about 38% from January 1999 to January 2000.

Check: We can verify that a 38% increase with a 91.3 initial amount equals an increase of 34.7. The amount of increase is 38% of 91.3.

$R = 38\% \cdot 91.3$

$R = 0.38(91.3)$

$R \approx 34.7$

$91.3 + 34.7 = 126$

63. **Understand:** We must calculate the percent of increase. We are given the price of gasoline in January 2004 and January 2006.

Plan: Because the percent of the increase is equal to the ratio of the amount of increase to the initial amount, we must first calculate the amount of increase. The amount can be calculated by subtracting the price of gasoline in 2004 from the price in 2006. We then can write the proportion and solve.

Execute:

amount of increase

= price gas Jan. 2006 – price gas Jan. 2004

$= 223.6 - 149.2$

$= 74.4$

What percent of 149.2 is 74.4?

$$\frac{P}{100} = \frac{74.4}{149.2}$$

$$149.2 \cdot P = 100 \cdot 74.4$$

$$149.2P = 7440$$

$$\frac{149.2P}{149.2} = \frac{7440}{149.2}$$

$$P \approx 49.9$$

Answer: The price of a gallon of gasoline increased by about 49.9% from January 2004 to January 2006.

Check: We can verify that a 49.9% increase with a 149.2 initial amount equals an increase of 74.4. The amount of increase is 49.9% of 149.2.

$R = 49.9\% \cdot 149.2$

$R = 0.499(149.2)$

$R \approx 74.4$

$149.2 + 74.4 = 223.6$

Review Exercises

1. $\left(\frac{1}{5}\right)^3 = \left(\frac{1}{5}\right)^3 = \frac{1}{125}$ 2. $(1 + 0.08)^2 = (1.08)^2$
$= 1.1664$

3. $\quad 20 = 9(y + 2) - 3y$

$\quad 20 = 9y + 18 - 3y$

$\quad 20 = 6y + 18$

$\quad \underline{-18 = \qquad -18}$

$\qquad 2 = 6y + 0$

$\qquad \dfrac{2}{6} = \dfrac{6y}{6}$

$\qquad \dfrac{1}{3} = y$

Check: $20 = 9\left(\dfrac{1}{3} + 2\right) - 3\left(\dfrac{1}{3}\right)$

$\qquad 20 = 3 + 18 - 1$

$\qquad 20 = 20$

4. $\quad 0.12x = 450 \qquad$ Check: $0.12(3750) = 450$

$\quad \dfrac{0.12x}{0.12} = \dfrac{450}{0.12} \qquad\qquad\qquad 450 = 450$

$\qquad\quad x = 3750$

5.

Categories	Value	Number	Amount
$5 bills	5	$16 - x$	$5(16 - x)$
$10 bills	10	x	$10x$

$\quad 5(16 - x) + 10x = 105$

$\quad\; 80 - 5x + 10x = 105$

$\qquad\quad 80 + 5x = 105$

$\qquad \underline{-80 \qquad\quad = -80}$

$\qquad\quad 0 + 5x = 25$

$\qquad\qquad \dfrac{5x}{5} = \dfrac{25}{5}$

$\qquad\qquad\; x = 5$

Will has 5 ten dollar bills and $16 - 5 = 11$ five dollar bills.

Exercise Set 8.6

1. Principal is an initial amount of money borrowed or invested.

3. $I = Prt$, where I represents simple interest, P represents principal, r represents the simple interest rate, and t represents the time.

5. $B = P\left(1 + \dfrac{r}{n}\right)^{nt}$, where B represents the final balance, P represents the principal, r represents the APR, n represents the number of compound periods per year, and t represents the time in years.

7. simple interest: $I = Prt$
$$I = (4000)(3\%)(1)$$
$$I = (4000)(0.03)(1)$$
$$I = \$120$$

final balance: $4000 + 120 = \$4120$

9. simple interest: $I = Prt$
$$I = (350)(5.5\%)(2)$$
$$I = (350)(0.055)(2)$$
$$I = \$38.50$$

final balance: $350 + 38.50 = \$388.50$

11. simple interest: $I = Prt$
$$I = (12,250)(12.9\%)(5)$$
$$I = (12,250)(0.129)(5)$$
$$I = \$7901.25$$

final balance: $12,250 + 7901.25 = \$20,151.25$.

13. simple interest: $I = Prt$
$$I = (2400)(8\%)\left(\frac{1}{2}\right)$$
$$I = \frac{2400}{1} \cdot \frac{0.08}{1} \cdot \frac{1}{2}$$
$$I = \frac{192}{2}$$
$$I = \$96$$

final balance: $2400 + 96 = \$2496$

15. simple interest: $I = Prt$
$$I = (2000)(6.9\%)\left(\frac{60}{365}\right)$$
$$I = \frac{2000}{1} \cdot \frac{0.069}{1} \cdot \frac{60}{365}$$
$$I = \frac{8280}{365}$$
$$I \approx \$22.68$$

final balance: $2000 + 22.68 = \$2022.68$

17. $$I = Prt$$
$$331.12 = P(0.069)(1)$$
$$331.12 = 0.069P$$
$$\frac{331.12}{0.069} = \frac{0.069P}{0.069}$$
$$\$4798.84 \approx P$$

The principal was $\$4798.84$.

19. $$I = Prt$$
$$15.57 = P(0.032)\left(\frac{3}{12}\right)$$
$$15.57 = P\left(\frac{0.032}{1} \cdot \frac{3}{12}\right)$$
$$15.57 = P\left(\frac{0.096}{12}\right)$$
$$15.57 = 0.008P$$
$$\frac{15.57}{0.008} = \frac{0.008P}{0.008}$$
$$\$1946.25 = P$$

The principal was $\$1946.25$.

21. $$B = P\left(1 + \frac{r}{n}\right)^{nt}$$
$$B = 5000\left(1 + \frac{0.08}{1}\right)^{1 \cdot 2}$$
$$B = 5000(1.08)^2$$
$$B = 5000(1.1664)$$
$$B = \$5832$$

The final balance is $\$5832$.

23. $B = P\left(1+\dfrac{r}{n}\right)^{nt}$

    $B = 840\left(1+\dfrac{0.06}{1}\right)^{1\cdot 2}$

    $B = 840\left(1.06\right)^{2}$

    $B = 840\left(1.1236\right)$

    $B \approx \$943.82$

    The final balance is \$943.82.

25. $B = P\left(1+\dfrac{r}{n}\right)^{nt}$

    $B = 14{,}000\left(1+\dfrac{0.08}{2}\right)^{2\cdot 1}$

    $B = 14{,}000\left(1+0.04\right)^{2}$

    $B = 14{,}000\left(1.04\right)^{2}$

    $B = 14{,}000\left(1.0816\right)$

    $B = \$15{,}142.40$

    The final balance is \$15,142.40.

27. $B = P\left(1+\dfrac{r}{n}\right)^{nt}$

    $B = 400\left(1+\dfrac{0.09}{1}\right)^{1\cdot 3}$

    $B = 400\left(1.09\right)^{3}$

    $B = 400\left(1.1295029\right)$

    $B \approx \$518.01$

    The final balance is \$518.01.

29. $B = P\left(1+\dfrac{r}{n}\right)^{nt}$

    $B = 1600\left(1+\dfrac{0.06}{2}\right)^{2\cdot 3}$

    $B = 1600\left(1+0.03\right)^{6}$

    $B = 1600\left(1.03\right)^{6}$

    $B \approx 1600\left(1.194052297\right)$

    $B \approx \$1910.48$

    The final balance is \$1910.48.

31. $B = P\left(1+\dfrac{r}{n}\right)^{nt}$

    $B = 290\left(1+\dfrac{0.09}{2}\right)^{2\cdot 4}$

    $B = 290\left(1+0.045\right)^{8}$

    $B = 290\left(1.045\right)^{8}$

    $B \approx 290\left(1.422100613\right)$

    $B \approx \$412.41$

    The final balance is \$412.41.

33. $B = P\left(1+\dfrac{r}{n}\right)^{nt}$

    $B = 1300\left(1+\dfrac{0.12}{4}\right)^{4\cdot 2}$

    $B = 1300\left(1+0.03\right)^{8}$

    $B = 1300\left(1.03\right)^{8}$

    $B \approx 1300\left(1.266770081\right)$

    $B \approx \$1646.80$

    The final balance is \$1646.80.

35. $B = P\left(1+\dfrac{r}{n}\right)^{nt}$

    $B = 450\left(1+\dfrac{0.06}{4}\right)^{4\cdot 4}$

    $B = 450\left(1+0.015\right)^{16}$

    $B = 450\left(1.015\right)^{16}$

    $B \approx 450\left(1.268985548\right)$

    $B \approx \$571.04$

    The final balance is \$571.04.

37. $B = P\left(1+\dfrac{r}{n}\right)^{nt}$

    $B = 860.20\left(1+\dfrac{0.169}{365}\right)^{365\cdot\frac{30}{365}}$

    $B \approx 860.20\left(1+0.000463014\right)^{30}$

    $B \approx 860.20\left(1.000463014\right)^{30}$

    $B \approx 860.20\left(1.013984071\right)$

    $B \approx \$872.23$

    The balance on the next bill will be \$872.23.

39. $B = P\left(1 + \dfrac{r}{n}\right)^{nt}$

$B = 694.75\left(1 + \dfrac{0.198}{365}\right)^{365 \cdot \frac{30}{365}}$

$B \approx 694.75\left(1 + 0.000542466\right)^{30}$

$B \approx 694.75\left(1.000542466\right)^{30}$

$B \approx 694.75\left(1.01640263\right)$

$B \approx \$706.15$

The balance on the next bill will be \$706.15.

41. Using the table, we can locate a principal amount of 7,000. Look under the 5 year column and read the corresponding monthly payment. The monthly payment is \$145.31.

43. We can make 17,000 by adding 15,000 + 2000. Look under the 4 year column and add the corresponding monthly payment amounts for each of these principals. The monthly payment will be 373.28 + 49.77 = \$423.05.

45. We can make 92,500 by adding 90,000 + 2000 + 500. Look under the 30 year column and add the corresponding monthly payment amount for each of these principals. The monthly payment will be 724.16 + 16.09 + 4.02 = \$744.27.

47. We can make 84,100 by adding 80,000 + 2 × 2000 + 100. Look under the 15 year column and add the corresponding monthly payment amount for each of these principals. The monthly payment will be 811.41 + (2 × 20.29) + 1.01 = 811.41 + 40.58 + 1.01 = \$853.00.

49. We can make 115,200 by adding 110,000 + 5000 + 200. Look under the 30 year column and add the corresponding monthly payment amount for each of these principals. The monthly payment will be 885.08 + 40.23 + 1.61 = \$926.92.

Review Exercises

1.

2. $\dfrac{5}{6} = 0.8\overline{3}$ 3. $\dfrac{65 + 90 + 72 + 84}{4} = \dfrac{311}{4}$

$= 77.75$

4. $2(-3) + 3(4) = -6 + 12$

$= 6$

5. $\dfrac{3}{8}x = -\dfrac{21}{30}$ Check: $\dfrac{3}{8}\left(-1\dfrac{13}{15}\right) = -\dfrac{21}{30}$

$\dfrac{\cancel{8}}{\cancel{3}} \cdot \dfrac{\cancel{3}}{\cancel{8}}x = -\dfrac{\overset{7}{\cancel{21}}}{\underset{15}{\cancel{30}}} \cdot \dfrac{\cancel{8}^{4}}{\cancel{3}_{1}}$ $\dfrac{\overset{1}{\cancel{3}}}{\underset{2}{\cancel{8}}}\left(-\dfrac{\overset{7}{\cancel{28}}}{\cancel{15}_{5}}\right) = -\dfrac{21}{30}$

$1x = -\dfrac{28}{15}$ $-\dfrac{7}{10} = -\dfrac{7}{10}$

$x = -1\dfrac{13}{15}$

Chapter 8 Review Exercises

1. false   2. false   3. false

4. false   5. true   6. true

7. 100   8. multiply

9. Select a variable for the unknown. Translate *is* to an equal sign. If *of* is preceded by the percent, translate it to multiplication. If *of* is preceded by a whole number, translate it to division.

10. Write the proportion in the form

percent = $\dfrac{\text{part}}{\text{whole}}$, where the percent is expressed as a ratio with a denominator of 100.

11. a) $40\% = \dfrac{40 \div 20}{100 \div 20} = \dfrac{2}{5}$

b) $26\% = \dfrac{26 \div 2}{100 \div 2} = \dfrac{13}{50}$

c) $6.5\% = \dfrac{6.5}{100}$ d) $24\dfrac{1}{2}\% = \dfrac{24\dfrac{1}{2}}{100}$

$= \dfrac{6.5(10)}{100(10)}$ $= 24\dfrac{1}{2} \div 100$

$= \dfrac{65}{1000}$ $= \dfrac{49}{2} \cdot \dfrac{1}{100}$

$= \dfrac{13}{200}$ $= \dfrac{49}{200}$

12. a) $16\% = 16 \div 100 = 0.16$

b) $150\% = 150 \div 100 = 1.5$

c) $3.2\% = 3.2 \div 100 = 0.032$

d) $40\dfrac{1}{3}\% = 40.\overline{3}\% = 40.\overline{3} \div 100 = 0.40\overline{3}$

13. $\dfrac{3}{8} = \dfrac{3}{\overset{25}{\cancel{8}}_2} \cdot \dfrac{\cancel{100}}{1} \%$ 14. $\dfrac{4}{9} = \dfrac{4}{9} \cdot \dfrac{100}{1} \%$

$\qquad = \dfrac{75}{2} \%$ $\qquad = \dfrac{400}{9} \%$

$\qquad = 37\dfrac{1}{2}\% \text{ or } 37.5\%$ $\qquad = 44\dfrac{4}{9}\% \text{ or } 44.\overline{4}\%$

15. $0.54 = (0.54)(100\%)$
$\qquad = 54\%$

16. $1.3 = (1.3)(100\%)$
$\qquad = 130\%$

17. $c = 15\% \cdot 90$ 18. $12.8\% \cdot w = 5.12$
$\quad c = (0.15)(90)$ $\quad 0.128w = 5.12$
$\quad c = 13.5$ $\quad \dfrac{0.128w}{0.128} = \dfrac{5.12}{0.128}$
$\qquad\qquad\qquad\qquad\qquad w = 40$

19. $\qquad 12.5 = p \cdot 20$
$\qquad 12.5 = 20p$
$\qquad \dfrac{12.5}{20} = \dfrac{20p}{20}$
$\qquad 0.625 = 1p$
$\quad 0.625(100\%) = p$
$\qquad 62.5\% = p$

20. $p = \dfrac{40}{150}$ 21. $\qquad \dfrac{40.5}{100} = \dfrac{c}{800}$
$\quad p = 0.2\overline{6}$ $\quad 800 \cdot (40.5) = 100 \cdot c$
$\quad p = (0.2\overline{6})(100\%)$ $\quad 32{,}400 = 100c$
$\quad p = 26.\overline{6}\%$ $\quad \dfrac{32{,}400}{100} = \dfrac{100c}{100}$
$\qquad\qquad\qquad\qquad\qquad 324 = c$

22. $\dfrac{15}{100} = \dfrac{10\frac{1}{2}}{w}$ 23. $\dfrac{P}{100} = \dfrac{8.1}{45}$
$\quad w \cdot 15 = 100 \cdot (10.5)$ $\quad 45 \cdot P = 100 \cdot 8.1$
$\quad 15w = 1050$ $\quad 45P = 810$
$\quad \dfrac{15w}{15} = \dfrac{1050}{15}$ $\quad \dfrac{45P}{45} = \dfrac{810}{45}$
$\qquad w = 70$ $\qquad P = 18\%$

24. $\dfrac{P}{100} = \dfrac{16}{30}$
$\quad 30 \cdot P = 100 \cdot 16$
$\quad 30P = 1600$
$\quad \dfrac{30P}{30} = \dfrac{1600}{30}$
$\qquad P = 53.\overline{3}$

25. a) Simple percent sentence:
65% of 680 is what number?

Word-for-Word Translation Method:
$\quad 65\% \cdot 680 = c$
$\quad 0.65(680) = c$
$\qquad 442 = c$

Proportion Method: $\qquad \dfrac{65}{100} = \dfrac{c}{680}$
$\qquad\qquad 680 \cdot 65 = 100 \cdot c$
$\qquad\qquad 44{,}200 = 100c$
$\qquad\qquad \dfrac{44{,}200}{100} = \dfrac{100c}{100}$
$\qquad\qquad 442 = c$

442 people agreed.

b) Simple percent sentence:
12.9% of 680 is what number?

Word-for-Word Translation Method:
$\quad 12.9\% \cdot 680 = c$
$\quad 0.129(680) = c$
$\qquad 88 \approx c$

Proportion Method: $\qquad \dfrac{12.9}{100} = \dfrac{c}{680}$
$\qquad\qquad 680 \cdot 12.9 = 100 \cdot c$
$\qquad\qquad 8772 = 100c$
$\qquad\qquad \dfrac{8772}{100} = \dfrac{100c}{100}$
$\qquad\qquad 88 \approx c$

88 people had no opinion.

26. Simple percent sentence:
What percent is 321 out of 554?

Word-for-Word Translation Method:
$\quad p = \dfrac{321}{554}$
$\quad p \approx 0.579$
$\quad p \approx 0.579(100\%)$
$\quad p \approx 57.9\%$

Proportion Method: $\qquad \dfrac{P}{100} = \dfrac{321}{554}$
$\qquad\qquad 554 \cdot P = 100 \cdot 321$
$\qquad\qquad 554P = 32{,}100$
$\qquad\qquad \dfrac{554P}{554} = \dfrac{32{,}100}{554}$
$\qquad\qquad P \approx 57.9$

57.9% in the survey supported raising the minimum wage.

27. Simple percent sentence:
180 is 35% or what number?

Word-for-Word Translation Method:
$$180 = 35\% \cdot w$$
$$180 = 0.35w$$
$$\frac{180}{0.35} = \frac{0.35w}{0.35}$$
$$514.29 \approx w$$

Proportion Method: $\dfrac{35}{100} = \dfrac{180}{w}$
$$w \cdot 35 = 100 \cdot 180$$
$$35w = 18,000$$
$$\frac{35w}{35} = \frac{18,000}{35}$$
$$w \approx 514.29$$

The total distance of the trip is about 514.29 mi.

28. The sales tax is 6% of the initial price. What is 6% of 285.75?

$$c = 6\% \cdot 285.75$$
$$c = 0.06(285.75)$$
$$c \approx 17.15$$

The sales tax is $17.15. Because the sales tax increases the initial price, we add the tax to the initial price to get the total amount.

total = 285.75 + 17.15 = $302.90

29. Simple percent sentence:
25% of 56.95 is what number?

Word-for-Word Translation Method:
$$25\% \cdot 56.95 = c$$
$$0.25(56.95) = c$$
$$14.24 = c$$

The discount is $14.24. To find the price after the discount, we must subtract the discount from the initial price. 56.95 − 14.24 = $42.71

30. Simple percent sentence:
1102.5 is 3.5% of what number?

Word-for-Word Translation Method:
$$1102.5 = 3.5\% \cdot w$$
$$1102.5 = 0.035w$$
$$\frac{1102.5}{0.035} = \frac{0.035w}{0.035}$$
$$31,500 = w$$

Proportion Method: $\dfrac{3.5}{100} = \dfrac{1102.5}{w}$
$$w \cdot 3.5 = 100 \cdot (1102.5)$$
$$3.5w = 110,250$$
$$\frac{3.5w}{3.5} = \frac{110,250}{3.5}$$
$$w = 31,500$$

Kat's former salary was $31,500.

31. Simple percent sentence:
What percent is 27.68 out of 86.50

Word-for-Word Translation Method:
$$p = \frac{27.68}{86.50}$$
$$p = 0.32$$
$$p = 0.32(100\%)$$
$$p = 32\%$$

Proportion Method: $\dfrac{P}{100} = \dfrac{27.68}{86.50}$
$$86.50 \cdot P = 100 \cdot 27.68$$
$$86.50P = 2768$$
$$\frac{86.50P}{86.50} = \frac{2768}{86.50}$$
$$P = 32$$

The power bill increased by 32%.

32. We must find the amount of increase. 15.50 − 12 = $3.50.

Simple percent sentence:
What percent is 3.50 out of 12?

$$\frac{P}{100} = \frac{3.50}{12}$$
$$12 \cdot P = 100 \cdot 3.50$$
$$12P = 350$$
$$\frac{12P}{12} = \frac{350}{12}$$
$$P = 29.1\overline{6}$$

Dave's hourly wage increased by about 29.2%.

33. simple interest: $I = Prt$
$$I = (480)(6\%)(1)$$
$$I = (480)(0.06)(1)$$
$$I = \$28.80$$

34. simple interest: $I = Prt$

$$I = (8000)(5.2\%)\left(\frac{3}{12}\right)$$

$$I = \frac{8000}{1} \cdot \frac{0.052}{1} \cdot \frac{3}{12}$$

$$I = \frac{1248}{12}$$

$$I = \$104$$

final balance: $8000 + 104 = \$8104$.

35. Since we know the ending balance, we know that there is 100% of the original deposit plus the 4.5% interest for a total of 104.5%.

$$I = Prt$$
$$627 = P(1.045)(1)$$
$$627 = 1.045P$$
$$\frac{627}{1.045} = \frac{1.045P}{1.045}$$
$$\$600 = P$$

The principal was \$600.

36. $B = P\left(1 + \frac{r}{n}\right)^{nt}$

$$B = 5000\left(1 + \frac{0.08}{1}\right)^{1 \cdot 3}$$

$$B = 5000(1.08)^3$$

$$B = 5000(1.259712)$$

$$B = \$6298.56$$

The final balance is \$6298.56.

37. $B = P\left(1 + \frac{r}{n}\right)^{nt}$

$$B = 1800\left(1 + \frac{0.124}{4}\right)^{4 \cdot 1}$$

$$B = 1800(1 + 0.031)^4$$

$$B = 1800(1.031)^4$$

$$B \approx 1800(1.129886088)$$

$$B \approx \$2033.79$$

The final balance is \$2033.79.

38. We can make 90,500 by adding $90,000 + 500$. Look under the 15 year column and add the corresponding monthly payment amount for each of these principals. The monthly payment will be $912.84 + 5.07 = \$917.91$.

Chapter 8 Practice Test

1. $24\% = \frac{24}{100} = \frac{6}{25}$

2. $40\frac{3}{4}\% = \frac{40\frac{3}{4}}{100}$
$$= \frac{163}{4} \div 100$$
$$= \frac{163}{4} \cdot \frac{1}{100}$$
$$= \frac{163}{400}$$

3. $4.2\% = 4.2 \div 100 = 0.042$

4. $12\frac{1}{2}\% = 12.5 \div 100 = 0.125$

5. $\frac{2}{5} = \frac{2}{5} \cdot \frac{\overset{20}{\cancel{100}}}{1}$
$$= 40\%$$

6. $\frac{5}{9} = \frac{5}{9} \cdot \frac{100}{1}\%$
$$= \frac{500}{9}\%$$
$$= 55\frac{5}{9}\% \text{ or } 55.\overline{5}\%$$

7. $0.26 = 0.26 \cdot 100$
$$= 26\%$$

8. $1.2 = 1.2(100\%)$
$$= 120\%$$

9. $c = 15\% \cdot 76$
$$c = 0.15(76)$$
$$c = 11.4$$

10. $6.5\% \cdot w = 8.32$
$$0.065w = 8.32$$
$$\frac{0.065w}{0.065} = \frac{8.32}{0.065}$$
$$w = 128$$

11. $\frac{P}{100} = \frac{14}{60}$
$$60 \cdot P = 100 \cdot 14$$
$$60P = 1400$$
$$\frac{60P}{60} = \frac{1400}{60}$$
$$P = 23\frac{1}{3}$$
Answer: 23.3%

12. $\frac{P}{100} = \frac{12}{32}$
$$32 \cdot P = 100 \cdot 12$$
$$32P = 1200$$
$$\frac{32P}{32} = \frac{1200}{32}$$
$$P = 37\frac{1}{2}$$
Answer: 37.5%

13. Simple percent sentence:
12% of 50 is what number?

Word-for-Word Translation Method:
$$12\% \cdot 50 = c$$
$$0.12 \cdot 50 = c$$
$$6 = c$$

Proportion Method: $\dfrac{12}{100} = \dfrac{c}{50}$
$$50 \cdot 12 = 100 \cdot c$$
$$600 = 100c$$
$$\dfrac{600}{100} = \dfrac{100c}{100}$$
$$6 = c$$

6 students earned an A.

14. Simple percent sentence:
25% of 1218 is what number?

Word-for-Word Translation Method:
$$25\% \cdot 1218 = c$$
$$0.25 \cdot 1218 = c$$
$$304.5 = c$$

Proportion Method: $\dfrac{25}{100} = \dfrac{c}{1218}$
$$1218 \cdot 25 = 100 \cdot c$$
$$30{,}450 = 100c$$
$$\dfrac{30{,}450}{100} = \dfrac{100c}{100}$$
$$304.5 = c$$

Carolyn's commission was $304.50.

15. Simple percent sentence:
What percent is 345.75 out of 2786.92?

Word-for-Word Translation Method:
$$p = \dfrac{345.75}{2786.92}$$
$$p \approx 0.124$$
$$p \approx 0.124(100\%)$$
$$p \approx 12.4\%$$

Proportion Method: $\dfrac{P}{100} = \dfrac{345.75}{2786.92}$
$$2786.92 \cdot P = 100 \cdot 345.75$$
$$2786.92P = 34{,}575$$
$$\dfrac{2786.92P}{2786.92} = \dfrac{34{,}575}{2786.92}$$
$$P \approx 12.4$$

About 12.4% of the Morgan family's net monthly income goes toward paying for the car.

16. The sales tax is 5% of the initial price. What is 5% of 295.75?

Word-for-Word Translation Method:
$$c = 5\% \cdot 295.75$$
$$c = 0.05(295.75)$$
$$c \approx 14.79$$

Proportion Method: $\dfrac{5}{100} = \dfrac{c}{295.75}$
$$295.75 \cdot 5 = 100 \cdot c$$
$$1478.75 = 100c$$
$$\dfrac{1478.75}{100} = \dfrac{100c}{100}$$
$$14.7875 = c$$
$$14.79 \approx c$$

The sales tax is $14.79. Because the sales tax increases the initial price, we add the tax to the initial price to get the total amount.

total $= 295.75 + 14.79 = \$310.54$

17. If the discount amount is 30% of the initial price, then the final price is 100% – 30% = 70% of the initial price. What amount is 70% of 84.95?

Word-for-Word Translation Method:
$$c = 70\% \cdot 84.95$$
$$c = 0.7(84.95)$$
$$c \approx 59.47$$

Proportion Method: $\dfrac{70}{100} = \dfrac{c}{84.95}$
$$70 \cdot 84.95 = 100 \cdot c$$
$$5946.5 = 100c$$
$$\dfrac{5946.5}{100} = \dfrac{100c}{100}$$
$$59.465 = c$$
$$59.47 \approx c$$

The final price is $59.47.

18. 2.5% of the former salary is the raise amount. 2.5% of what amount is 586.25?

Word-for-Word Translation Method:
$$2.5\% \cdot f = 586.25$$
$$0.025f = 586.25$$
$$\dfrac{0.025f}{0.025} = \dfrac{586.25}{0.025}$$
$$f = 23{,}450$$

Proportion Method:

$$\frac{2.5}{100} = \frac{586.25}{f}$$

$$2.5 \cdot f = 100 \cdot 586.25$$

$$2.5f = 58,625$$

$$\frac{2.5f}{2.5} = \frac{58,625}{2.5}$$

$$f = 23,450$$

Barbara's former salary was $23,450.

19. amount of increase
   = new hourly wage − initial hourly wage
   = 10.50 − 8.75
   = $1.75

   What percent of 8.75 is 1.75?

   $$\frac{P}{100} = \frac{1.75}{8.75}$$

   $$8.75 \cdot P = 100 \cdot (1.75)$$

   $$8.75P = 175$$

   $$\frac{8.75P}{8.75} = \frac{175}{8.75}$$

   $$P = 20$$

   Andre received a 20% raise.

20. simple interest: $I = Prt$

   $$I = (5000)(6\%)(1)$$

   $$I = (5000)(0.06)(1)$$

   $$I = \$300$$

21. simple interest: $I = Prt$

   $$I = (800)(4\%)\left(\frac{6}{12}\right)$$

   $$I = \frac{800}{1} \cdot \frac{0.04}{1} \cdot \frac{6}{12}$$

   $$I = \frac{192}{12}$$

   $$I = \$16$$

   final balance: $800 + 16 = \$816$

22. Since we know the ending balance, we know that there is 100% of the original deposit plus the 6% interest for a total of 106%.

   $$I = Prt$$

   $$2544 = P(1.06)(1)$$

   $$2544 = 1.06P$$

   $$\frac{2544}{1.06} = \frac{1.06P}{1.06}$$

   $$\$2400 = P$$

   The principle was $2400.

23. $B = P\left(1+\dfrac{r}{n}\right)^{nt}$

   $$B = 2000\left(1+\frac{0.09}{1}\right)^{1 \cdot 3}$$

   $$B = 2000(1.09)^3$$

   $$B = 2000(1.295029)$$

   $$B \approx \$2590.06$$

   The final balance is $2590.06.

24. $B = P\left(1+\dfrac{r}{n}\right)^{nt}$

   $$B = 1200\left(1+\frac{0.04}{4}\right)^{4 \cdot 1}$$

   $$B = 1200(1+0.01)^4$$

   $$B = 1200(1.01)^4$$

   $$B \approx 1200(1.04060401)$$

   $$B \approx \$1248.72$$

   The final balance is $1248.72.

25. We can make 112,000 by adding 110,000 + 2000. Look under the 30 year column and add the corresponding monthly payment amount for each of these principles. The monthly payment will be 885.08 + 16.09 = $901.17.

Chapter 8 Cumulative Review Exercises

1. true 2. true 3. true

4. true 5. false 6. false

7. Multiply each term in the second polynomial by each term in the first polynomial. Then combine like terms.

8. Find a common denominator. Rewrite. Add numerators and keep the common denominator. Simplify.

9. Multiply both sides by an appropriate power of 10 as determined by the decimal number with the most decimal places.

10. Multiply the given measurement by unit fractions so that the undesired units divide out.

11. 2.75×10^{10}

12.

13. $7800 \cdot 200 = 1,560,000$: Think $78 \times 2 = 156$ and add 4 zeros.

14. 1

15. $[12+4(6-8)]-(-2)^3+\sqrt{100-36}$

$=[12+4(-2)]-(-2)^3+\sqrt{64}$

$=[12+(-8)]-(-2)^3+\sqrt{64}$

$=4-(-2)^3+\sqrt{64}$

$=4-(-8)+8$

$=4+8+8$

$=20$

16. $8\frac{3}{4}-\left(-5\frac{2}{3}\right)=8\frac{3}{4}+5\frac{2}{3}$

$=8\frac{3\cdot3}{4\cdot3}+5\frac{2\cdot4}{3\cdot4}$

$=8\frac{9}{12}+5\frac{8}{12}$

$=13\frac{17}{12}$

$=13+1\frac{5}{12}$

$=14\frac{5}{12}$

17. $10\frac{5}{8}\div\left(-\frac{3}{16}\right)-\frac{1}{2}=\frac{85}{\cancel{8}}\cdot\left(-\frac{\cancel{16}^2}{3}\right)-\frac{1}{2}$

$=-\frac{170}{3}-\frac{1}{2}$

$=-\frac{170\cdot2}{3\cdot2}-\frac{1\cdot3}{2\cdot3}$

$=-\frac{340}{6}-\frac{3}{6}$

$=-\frac{343}{6}$

$=-57\frac{1}{6}$

18. $4.86\div0.8+58.9=6.075+58.9$

$=64.975$

19. $\left(4\frac{3}{8}\right)(-5.6)=4.375(-5.6)$

$=-24.5$

20. $\frac{3}{4}-(0.6)^2=0.75-0.36$

$=0.39$

21. $\sqrt{(9)(0.25)}=\sqrt{2.25}=1.5$

22. $\sqrt{72}\approx8.49$

23. $12\frac{1}{2}\%=12.5\%$

$=12.5\div100$

$=0.125$

24. $0.71=0.71(100\%)=71\%$

25. $\left(x^4-\frac{1}{3}x^2+9.6\right)-\left(4x^3+\frac{3}{5}x^2-14.6\right)$

$=\left(x^4-\frac{1}{3}x^2+9.6\right)+\left(-4x^3-\frac{3}{5}x^2+14.6\right)$

$=x^4-4x^3-\frac{1}{3}x^2-\frac{3}{5}x^2+9.6+14.6$

$=x^4-4x^3-\frac{5}{15}x^2-\frac{9}{15}x^2+24.2$

$=x^4-4x^3-\frac{14}{15}x^2+24.2$

26. $(7.8x^6)(5.9x^8)=(7.8)(5.9)x^{6+8}$

$=46.02x^{14}$

27. $(6.2x-1)(4x+5)$

$=6.2x\cdot4x+6.2x\cdot5-1\cdot4x-1\cdot5$

$=24.8x^2+31x-4x-5$

$=24.8x^2+27x-5$

28. $540=2^2\cdot3^3\cdot5$

29. $\qquad 24y^2=2^3\cdot3\cdot y^2$

$\qquad 30x=2\cdot3\cdot5\cdot x$

$\text{LCM}(24y^2,30x)=2^3\cdot3\cdot5\cdot xy^2=120xy^2$

30. $\frac{18x^6}{3x^2}=\frac{18}{3}x^{6-2}$

$=6x^4$

31. $32m^4+24m^2-16m$

$=8m\left(\frac{32m^4+24m^2-16m}{8m}\right)$

$=8m\left(\frac{32m^4}{8m}+\frac{24m^2}{8m}-\frac{16m}{8m}\right)$

$=8m(4m^3+3m-2)$

32. $\frac{10m}{9n^2}\div\frac{5}{12n}=\frac{10m}{9n^2}\cdot\frac{12n}{5}$

$=\frac{2\cdot\cancel{5}\cdot m}{\cancel{3}\cdot3\cdot\cancel{n}\cdot n}\cdot\frac{2\cdot2\cdot\cancel{3}\cdot\cancel{n}}{\cancel{5}}$

$=\frac{8m}{3n}$

33. $\dfrac{3}{5} - \dfrac{y}{4} = \dfrac{3 \cdot 4}{5 \cdot 4} - \dfrac{y \cdot 5}{4 \cdot 5}$

$= \dfrac{12}{20} - \dfrac{5y}{20}$

$= \dfrac{12 - 5y}{20}$

34. $\dfrac{5}{8}y - 3 = \dfrac{3}{4}$

$8 \cdot \left(\dfrac{5}{8}y - 3 \right) = 8 \cdot \left(\dfrac{3}{4} \right)$

$\dfrac{\cancel{8}}{1} \cdot \dfrac{5}{\cancel{8}}y - 8 \cdot 3 = \dfrac{\overset{2}{\cancel{8}}}{1} \cdot \dfrac{3}{\cancel{4}}$

$5y - 24 = 6$

$\underline{+24 = +24}$

$5y = 30$

$\dfrac{5y}{5} = \dfrac{30}{5}$

$y = 6$

Check: $\dfrac{5}{\underset{4}{\cancel{8}}} \cdot \dfrac{\overset{3}{\cancel{6}}}{1} - 3 = \dfrac{3}{4}$

$\dfrac{15}{4} - \dfrac{12}{4} = \dfrac{3}{4}$

$\dfrac{3}{4} = \dfrac{3}{4}$

35. $3.5x - 12.1 = 6.8x + 3.08$

$100(3.5x - 12.1) = 100(6.8x + 3.08)$

$100 \cdot 3.5x - 100 \cdot 12.1 = 100 \cdot 6.8x + 100 \cdot 3.08$

$350x - 1210 = 680x + 308$

$\underline{-350x \qquad = -350x}$

$0 - 1210 = 330x + 308$

$\underline{\qquad -308 = \qquad -308}$

$-1518 = 330x + 0$

$\dfrac{-1518}{330} = \dfrac{330x}{330}$

$-4.6 = x$

Check: $3.5(-4.6) - 12.1 = 6.8(-4.6) + 3.08$

$-16.1 - 12.1 = -31.28 + 3.08$

$-28.2 = -28.2$

36. $\dfrac{4\frac{1}{3}}{9} = \dfrac{1\frac{2}{3}}{n}$ Check:

$n \cdot 4\dfrac{1}{3} = 9 \cdot 1\dfrac{2}{3}$

$\dfrac{13}{3}n = \dfrac{\overset{3}{\cancel{9}}}{1} \cdot \dfrac{5}{\cancel{3}}$

$\dfrac{13}{3}n = 15$

$\dfrac{3}{13} \cdot \dfrac{13}{3}n = \dfrac{15}{1} \cdot \dfrac{3}{13}$

$1n = \dfrac{45}{13}$

$n = 3\dfrac{6}{13}$

Check: $\dfrac{4\frac{1}{3}}{9} = \dfrac{1\frac{2}{3}}{3\frac{6}{13}}$

$3\dfrac{6}{13} \cdot 4\dfrac{1}{3} = 9 \cdot 1\dfrac{2}{3}$

$\dfrac{\overset{15}{\cancel{45}}}{\cancel{13}} \cdot \dfrac{\cancel{13}}{\cancel{3}} = \dfrac{\overset{3}{\cancel{9}}}{1} \cdot \dfrac{5}{\cancel{3}}$

$15 = 15$

37. $9.5 \text{ lb.} = \dfrac{9.5 \ \cancel{\text{lb.}}}{1} \cdot \dfrac{16 \text{ oz.}}{1 \ \cancel{\text{lb.}}} = 152 \text{ oz.}$

38. Because l is 3 units to the left of ml, we move the decimal point 3 places to the left in 80.

80 ml = 0.08 l

39. $C = \dfrac{5}{9}(F - 32)$

$C = \dfrac{5}{9}(40 - 32)$

$C = \dfrac{5}{9} \cdot \dfrac{8}{1}$

$C = \dfrac{40}{9}$

$C = 4.\overline{4}°$

40. a) $\dfrac{\left(\begin{array}{c} 67 + 38 + 29 + 29 + 29 + 27 \\ +25 + 25 + 25 + 23 + 22 + 21 \end{array} \right)}{12} = \dfrac{360}{12} = 30$

The mean income of the actors is $30,000,000.

b) Put the incomes in order from least to greatest.

21, 22, 23, 25, 25, 25, 27, 29, 29, 29, 38, 67

Find the mean of the two middle scores, which are 25 and 27.

$\dfrac{25 + 27}{2} = \dfrac{52}{2} = 26$

The median income of the actors is $26,000,000.

41. $A = bh$

$$A = (4x+1)(3x-5)$$
$$A = 4x(3x) + 4x(-5) + 1(3x) + 1(-5)$$
$$A = 12x^2 - 20x + 3x - 5$$
$$A = 12x^2 - 17x - 5$$

Let $x = 3$.
$$A = 12x^2 - 17x - 5$$
$$A = 12(3^2) - 17(3) - 5$$
$$A = 12(9) - 17(3) - 5$$
$$A = 108 - 51 - 5$$
$$A = 52 \text{ in.}^2$$

42.   one integer $= 3 + 2x$
     other integer $= x$

$$(3+2x) + x = 123$$
$$3 + 3x = 123$$
$$\underline{-3 \qquad\quad = -3}$$
$$0 + 3x = 120$$
$$\frac{3x}{3} = \frac{120}{3}$$
$$x = 40$$

One integer is $3 + 2(40) = 83$ and the other integer is 40.

43. $A = bh$
$$201.3 = 16.5 \cdot h$$
$$201.3 = 16.5h$$
$$\frac{201.3}{16.5} = \frac{16.5h}{16.5}$$
$$12.2 \text{ ft.} = h$$

$$12.2 \text{ ft.} \approx \frac{12.2 \;\cancel{\text{ft.}}}{1} \cdot \frac{1 \;\cancel{\text{yd.}}}{3 \;\cancel{\text{ft.}}} \cdot \frac{0.914 \text{ m}}{1 \;\cancel{\text{yd.}}}$$
$$\approx \frac{11.1508}{3} \text{ m}$$
$$\approx 3.7 \text{ m}$$

44.

Categories	Rate	Time	Distance
Liam	6	t	$6t$
Starnell	4	t	$4t$

$$6t + 4t = \frac{1}{4}$$
$$10t = \frac{1}{4}$$
$$\frac{10t}{10} = \frac{\frac{1}{4}}{10}$$
$$t = \frac{1}{4} \div 10$$
$$t = \frac{1}{4} \cdot \frac{1}{10}$$
$$t = \frac{1}{40} \text{ hours}$$

To convert hours to minutes, we must multiply by

60. $\dfrac{1}{40} \times 60 = \dfrac{60}{40} = \dfrac{3}{2} = 1\dfrac{1}{2}$ minutes

They will meet in $1\dfrac{1}{2}$ minutes.

45.   We must subtract the area of the circle from the area of the trapezoid.

$$A = \frac{1}{2} \cdot h \cdot (a+b) - \pi r^2$$
$$A = \frac{1}{2} \cdot 11 \cdot (15+9) - \pi \cdot 4^2$$
$$A = 0.5(11)(24) - \pi \cdot 16$$
$$A = 132 - 16\pi$$
$$A \approx 132 - 16(3.14)$$
$$A \approx 132 - 50.24$$
$$A \approx 81.76 \text{ cm}^2$$

46. $V = \pi r^2 h$
$$V = \pi \cdot (1.2)^2 \cdot 4$$
$$V = 1.44 \cdot 4\pi$$
$$V = 5.76\pi$$
$$V \approx 5.76(3.14)$$
$$V \approx 18.1 \text{ in.}^3$$

47.   There are 12 favorable outcomes out of 480 possible outcomes. $P = \dfrac{12}{480} = \dfrac{1}{40}$

48.

Smaller triangle		Larger triangle
7 cm	$\leftrightarrow$	9 cm
12.5 cm	$\leftrightarrow$	x
15 cm	$\leftrightarrow$	y

$$\frac{7}{9} = \frac{12.5}{x} \qquad \frac{7}{9} = \frac{15}{y}$$

$$x \cdot 7 = 9 \cdot (12.5) \qquad y \cdot 7 = 9 \cdot 15$$

$$7x = 112.5 \qquad 7y = 135$$

$$\frac{7x}{7} = \frac{112.5}{7} \qquad \frac{7y}{7} = \frac{135}{7}$$

$$x \approx 16.1 \text{ cm} \qquad y \approx 19.3 \text{ cm}$$

49. Basic percent sentence:
What percent is 384 out of 450?

$$p = \frac{384}{450}$$

$$p = 0.85\overline{3}$$

$$p = (0.85\overline{3})(100\%)$$

$$p = 85\frac{1}{3}\% \text{ or } 85.\overline{3}\%$$

$85.\overline{3}\%$ of the people in attendance are over 50.

50. $B = P\left(1 + \dfrac{r}{n}\right)^{nt}$

$$B = 3200\left(1 + \frac{0.06}{2}\right)^{2 \cdot 2}$$

$$B = 3200(1 + 0.03)^4$$

$$B = 3200(1.03)^4$$

$$B = 3200(1.12550881)$$

$$B \approx \$3601.63$$

The balance will be \$3601.63

Chapter 9 More with Geometry and Graphs

Exercise Set 9.1

1. A point is a position in space having no length or width.

3. The measure of an acute angle is between 0° and 90°, whereas the measure of an obtuse angle is between 90° and 180°.

5. The consecutive interior angles are supplementary.

7. line $\overleftrightarrow{XY}$

9. ray $\overrightarrow{MN}$

11. point J

13. plane R

15. line segment $\overline{TU}$

17. parallel lines a and b, or $a \parallel b$

19. The angle is obtuse because its measure is between 90° and 180°.

21. The angle is straight because its measure is 180°.

23. The angle is right because its measure is 90°.

25. The angle is acute because its measure is between 0° and 90°.

27. congruent; They are vertical angles.

29. congruent; They are alternate interior angles.

31. supplementary: They are consecutive exterior angles.

33. supplementary; They form a straight angle.

35. congruent; They are alternate exterior angles.

37. supplementary: They are consecutive interior angles.

39. congruent; They are corresponding angles.

41. a) Because $\angle 1$ and $\angle 3$ form a straight angle, they are supplementary, so $\angle 1 + \angle 3 = 180°$.

$$52° + \angle 3 = 180°$$
$$\underline{-52°\qquad\quad -52°}$$
$$0 + \angle 3 = 128°$$
$$\angle 3 = 128°$$

   b) Because $\angle 1$ and $\angle 4$ are vertical, they are congruent, so $\angle 1 \cong \angle 4 = 52°$.

c) Because $\angle 1$ and $\angle 5$ are corresponding angles, they are congruent, so $\angle 1 \cong \angle 5 = 52°$.

d) Because $\angle 5$ and $\angle 6$ form a straight angle, they are supplementary, so $\angle 5 + \angle 6 = 180°$.

$$52° + \angle 6 = 180°$$
$$\underline{-52°\qquad\quad -52°}$$
$$0 + \angle 6 = 128°$$
$$\angle 6 = 128°$$

43. a) Because $\angle 2$ and $\angle 4$ form a straight angle, they are supplementary, so $\angle 2 + \angle 4 = 180°$.

$$110° + \angle 4 = 180°$$
$$\underline{-110°\qquad\quad -110°}$$
$$0 + \angle 4 = 70°$$
$$\angle 4 = 70°$$

   b) Because $\angle 4$ and $\angle 5$ are alternate interior angles, they are congruent, so $\angle 4 \cong \angle 5 = 70°$.

c) Because $\angle 2$ and $\angle 7$ are alternate exterior angles, they are congruent, so $\angle 2 \cong \angle 7 = 110°$.

d) Because $\angle 2$ and $\angle 8$ form a consecutive exterior angles, they are supplementary, so $\angle 2 + \angle 8 = 180°$.

$$110° + \angle 8 = 180°$$
$$\underline{-110°\qquad\quad -110°}$$
$$0 + \angle 8 = 70°$$
$$\angle 8 = 70°$$

45. **Understand:** Because the two angles in the variable expressions form a straight angle, they are supplementary, so the sum of their measures is 180°. Also, because the two intersecting lines form two pairs of vertical angles, the measures of the other two angles will be equal to x and $3x + 32$ as well.

Plan: Translate to an equation, then solve. The sum of x and $3x + 32$ is 180.

Execute: $x + 3x + 32 = 180$
$$4x + 32 = 180$$
$$4x = 148$$
$$x = 37$$

Answer: The measure of the smaller angle is 37°. To find the measure of the larger angle, we replace x in $3x + 32$ with 37.

$3(37) + 32 = 111 + 32 = 143$

The two angles are 37° and 143°. Because the two intersecting lines form two pair of vertical angles, the measures of the other two angles are also 37° and 143°.

Check: The sum of the measures of the supplementary angles is 37° + 143° = 180°.

47. **Understand:** Because the two angles in the variable expressions form a straight angle, they are supplementary, so the sum of their measures is 180°. Also, because the two intersecting lines form two pairs of vertical angles, the measures of the other two angles will be equal to y and $0.5y - 9$ as well.

    Plan: Translate to an equation, then solve. The sum of y and $0.5y - 9$ is 180.

    Execute: $y + 0.5y - 9 = 180$
    $$1.5y - 9 = 180$$
    $$1.5y = 189$$
    $$y = 126$$

    Answer: The measure of the larger angle is 126°. To find the measure of the smaller angle, we replace y in $0.5y - 9$ with 126.
    $0.5(126) - 9 = 63 - 9 = 54$

    The two angles are 126° and 54°. Because the two intersecting lines form two pair of vertical angles, the measures of the other two angles are also 126° and 54°.

    Check: The sum of the measures of the supplementary angles is 126° + 54° = 180°.

49. **Understand:** The variable expressions represent the measures of angles that are consecutive interior angles, so they are supplementary.

    Plan: Translate to an equation, then solve. The sum of x and $2x - 21$ is 180.

    Execute: $x + 2x - 21 = 180$
    $$3x - 21 = 180$$
    $$3x = 201$$
    $$x = 67$$

    Answer: The measure of the smaller angle is 67°. To find the measure of the larger angle, we replace x in $2x - 21$ with 67.
    $2(67) - 21 = 134 + 21 = 113$

    The two angles are 67° and 113°. We can find all the other angles using the rules concerning angles formed by a transversal intersecting two parallel lines.

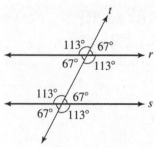

Check: The sum of the measures of two consecutive interior angles is 67° + 113° = 180°.

51. **Understand:** The variable expressions represent the measures of angles that are consecutive exterior angles, so they are supplementary.

    Plan: Translate to an equation, then solve. The sum of y and $0.4y + 15.5$ is 180.

    Execute: $y + 0.4y + 15.5 = 180$
    $$1.4y + 15.5 = 180$$
    $$1.4y = 164.5$$
    $$y = 117.5$$

    Answer: The measure of the larger angle is 117.5°. To find the measure of the larger angle, we replace y in $0.4y + 15.5$ with 117.5.
    $0.4(117.5) + 15.5 = 47 + 15.5 = 62.5$

    The two angles are 117.5° and 62.5°. We can find all the other angles using the rules concerning angles formed by a transversal intersecting two parallel lines.

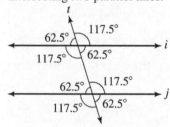

Check: The sum of the measures of two consecutive interior angles is 117.5° + 62.5° = 180°.

53. acute triangle; The measure of each angle is less than 90°.

55. obtuse triangle; The measure of one angle is greater than 90°.

57. right triangle; The measure of one angle is 90°.

59. **Understand:** The sum of the measures of the angles is 180°.

    Plan: Translate to an equation, then solve. The sum of x, $0.5x + 16$, and $x - 46$ is 180.

Execute: $x + 0.5x + 16 + x - 46 = 180$

$$2.5x - 30 = 180$$
$$2.5x = 210$$
$$x = 84$$

Answer: The measure of the angle labeled x is 84°. To find the measure of the other two angles, we replace x in $0.5x + 16$ and $x - 46$ with 84.

$0.5(84) + 16 = 42 + 16 = 58$

$84 - 46 = 38$

The measures of the angles are 84°, 58°, and 38°.

Check: The sum of the measures of the angles in the triangle is 84° + 58° + 38° = 180°.

61. **Understand:** The sum of the measures of the angles is 180°.

Plan: Translate to an equation, then solve. The sum of n, $0.2n - 0.6$, and $0.3n - 11.4$ is 180.

Execute: $n + 0.2n - 0.6 + 0.3n - 11.4 = 180$

$$1.5n - 12 = 180$$
$$1.5n = 192$$
$$n = 128$$

Answer: The measure of the angle labeled n is 128°. To find the measure of the other two angles, we replace n in $0.2n - 0.6$ and $0.3n - 11.4$ with 128.

$0.2(128) - 0.6 = 25.6 - 0.6 = 25$

$0.3(128) - 11.4 = 38.4 - 11.4 = 27$

The measures of the angles are 128°, 25°, and 27°.

Check: The sum of the measures of the angles in the triangle is 128° + 25° + 27° = 180°.

Review Exercises

1. Mean: $\dfrac{(84 + 98 + 46 + 72 + 88 + 98)}{6} = \dfrac{486}{6} = 81$

Median: Arrange scores in order from least to greatest, then take the mean of the middle two scores which are 84 and 88.

46, 72, 84, 88, 98, 98

$\dfrac{84 + 88}{2} = \dfrac{172}{2} = 86$

2. ![number line from -4 to 4 with point at 3]

3. ![number line from -4 to 4 with point at -4]

4. $x + 2y = 3 + 2(-5) = 3 - 10 = -7$

5. $x + 2y = -1 + 2(-2) = -1 - 4 = -5$

Exercise Set 9.2

1.

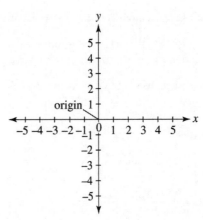

3. Beginning at the origin, move to the left or right along the x-axis the amount indicated by the first coordinate. From that position on the x-axis, move up or down by the amount indicated by the second coordinate.

5. Both coordinates are negative.

7. A: $(2, 3)$ B: $(-3, 0)$ C: $(-1, -3)$ D: $(3, -4)$

9. A: $(-4, 1)$ B: $(0, 3)$ C: $(-3, -3)$ D: $(3, -5)$

11.

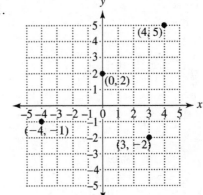

13.

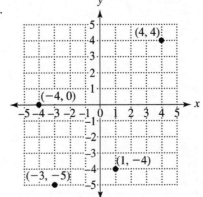

15. Quadrant II because the first coordinate is negative and the second coordinate is positive.

17. Quadrant I because both coordinates are positive.

19. Quadrant IV because the first coordinate is positive and the second coordinate is negative.

21. Quadrant III because both coordinates are negative.

23. $(0, -34)$ is not in a quadrant, it is on the y-axis.

25. $(39, 0)$ is not in a quadrant, it is on the x-axis.

27. midpoint $= \left(\dfrac{x_1 + x_2}{2}, \dfrac{y_1 + y_2}{2} \right)$

$= \left(\dfrac{0 + 8}{2}, \dfrac{6 + 0}{2} \right)$

$= (4, 3)$

29. midpoint $= \left(\dfrac{x_1 + x_2}{2}, \dfrac{y_1 + y_2}{2} \right)$

$= \left(\dfrac{3 + 7}{2}, \dfrac{-1 + 2}{2} \right)$

$= \left(5, \dfrac{1}{2} \right)$

31. midpoint $= \left(\dfrac{x_1 + x_2}{2}, \dfrac{y_1 + y_2}{2} \right)$

$= \left(\dfrac{2 + (-3)}{2}, \dfrac{-6 + 8}{2} \right)$

$= \left(-\dfrac{1}{2}, 1 \right)$

33. midpoint $= \left(\dfrac{x_1 + x_2}{2}, \dfrac{y_1 + y_2}{2} \right)$

$= \left(\dfrac{0 + (-7)}{2}, \dfrac{-8 + 11}{2} \right)$

$= \left(-\dfrac{7}{2}, \dfrac{3}{2} \right)$

35. midpoint $= \left(\dfrac{x_1 + x_2}{2}, \dfrac{y_1 + y_2}{2} \right)$

$= \left(\dfrac{-9 + (-5)}{2}, \dfrac{-3 + (-2)}{2} \right)$

$= \left(-7, -\dfrac{5}{2} \right)$

37. midpoint $= \left(\dfrac{x_1 + x_2}{2}, \dfrac{y_1 + y_2}{2} \right)$

$= \left(\dfrac{1.5 + (-4.5)}{2}, \dfrac{-9 + 2.2}{2} \right)$

$= (-1.5, -3.4)$

39. midpoint $= \left(\dfrac{x_1 + x_2}{2}, \dfrac{y_1 + y_2}{2} \right)$

$= \left(\dfrac{3\frac{1}{4} + 5}{2}, \dfrac{8 + 2\frac{1}{2}}{2} \right)$

$= \left(4\dfrac{1}{8}, 5\dfrac{1}{4} \right)$

Review Exercises

1.

2. $2x - 3y = 2 \cdot 4 - 3 \cdot 2 = 8 - 6 = 2$

3. $-4x^2 - 5y = -4(-3)^2 - 5 \cdot 0$

$= -4(9) - 0$

$= -36 - 0$

$= -36$

4. $2x - 8 = 10$ Check: $2(9) - 8 = 10$

$\underline{+8 = +8}$ $18 - 8 = 10$

$2x + 0 = 18$ $10 = 10$

$\dfrac{2x}{2} = \dfrac{18}{2}$

$x = 9$

5. $-\dfrac{5}{6} y = \dfrac{3}{4}$ Check: $-\dfrac{\cancel{6}^{\,1}}{\cancel{2}\cancel{6}} \cdot -\dfrac{\cancel{9}^{\,3}}{\cancel{10}_{\,2}} = \dfrac{3}{4}$

$-\dfrac{\cancel{6}}{\cancel{5}} \cdot -\dfrac{\cancel{5}}{\cancel{6}} y = \dfrac{3}{\cancel{4}_{\,2}} \cdot -\dfrac{\cancel{6}^{\,3}}{5}$ $\dfrac{3}{4} = \dfrac{3}{4}$

$y = -\dfrac{9}{10}$

Exercise Set 9.3

1. Answers may vary, but $x + y = 5$ is one possibility.

3. 1. Replace one of the corresponding variables with a value (any value).
    2. Solve the equation for the other variable.

5. a straight line

7. a point where a graph intersects the x-axis

9. 0

11. Replace x with 2 and y with 3, and see if the equation is true.

$x + 2y = 8$ This checks which means $(2,3)$
$2 + 2(3)?8$ is a solution for $x + 2y = 8$.
$2 + 6?8$
$8 = 8$

13. Replace x with -5 and y with 2, and see if the equation is true.

$y - 4x = 3$ Because the equation is not true,
$2 - 4(-5)?3$ $(-5,2)$ is not a solution for
$2 - (-20)?3$ $y = 2x - 1$
$2 + 20?3$
$22 \neq 3$

15. Replace x with 9 and y with 0, and see if the equation is true.

$y = -2x + 18$ This checks which means $(9,0)$
$0? - 2 \cdot 9 + 18$ is a solution for $y = -2x + 18$.
$0? - 18 + 18$
$0 = 0$

17. Replace x with 6 and y with -2, and see if the equation is true.

$y = -\dfrac{2}{3}x$ Because the equation is not true,
$-2? - \dfrac{2}{\cancel{3}} \cdot \dfrac{\cancel{6}^2}{1}$ $(6, -2)$ is not a solution for
$-2 \neq -4$ $y = -\dfrac{2}{3}x$.

19. Replace x with $-1\dfrac{2}{5}$ and y with 0, and see if the equation is true.

$y - 3x = 5$ Because the equation is not
$0 - 3\left(-1\dfrac{2}{5}\right)?5$ true, $\left(-1\dfrac{2}{5}, 0\right)$ is not a
$0 - \left(\dfrac{3}{1} \cdot \dfrac{-7}{5}\right)?5$ solution for $y - 3x = 5$.
$0 - \left(-\dfrac{21}{5}\right)?5$
$0 + \dfrac{21}{5}?5$
$\dfrac{21}{5} \neq 5$

21. Replace x with 2.2 and y with -11.2, and see if the equation is true.

$y + 6x = -2$
$-11.2 + 6(2.2)? - 2$
$-11.2 + 13.2? - 2$
$2 \neq -2$

Because the equation is not true, $(2.2, -11.2)$ is not a solution for $y + 6x = -2$.

23. Individual solution points may vary but should all lie on the same line. Some possible solutions are $(2, -6), (4, -4),$ and $(5, -3)$.

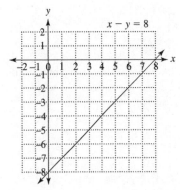

25. Individual solution points may vary but should all lie on the same line. Some possible solutions are $(1,4), (2,2),$ and $(3,0)$.

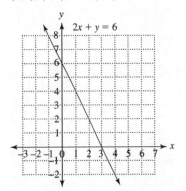

27. Individual solution points may vary but should all lie on the same line. Some possible solutions are $(0, 0), (1, 1),$ and $(2, 2)$.

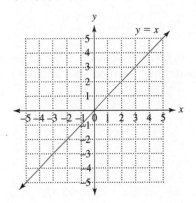

29. Individual solution points may vary but should all lie on the same line. Some possible solutions are (0, 0), (1, 2), and (2, 4).

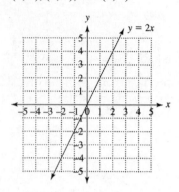

31. Individual solution points may vary but should all lie on the same line. Some possible solutions are (−1, 5), (0, 0), and (1, −5).

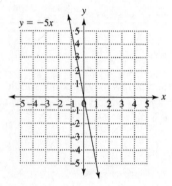

33. Individual solution points may vary but should all lie on the same line. Some possible solutions are (0, −3), (1, −2), and (2, −1).

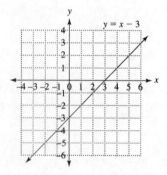

35. Individual solution points may vary but should all lie on the same line. Some possible solutions are (0, 4), (1, 2), and (2, 0).

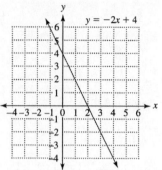

37. Individual solution points may vary but should all lie on the same line. Some possible solutions are (−1, −1), (0, 2), and (1, 5).

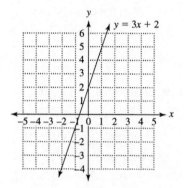

39. Individual solution points may vary but should all lie on the same line. Some possible solutions are (0, 0), (2, 1), and (4, 2).

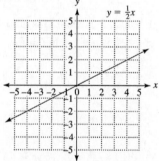

41. Individual solution points may vary but should all lie on the same line. Some possible solutions are (0, 4), (3, 2), and (6, 0).

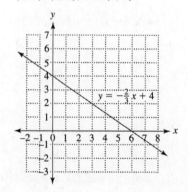

43. Individual solution points may vary but should all lie on the same line. Some possible solutions are $(1, -4)$, $(2, 0)$, and $(3, 4)$.

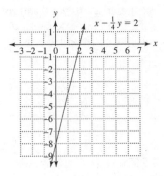

45. Individual solution points may vary but should all lie on the same line. Some possible solutions are $(-5, -4.5)$, $(0, -2.5)$, and $(5, -0.5)$.

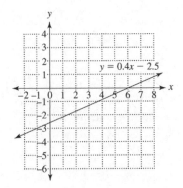

47. Individual solution points may vary but should all lie on the same line. Some possible solutions are $(0, -5)$, $(1, -5)$, and $(4, -5)$.

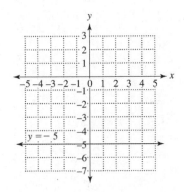

49. Individual solution points may vary but should all lie on the same line. Some possible solutions are $(7, 0)$, $(7, 2)$, and $(7, 5)$.

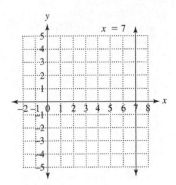

51. To find the x-intercept, replace y with 0 and solve the equation for x. To find the y-intercept, replace the x with 0 and solve the equation for y.

x-intercept: $x + 2y = 12$ $(12, 0)$
$$x + 2 \cdot 0 = 12$$
$$x + 0 = 12$$
$$x = 12$$

y-intercept: $x + 2y = 12$ $(0, 6)$
$$0 + 2y = 12$$
$$2y = 12$$
$$\frac{2y}{2} = \frac{12}{2}$$
$$y = 6$$

53. To find the x-intercept, replace y with 0 and solve the equation for x. To find the y-intercept, replace the x with 0 and solve the equation for y.

x-intercept: $5x - 4y = 20$ $(4, 0)$
$$5x - 4 \cdot 0 = 20$$
$$5x - 0 = 20$$
$$5x = 20$$
$$\frac{5x}{5} = \frac{20}{5}$$
$$x = 4$$

y-intercept: $5x - 4y = 20$ $(0, -5)$
$$5 \cdot 0 - 4y = 20$$
$$0 - 4y = 20$$
$$-4y = 20$$
$$\frac{-4y}{-4} = \frac{20}{-4}$$
$$y = -5$$

55. To find the x-intercept, replace y with 0 and solve the equation for x. To find the y-intercept, replace the x with 0 and solve the equation for y.

x-intercept: $\dfrac{3}{4}x - y = -6$ $(-8,0)$

$$\dfrac{3}{4}x - 0 = -6$$

$$\dfrac{3}{4}x = -6$$

$$\dfrac{4}{3} \cdot \dfrac{3}{4}x = \dfrac{-{}^2\cancel{6}}{1} \cdot \dfrac{4}{\cancel{3}}$$

$$x = -8$$

y-intercept: $\dfrac{3}{4}x - y = -6$ $(0,6)$

$$\dfrac{3}{4} \cdot 0 - y = -6$$

$$0 - y = -6$$

$$-1y = -6$$

$$\dfrac{-1y}{-1} = \dfrac{-6}{-1}$$

$$y = 6$$

57. To find the x-intercept, replace y with 0 and solve the equation for x. To find the y-intercept, replace the x with 0 and solve the equation for y.

x-intercept: $6.5x + 2y = 1.3$ $(0.2,0)$

$$6.5x + 2 \cdot 0 = 1.3$$

$$6.5x + 0 = 1.3$$

$$6.5x = 1.3$$

$$\dfrac{6.5x}{6.5} = \dfrac{1.3}{6.5}$$

$$x = 0.2$$

y-intercept: $6.5x + 2y = 1.3$ $(0,0.65)$

$$6.5 \cdot 0 + 2y = 1.3$$

$$0 + 2y = 1.3$$

$$2y = 1.3$$

$$\dfrac{2y}{2} = \dfrac{1.3}{2}$$

$$y = 0.65$$

59. To find the x-intercept, replace y with 0 and solve the equation for x.

x-intercept: $y = 6x$ $(0,0)$

$$0 = 6x$$

$$\dfrac{0}{6} = \dfrac{6x}{6}$$

$$0 = x$$

Since the x-intercept is the origin, $(0,0)$. In fact the origin is the only point in the rectangular coordinate system where a line can pass through both the x-axis and y-axis simultaneously.

Therefore both the x- and y-intercept coordinates are $(0,0)$.

61. To find the x-intercept, replace y with 0 and solve the equation for x. To find the y-intercept, replace the x with 0 and solve the equation for y.

x-intercept: $y = -4x + 1$ $\left(\dfrac{1}{4},0\right)$

$$0 = -4x + 1$$

$$\dfrac{-1}{} \qquad \dfrac{-1}{}$$

$$-1 = -4x + 0$$

$$\dfrac{-1}{-4} = \dfrac{-4x}{-4}$$

$$\dfrac{1}{4} = x$$

y-intercept: $y = -4x + 1$ $(0,1)$

$$y = -4 \cdot 0 + 1$$

$$y = 0 + 1$$

$$y = 1$$

63. To find the x-intercept, replace y with 0 and solve the equation for x. To find the y-intercept, replace the x with 0 and solve the equation for y.

x-intercept: $y = \dfrac{1}{5}x - 5$ $(25,0)$

$$0 = 0.2x - 5$$

$$\underline{+5} \qquad \underline{+5}$$

$$5 = 0.2x + 0$$

$$5 = 0.2x$$

$$\dfrac{5}{0.2} = \dfrac{0.2x}{0.2}$$

$$25 = x$$

y-intercept: $y = \dfrac{1}{5}x - 5$ $(0,-5)$

$$y = \dfrac{1}{5} \cdot 0 - 5$$

$$y = 0 - 5$$

$$y = -5$$

65. The graph of $x = -9$ is a vertical line parallel to the y-axis that passes through the x-axis at the point $(-9,0)$. Notice that this point is the x-intercept. Because the line is parallel to the y-axis, it will never intersect the y-axis. Therefore there is no y-intercept.

67. The graph of $y = 4$ is a horizontal line parallel to the x-axis that passes through the y-axis at the point $(0,4)$. Notice that this point is the y-intercept. Because the line is parallel to the x-

axis, it will never intersect the *x*-axis. Therefore there is no *x*-intercept.

69. The graph of $y = -3$ is a horizontal line parallel to the *x*-axis that passes through the *y*-axis at the point $(0, -3)$. Notice that this point is the *y*-intercept. Because the line is parallel to the *x*-axis, it will never intersect the *x*-axis. Therefore there is no *x*-intercept.

Review Exercises

1. $-32t + 70 = -32(4.2) + 70$
    $$= -134.4 + 70$$
    $$= -64.4$$

2. To find the perimeter, we must add all the sides.
    $10.4 + 8.2 + 6.1 + 6.3 + 9 = 40$ m

3. $A = bh$
    $$A = \left(6\frac{1}{2}\right)(7)$$
    $$A = \frac{13}{2} \cdot \frac{7}{1}$$
    $$A = \frac{91}{2}$$
    $$A = 45\frac{1}{2} \text{ ft.}^2$$

4. $A = \frac{1}{2}bh$
    $$A = 0.5(4.2)(2.7)$$
    $$A = 5.67 \text{ in.}^2$$

5. $A = \frac{1}{2}h(a+b)$
    $$A = \frac{1}{2} \cdot 6 \cdot (11 + 7.5)$$
    $$A = \frac{1}{2} \cdot \frac{6}{1} \cdot \frac{18.5}{1}$$
    $$A = \frac{111}{2}$$
    $$A = 55.5 \text{ ft.}^2$$

Exercise Set 9.4

1. When the object stops, its velocity is 0. Therefore, we replace *v* with 0 and solve for *t*.

3. a square

5. a) Initial velocity means the velocity at the time the rocket is released, which is 0 seconds. Let $t = 0$.

    $$v = -32.2t + 600$$
    $$v = -32.2(0) + 600$$
    $$v = 0 + 600$$
    $$v = 600 \text{ ft./sec.}$$

    b) We are given a time of 3 seconds. Let $t = 3$.

$$v = -32.2t + 600$$
$$v = -32.2(3) + 600$$
$$v = -96.6 + 600$$
$$v = 503.4 \text{ ft./sec.}$$

c) We must calculate the time at which the rocket stops in midair. After stopping, the rocket will then reverse direction and begin falling back toward the ground. When an object is at a stop, it is not moving and therefore has no velocity. We must solve for *t* when the velocity, *v* is 0. Let $v = 0$ and solve for *t*.

$$v = -32.2t + 600$$
$$0 = -32.2t + 600$$
$$\underline{-600 = \qquad\qquad -600}$$
$$-600 = -32.2t + 0$$
$$\frac{-600}{-32.2} = \frac{-32.2t}{-32.2}$$
$$18.6 \text{ sec.} \approx t$$

d)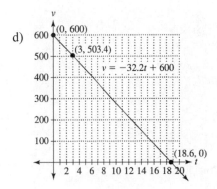

7. a) The revenue is \$250,000, so let $r = 250,000$.

    $$p = 0.24r - 45,000$$
    $$p = 0.24 \cdot 250,000 - 45,000$$
    $$p = 60,000 - 45,000$$
    $$p = \$15,000$$

    b) Let $p = 0$ and solve for *r*.

    $$p = 0.24r - 45,000$$
    $$0 = 0.24r - 45,000$$
    $$\underline{+45,000 = \qquad\qquad +45,000}$$
    $$45,000 = 0.24r + 0$$
    $$\frac{45,000}{0.24} = \frac{0.24r}{0.24}$$
    $$\$187,500 = r$$

c)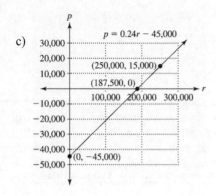

9. a) The initial balance is the amount you invest. Let $t = 0$.

$b = 13.5t + 300$
$b = 13.5 \cdot 0 + 300$
$b = 0 + 300$
$b = \$300$

b) Let $t = 5$.

$b = 13.5t + 300$
$b = 13.5 \cdot 5 + 300$
$b = 67.5 + 300$
$b = \$367.50$

c) Let $t = 20$.

$b = 13.5t + 300$
$b = 13.5 \cdot 20 + 300$
$b = 270 + 300$
$b = \$570$

d)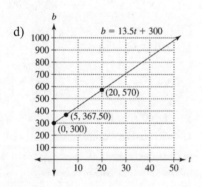

11. a) To find where the graph crosses the F-axis, we must let $C = 0$ and solve for F.

$F = \dfrac{9}{5}C + 32$

$F = \dfrac{9}{5} \cdot 0 + 32$

$F = 0 + 32$

$F = 32$

The F-intercept is $(0, 32)$.

b) To find where the graph crosses the C-axis, we must let $F = 0$ and solve for C.

$$F = \frac{9}{5}C + 32$$

$$0 = \frac{9}{5}C + 32$$

$$\underline{-32} \qquad \underline{-32}$$

$$-32 = \frac{9}{5}C + 0$$

$$\frac{5}{9} \cdot \frac{-32}{1} = \frac{9}{5}C \cdot \frac{5}{9}$$

$$-\frac{160}{9} = 1C$$

$$-17\frac{7}{9} = C$$

The C-intercept is $\left(-17\dfrac{7}{9}, 0\right)$.

c) Let $C = 40$.

$$F = \frac{9}{5}C + 32$$

$$F = \frac{9}{\cancel{5}} \cdot \frac{\overset{8}{\cancel{40}}}{1} + 32$$

$$F = 72 + 32$$

$$F = 104°$$

d)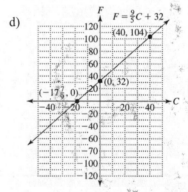

13. a) The labor cost would be the hourly rate times the number of hours ($40h$). The total cost T is the sum of labor cost and flat fee for the visit.

$T = 40h + 75$

b) Let $h = 2.5$

$T = 40h + 75$
$T = 40(2.5) + 75$
$T = 100 + 75$
$T = \$175$

c) Let $T = 200$.

$$T = 40h + 75$$
$$200 = 40h + 75$$
$$\underline{-75 \qquad\quad -75}$$
$$125 = 40h + \ 0$$
$$\frac{125}{40} = \frac{40h}{40}$$
$$3\frac{1}{8} \text{ hr.} = h$$

d)

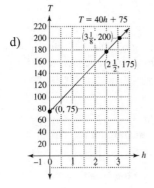

15.

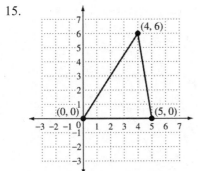

$$\text{centroid} = \left(\frac{x_1 + x_2 + x_3}{3}, \frac{y_1 + y_2 + y_3}{3} \right)$$
$$= \left(\frac{0 + 5 + 4}{3}, \frac{0 + 0 + 6}{3} \right)$$
$$= (3, 2)$$

The figure is a triangle. The formula for the area of a triangle is $A = \frac{1}{2}bh$. For the base, we can subtract the x-coordinates.

base $= |5 - 0| = |5| = 5$

For the height, we can subtract the y-coordinates.

height $= |6 - 0| = |6| = 6$

Now we can calculate the area.

$$A = \frac{1}{2}bh$$
$$= \frac{1}{2}(5)(6)$$
$$= \frac{1}{2}(30)$$
$$= 15 \text{ square units}$$

17.

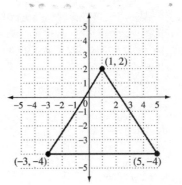

$$\text{centroid} = \left(\frac{x_1 + x_2 + x_3}{3}, \frac{y_1 + y_2 + y_3}{3} \right)$$
$$= \left(\frac{-3 + 5 + 1}{3}, \frac{-4 + (-4) + 2}{3} \right)$$
$$= (-1, -2)$$

The figure is a triangle. The formula for the area of a triangle is $A = \frac{1}{2}bh$. For the base, we can subtract the x-coordinates.

base $= |5 - (-3)| = |5 + 3| = |8| = 8$

For the height, we can subtract the y-coordinates.

height $= |2 - (-4)| = |2 + 4| = |6| = 6$

Now we can calculate the area.

$$A = \frac{1}{2}bh$$
$$= \frac{1}{2}(8)(6)$$
$$= \frac{1}{2}(48)$$
$$= 24 \text{ square units}$$

19.

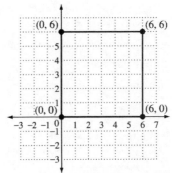

$$\text{centroid} = \left(\frac{x_1 + x_2 + x_3 + x_4}{4}, \frac{y_1 + y_2 + y_3 + y_4}{4} \right)$$
$$= \left(\frac{0 + 0 + 6 + 6}{4}, \frac{0 + 6 + 6 + 0}{4} \right)$$
$$= (3, 3)$$

The figure is a square. The formula for area of a square is $A = lw$, so we need the length and width. For the length, we can subtract the x-coordinates.

length $= |6 - 0| = |6| = 6$

For the width, we can subtract the y-coordinates.

width $= |6 - 0| = |6| = 6$

Now we can calculate the area.

$$\text{area} = lw$$
$$= 6 \cdot 6$$
$$= 36 \text{ square units}$$

21.

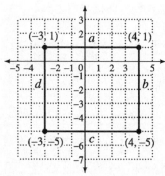

$$\text{centroid} = \left(\frac{x_1 + x_2 + x_3 + x_4}{4}, \frac{y_1 + y_2 + y_3 + y_4}{4} \right)$$
$$= \left(\frac{-3 + 4 + 4 + (-3)}{4}, \frac{1 + 1 + (-5) + (-5)}{4} \right)$$
$$= \left(\frac{2}{4}, \frac{-8}{4} \right)$$
$$= \left(\frac{1}{2}, -2 \right)$$

The figure is a rectangle. The formula for area of a rectangle is $A = lw$, so we need the length and width. For the length, we can subtract the x-coordinates.

length $= |4 - (-3)| = |4 + 3| = |7| = 7$

For the width, we can subtract the y-coordinates.

width $= |1 - (-5)| = |1 + 5| = |6| = 6$

Now we can calculate the area.

$$\text{area} = lw$$
$$= 7 \cdot 6$$
$$= 42 \text{ square units}$$

23.

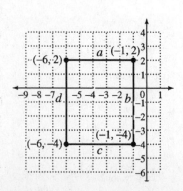

centroid

$$= \left(\frac{x_1 + x_2 + x_3 + x_4}{4}, \frac{y_1 + y_2 + y_3 + y_4}{4} \right)$$
$$= \left(\frac{-1 + (-6) + (-6) + (-1)}{4}, \frac{2 + 2 + (-4) + (-4)}{4} \right)$$
$$= \left(\frac{-14}{4}, \frac{-4}{4} \right)$$
$$= \left(-\frac{7}{2}, -1 \right)$$

The figure is a rectangle. The formula for area of a rectangle is $A = lw$, so we need the length and width. For the length, we can subtract the x-coordinates.

length $= |-1 - (-6)| = |-1 + 6| = |5| = 5$

For the width, we can subtract the y-coordinates.

width $= |2 - (-4)| = |2 + 4| = |6| = 6$

Now we can calculate the area.

$$\text{area} = lw$$
$$= 5 \cdot 6$$
$$= 30 \text{ square units}$$

25.

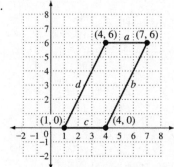

$$\text{centroid} = \left(\frac{x_1 + x_2 + x_3 + x_4}{4}, \frac{y_1 + y_2 + y_3 + y_4}{4} \right)$$
$$= \left(\frac{1 + 4 + 7 + 4}{4}, \frac{0 + 6 + 6 + 0}{4} \right)$$
$$= \left(\frac{16}{4}, \frac{12}{4} \right)$$
$$= (4, 3)$$

The figure is a parallelogram. The formula for the area of a parallelogram is $A = bh$, so we need the base and height. Because the base is on the x-axis, we can subtract the x-coordinates.

base = side c $= |4 - 1| = |3| = 3$

The height is the vertical measurement from the base to the top of the shape. The line along which height is measured is parallel to the y-axis. Therefore, we can subtract the y-coordinates.

height $= |6 - 0| = |6| = 6$

Now we can calculate the area.

area $= bh$

$= 3 \cdot 6$

$= 18$ square units

27.

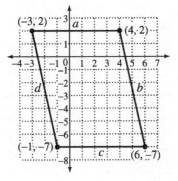

centroid

$= \left(\dfrac{x_1 + x_2 + x_3 + x_4}{4} , \dfrac{y_1 + y_2 + y_3 + y_4}{4} \right)$

$= \left(\dfrac{(-3) + 4 + 6 + (-1)}{4} , \dfrac{2 + 2 + (-7) + (-7)}{4} \right)$

$= \left(\dfrac{6}{4} , \dfrac{-10}{4} \right)$

$= \left(-\dfrac{3}{2} , -\dfrac{5}{2} \right)$

The figure is a parallelogram. The formula for the area of a parallelogram is $A = bh$, so we need the base and height. Because the base is parallel to the x-axis, we can subtract the x-coordinates.

base $=$ side c $= \left| 6 - (-1) \right| = \left| 6 + 1 \right| = \left| 7 \right| = 7$

The height is the vertical measurement from the base to the top of the shape. The line along which height is measured is parallel to the y-axis. Therefore, we can subtract the y-coordinates.

height $= \left| 2 - (-7) \right| = \left| 2 + 7 \right| = \left| 9 \right| = 9$

Now we can calculate the area.

area $= bh$

$= 7 \cdot 9$

$= 63$ square units

29.

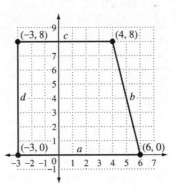

centroid $= \left(\dfrac{x_1 + x_2 + x_3 + x_4}{4} , \dfrac{y_1 + y_2 + y_3 + y_4}{4} \right)$

$= \left(\dfrac{-3 + (-3) + 4 + 6}{4} , \dfrac{0 + 8 + 8 + 0}{4} \right)$

$= \left(\dfrac{4}{4} , \dfrac{16}{4} \right)$

$= (1, 4)$

The figure is a trapezoid. The formula for the area of a trapezoid is $A = \dfrac{1}{2} h(a + b)$, so we need the lengths of the bases and the height. Because the base is parallel to the x-axis, we can subtract the x-coordinates.

base $=$ side a $= \left| 6 - (-3) \right| = \left| 6 + 3 \right| = \left| 9 \right| = 9$

base $=$ side c $= \left| 4 - (-3) \right| = \left| 4 + 3 \right| = \left| 7 \right| = 7$

The height is the vertical measurement from the base to the top of the shape. The line along which height is measured is parallel to the y-axis. Therefore, we can subtract the y-coordinates.

height $= \left| 8 - 0 \right| = \left| 8 \right| = 8$

Now we can calculate the area.

area $= \dfrac{1}{2} h(\text{side } a + \text{side } c)$

$= \dfrac{1}{2} \cdot 8 \cdot (9 + 7)$

$= \dfrac{1}{2} \cdot 8 \cdot 16$

$= 64$ square units

31.

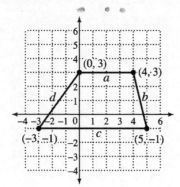

$$\text{centroid} = \left(\frac{x_1 + x_2 + x_3 + x_4}{4}, \frac{y_1 + y_2 + y_3 + y_4}{4} \right)$$

$$= \left(\frac{0 + 4 + 5 + (-3)}{4}, \frac{3 + 3 + (-1) + (-1)}{4} \right)$$

$$= \left(\frac{6}{4}, \frac{4}{4} \right)$$

$$= \left(\frac{3}{2}, 1 \right)$$

The figure is a trapezoid. The formula for the area of a trapezoid is $A = \frac{1}{2} h(a+b)$, so we need the lengths of the bases and the height. Because the base is parallel to the x-axis, we can subtract the x-coordinates.

$$\text{base} = \text{side a} = \left| 4 - 0 \right| = \left| 4 \right| = 4$$

$$\text{base} = \text{side c} = \left| 5 - (-3) \right| = \left| 5 + 3 \right| = \left| 8 \right| = 8$$

The height is the vertical measurement from the base to the top of the shape. The line along which height is measured is parallel to the y-axis. Therefore, we can subtract the y-coordinates.

$$\text{height} = \left| 3 - (-1) \right| = \left| 3 + 1 \right| = \left| 4 \right| = 4$$

Now we can calculate the area.

$$\text{area} = \frac{1}{2} h(\text{side } a + \text{side } c)$$

$$= \frac{1}{2} \cdot 4 \cdot (4+8)$$

$$= 0.5 \cdot 4 \cdot 12$$

$$= 24 \text{ square units}$$

33.

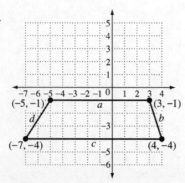

centroid

$$= \left(\frac{x_1 + x_2 + x_3 + x_4}{4}, \frac{y_1 + y_2 + y_3 + y_4}{4} \right)$$

$$= \left(\frac{-5 + (-7) + 3 + 4}{4}, \frac{-1 + (-4) + (-1) + (-4)}{4} \right)$$

$$= \left(\frac{-5}{4}, \frac{-10}{4} \right)$$

$$= \left(-\frac{5}{4}, -\frac{5}{2} \right)$$

The figure is a trapezoid. The formula for the area of a trapezoid is $A = \frac{1}{2} h(a+b)$, so we need the lengths of the bases and the height. Because the base is parallel to the x-axis, we can subtract the x-coordinates.

$$\text{base} = \text{side a} = \left| 3 - (-5) \right| = \left| 3 + 8 \right| = \left| 8 \right| = 8$$

$$\text{base} = \text{side c} = \left| 4 - (-7) \right| = \left| 4 + 7 \right| = \left| 11 \right| = 11$$

The height is the vertical measurement from the base to the top of the shape. The line along which height is measured is parallel to the y-axis. Therefore, we can subtract the y-coordinates.

$$\text{height} = \left| -1 - (-4) \right| = \left| -1 + 4 \right| = \left| 3 \right| = 3$$

Now we can calculate the area.

$$\text{area} = \frac{1}{2} h(\text{side } a + \text{side } c)$$

$$= \frac{1}{2} \cdot 3 \cdot (8 + 11)$$

$$= \frac{1}{2} \cdot 3 \cdot 19$$

$$= \frac{57}{2}$$

$$= 28\frac{1}{2} \text{ square units}$$

Review Exercises

1. one number $= 3x + 6$
    other number $= x$

$$(3x + 6) + x = 106$$

$$4x + 6 = 106$$

$$\underline{-6 = -6}$$

$$4x + 0 = 100$$

$$\frac{4x}{4} = \frac{10}{4}$$

$$x = 25$$

One number is $3(25) + 6 = 81$ and the other is 25.

2. $\dfrac{\cancel{3}}{{}_2\cancel{4}} \cdot \dfrac{\cancel{2}}{\cancel{3}} = \dfrac{1}{2}$ of the company's total cost goes toward non-salary wages.

3. Write the data in order from smallest to largest: 56, 65, 70, 70, 71, 72, 82, 84, 87, 88, 88, 90, 91, 95.

The mean is the sum of the test scores divided by the number of test scores.

$$\text{mean} = \dfrac{1109}{14} \approx 79.2$$

$$\text{median} = \dfrac{82+84}{2} = \dfrac{166}{2} = 83$$

4. 20% of 45 is what number?

$$20\% \cdot 45 = c$$
$$0.2 \cdot 45 = c$$
$$9 = c$$

9 people received an A.

5. $B = P\left(1 + \dfrac{r}{n}\right)^{nt}$

$$B = 400\left(1 + \dfrac{0.048}{4}\right)^{4 \cdot 2}$$
$$B = 400\left(1 + 0.012\right)^{8}$$
$$B = 400\left(1.012\right)^{8}$$
$$B = 400\left(1.100130234\right)$$
$$B \approx 440.05$$

The balance will be $440.05.

Chapter 9 Review Exercises

1. false   2. false   3. true

4. true   5. false   6. true

7. true   8. true

9. 1. Find at least two solutions.
2. Plot those solutions as points on the rectangular coordinate system.
3. Draw a straight line through the points.

10. To find the x-intercept, replace y with 0, then solve for x. To find the y-intercept, replace x with 0, then solve for y.

11. a) point P   b) ray $\overline{RT}$

c) parallel lines h and k, or $h \parallel k$

12. a) The angle is right because its measure is 90°.
b) The angle is obtuse because its measure is between 90° and 180°.
c) The angle is straight because its measure is 180°.
d) The angle is acute because its measure is between 0° and 90°.

13. a) congruent; They are vertical angles.
b) congruent; They are alternate interior angles.
c) supplementary: They are consecutive exterior angles.
d) supplementary; They form a straight angle.

14. a) Because $\angle 1$ and $\angle 3$ form a straight angle, they are supplementary, so $\angle 1 + \angle 3 = 180°$.

$$57° + \angle 3 = 180°$$
$$\underline{-57°\qquad\quad -57°}$$
$$0 + \angle 3 = 123°$$
$$\angle 3 = 123°$$

b) Because $\angle 1$ and $\angle 4$ are vertical, they are congruent, so $\angle 1 \cong \angle 4 = 57°$.

c) Because $\angle 1$ and $\angle 5$ are corresponding angles, they are congruent, so $\angle 1 \cong \angle 5 = 57°$.

d) Because $\angle 1$ and $\angle 7$ form consecutive exterior angles, they are supplementary, so $\angle 1 + \angle 7 = 180°$.

$$57° + \angle 7 = 180°$$
$$\underline{-57°\qquad\quad -57°}$$
$$0 + \angle 7 = 123°$$
$$\angle 7 = 123°$$

15. **Understand:** Because the two angles in the variable expressions form a straight angle, they are supplementary, so the sum of their measures is 180°. Also, because the two intersecting lines form two pairs of vertical angles, the measures of the other two angles will be equal to x and $4x + 25$ as well.

Plan: Translate to an equation, then solve. The sum of x and $4x + 25$ is 180.

Execute: $x + 4x + 25 = 180$
$$5x + 25 = 180$$
$$5x = 155$$
$$x = 31$$

Answer: The measure of the smaller angle is 31°. To find the measure of the larger angle, we replace x in $4x + 25$ with 31.
$$4(31) + 25 = 124 + 25 = 149$$
The two angles are 31° and 149°. Because the

two intersecting lines form two pair of vertical angles, the measures of the other two angles are also 31° and 149°.

Check: The sum of the measures of the supplementary angles is 31° + 149° = 180°.

16. **Understand:** The variable expressions represent the measures of angles that are consecutive exterior angles, so they are supplementary.

Plan: Translate to an equation, then solve. The sum of y and $0.6y - 19.2$ is 180.

Execute:
$$y + 0.6y - 19.2 = 180$$
$$1.6y - 19.2 = 180$$
$$1.6y = 199.2$$
$$y = 124.5$$

Answer: The measure of the larger angle is 124.5°. To find the measure of the smaller angle, we replace y in $0.6y - 19.2$ with 124.5.

$$0.6(124.5) - 19.2 = 74.7 - 19.2 = 55.5$$

The two angles are 124.5° and 55.5°. We can find all the other angles using the rules concerning angles formed by a transversal intersecting two parallel lines.

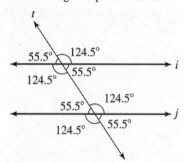

Check: The sum of the measures of two consecutive exterior angles is 124.5° + 55.5° = 180°.

17. a) acute triangle; The measure of each angle is less than 90°.

b) obtuse triangle; The measure of one angle is greater than 90°.

18. **Understand:** The sum of the measures of the angles is 180°.

Plan: Translate to an equation, then solve. The sum of x, $0.5x + 17$, and $x - 42$ is 180.

Execute:
$$x + 0.5x + 17 + x - 42 = 180$$
$$2.5x - 25 = 180$$
$$2.5x = 205$$
$$x = 82$$

Answer: The measure of the angle labeled x is 82°. To find the measure of the other two angles, we replace x in $0.5x + 17$ and $x - 42$ with 82.

$$0.5x + 17 = 0.5(82) + 17 = 41 + 17 = 58$$

$$x - 42 = 82 - 42 = 40$$

The measures of the angles are 82°, 58°, and 40°.

Check: The sum of the measures of the angles is 82° + 58° + 40° = 180°.

19. A: $(3, 4)$ B: $(-3, 2)$ C: $(-2, 0)$

D: $(-2, -4)$ E: $(0, -5)$ F: $(4, -2)$

20.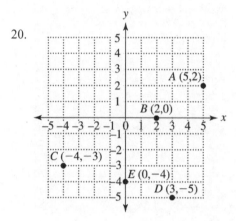

A: $(5, 2)$ B: $(2, 0)$ C: $(-4, -3)$

D: $(3, -5)$ E: $(0, -4)$

21. a) Quadrant II because the first coordinate is negative and the second coordinate is positive.

b) Quadrant III because both coordinates are negative.

c) Quadrant I because both coordinates are positive.

d) on the x-axis

22. midpoint $= \left(\dfrac{x_1 + x_2}{2}, \dfrac{y_1 + y_2}{2} \right)$

$= \left(\dfrac{2 + 6}{2}, \dfrac{9 + 1}{2} \right)$

$= (4, 5)$

23. midpoint $= \left(\dfrac{x_1 + x_2}{2}, \dfrac{y_1 + y_2}{2} \right)$

$= \left(\dfrac{-4 + 6}{2}, \dfrac{2 + (-3)}{2} \right)$

$= \left(1, -\dfrac{1}{2} \right)$

24. Replace x with 2 and y with 7, and see if the equation is true.

$$3x - y = -4$$
$$3 \cdot 2 - 7 ? - 4$$
$$6 - 7 ? - 4$$
$$-1 \neq -4$$

This does not check which means $(2,7)$ is not a solution for $3x - y = -4$.

25. Replace x with 2.5 and y with -0.5, and see if the equation is true.

$$y = -x + 2$$
$$-0.5 ? - 2.5 + 2$$
$$-0.5 = -0.5$$

This checks which means $(2.5, -0.5)$ is a solution for $y = -x + 2$.

26. Replace x with $-4\dfrac{3}{4} = -4.75$ and y with $-1\dfrac{9}{10} = -$
1.9, and see if the equation is true.

$$y = \frac{2}{5}x$$
$$-1.9 ? 0.4(-4.75)$$
$$-1.9 = -1.9$$

This checks which means $\left(-4\dfrac{3}{4}, -1\dfrac{9}{10}\right)$ is a solution for $y = \dfrac{2}{5}x$.

27. Individual solution points may vary but should all lie on the same line. Some possible solutions are (1, 4), (2, 2), and (3, 0).

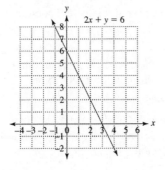

28. Individual solution points may vary but should all lie on the same line. Some possible solutions are (0, −2), (2, −1), and (4, 0).

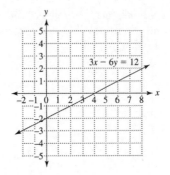

29. Individual solution points may vary but should all lie on the same line. Some possible solutions are (−1, 2), (0, 3), and (1, 4).

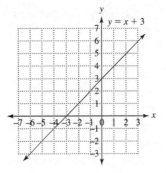

30. Individual solution points may vary but should all lie on the same line. Some possible solutions are (− 1, 3), (0, 0), and (1, −3).

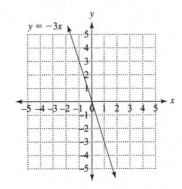

31. Individual solution points may vary but should all lie on the same line. Some possible solutions are (−3, −2), (0, 0), and (3, 2).

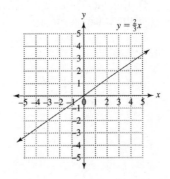

32. Individual solution points may vary but should all lie on the same line. Some possible solutions are $(-2, -2)$, $(-1, -3)$, and $(0, -4)$.

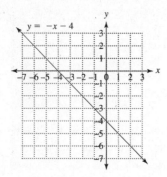

33. Individual solution points may vary but should all lie on the same line. Some possible solutions are $(0, 7)$, $(2, 7)$, and $(5, 7)$.

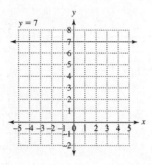

34. Individual solution points may vary but should all lie on the same line. Some possible solutions are $(-4, 0)$, $(-4, 4)$, and $(-4, 5)$.

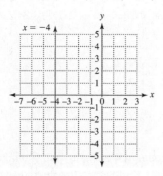

35. To find the x-intercept, replace y with 0 and solve the equation for x. To find the y-intercept, replace the x with 0 and solve the equation for y.

x-intercept: $5x + y = 10$ $(2, 0)$
$5x + 0 = 10$
$5x = 10$
$\dfrac{5x}{5} = \dfrac{10}{5}$
$x = 2$

y-intercept: $5x + y = 10$ $(0, 10)$
$5 \cdot 0 + y = 10$
$0 + y = 10$
$y = 10$

36. To find the x-intercept, replace y with 0 and solve the equation for x. To find the y-intercept, replace the x with 0 and solve the equation for y.

x-intercept: $y = 4x - 1$ $\left(\dfrac{1}{4}, 0\right)$
$0 = 4x - 1$
$\underline{+1 =} \quad \underline{+1}$
$1 = 4x + 0$
$\dfrac{1}{4} = \dfrac{4x}{4}$
$\dfrac{1}{4} = x$

y-intercept: $y = 4x - 1$ $(0, -1)$
$y = 4 \cdot 0 - 1$
$y = 0 - 1$
$y = -1$

37. To find the x-intercept, we replace y with 0 and solve the equation for x.

x-intercept: $y = \dfrac{1}{5}x$ $(0, 0)$
$0 = \dfrac{1}{5}x$
$5 \cdot 0 = \dfrac{1}{5}x \cdot 5$
$0 = x$

Since the x-intercept is the origin, $(0, 0)$. The origin is the only point in the rectangular coordinate system where a line can pass through both the x-axis and y-axis simultaneously. Therefore both the x- and y-intercept coordinates are $(0, 0)$.

38. The graph of $x = 6$ is a vertical line parallel to the y-axis that passes through the x-axis at the point $(6, 0)$. Notice that this point is the x-intercept. Because the line is parallel to the y-axis, it will never intersect the y-axis. Therefore there is no y-intercept.

39. The graph of $y = -2$ is a horizontal line parallel to the x-axis that passes through the y-axis at the point $(0, -2)$. Notice that this point is the y-intercept. Because the line is parallel to the x-axis, it will never intersect the x-axis. Therefore there is no x-intercept.

40. a) We are given a revenue of $100,000. Let $r =$ 100,000.

$$p = 0.4r - 12,000$$
$$p = 0.4 \cdot 100,000 - 12,000$$
$$p = 40,000 - 12,000$$
$$p = \$28,000$$

b) Let $p = 0$ and solve for r.

$$p = 0.4r - 12,000$$
$$0 = 0.4r - 12,000$$
$$\underline{+12,000 = \qquad +12,000}$$
$$12,000 = 0.4r + 0$$
$$\frac{12,000}{0.4} = \frac{0.4r}{0.4}$$
$$\$30,000 = r$$

c)

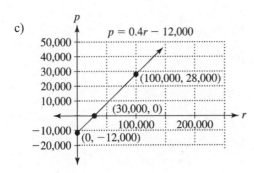

41.

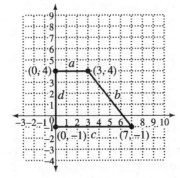

$$\text{centroid} = \left(\frac{x_1 + x_2 + x_3 + x_4}{4}, \frac{y_1 + y_2 + y_3 + y_4}{4} \right)$$
$$= \left(\frac{0 + 3 + 7 + 0}{4}, \frac{4 + 4 + (-1) + (-1)}{4} \right)$$
$$= \left(\frac{10}{4}, \frac{6}{4} \right)$$
$$= (2.5, 1.5)$$

42.

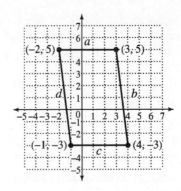

The figure is a parallelogram. The formula for the area of a parallelogram is $A = bh$, so we need the base and height. Because the base is parallel to the x-axis, we can subtract the x-coordinates.

$$\text{side a} = |3 - (-2)| = |3 + 2| = |5| = 5$$

$$\text{base} = \text{side c} = |4 - (-1)| = |4 + 1| = |5| = 5$$

The height is the vertical measurement from the base to the top of the shape. The line along which height is measured is parallel to the y-axis. Therefore, we can subtract the y-coordinates.

$$\text{height} = |5 - (-3)| = |5 + 3| = |8| = 8$$

Now we can calculate the area.

$$\text{area} = bh$$
$$= 5 \cdot 8$$
$$= 40 \text{ square units}$$

Chapter 9 Practice Test

1. a) line $\overleftrightarrow{AB}$ b) ray $\overrightarrow{RS}$

2. a) The angle is right because its measure is 90°.

   b) The angle is obtuse because its measure is between 90° and 180°.

3. **Understand:** Because the two angles in the variable expressions form a straight angle, they are supplementary, so the sum of their measures is 180°. Also, because the two intersecting lines form two pairs of vertical angles, the measures of the other two angles will be equal to x and $2x + 9$ as well.

   Plan: Translate to an equation, then solve. The sum of x and $2x + 9$ is 180.

   Execute: $x + 2x + 9 = 180$
   $$3x + 9 = 180$$
   $$3x = 171$$
   $$x = 57$$

   Answer: The measure of the smaller angle is 57°. To find the measure of the larger angle, we replace x in $2x + 9$ with 57.
   $$2(57) + 9 = 114 + 9 = 123$$

The two angles are 57° and 123°. Because the two intersecting lines form two pair of vertical angles, the measures of the other two angles are also 57° and 123°.

Check: The sum of the measures of the supplementary angles is 57° + 123° = 180°.

4. **Understand:** The variable expressions represent the measures of angles that are consecutive interior angles, so they are supplementary.

   Plan: Translate to an equation, then solve. The sum of x and $0.6x + 8$ is 180.

   Execute: $x + 0.6x + 8 = 180$

   $$1.6x + 8 = 180$$
   $$1.6x = 172$$
   $$x = 107.5$$

   Answer: The measure of the larger angle is 107.5°. To find the measure of the smaller angle, we replace x in $0.6x + 8$ with 107.5.
   $$0.6(107.5) + 8 = 64.5 + 8 = 72.5$$

   The two angles are 107.5° and 72.5°. We can find all the other angles using the rules concerning angles formed by a transversal intersecting two parallel lines.

   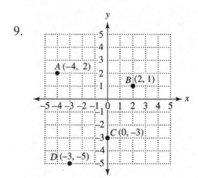

   Check: The sum of the measures of two consecutive interior angles is 107.5° + 72.5° = 180°.

5. **Understand:** The sum of the measures of the angles is 180°.

   Plan: Translate to an equation, then solve. The sum of x, $0.5x - 11.5$, and $x - 51$ is 180.

   Execute: $x + 0.5x - 11.5 + x - 51 = 180$

   $$2.5x - 62.5 = 180$$
   $$2.5x = 242.5$$
   $$x = 97$$

   Answer: The measure of the angle labeled x is 97°. To find the measure of the other two angles, we replace x in $0.5x - 11.5$ and $x - 51$ with 97.
   $$0.5x - 11.5 = 0.5(97) - 11.5 = 48.5 - 11.5 = 37$$
   $$x - 51 = 97 - 51 = 46$$
   The measures of the angles are 97°, 37°, and 46°.

   Check: The sum of the measures of the angles is 97° + 37° + 46° = 180°.

6. A: $(1,2)$ B: $(-3,0)$

C: $(-2,-5)$ D: $(4,-4)$

7. Quadrant II because the first coordinate is negative and the second coordinate is positive

8. $(0, -4)$ is not in a quadrant, it is on the y-axis.

9.

$A(-4,2)$ $B(2,1)$

$C(0,-3)$ $D(-3,-5)$

10. midpoint $= \left(\dfrac{x_1 + x_2}{2}, \dfrac{y_1 + y_2}{2} \right)$

    $$= \left(\dfrac{5+1}{2}, \dfrac{3+1}{2} \right)$$
    $$= (3,2)$$

11. midpoint $= \left(\dfrac{x_1 + x_2}{2}, \dfrac{y_1 + y_2}{2} \right)$

    $$= \left(\dfrac{4+(-3)}{2}, \dfrac{-2+(-7)}{2} \right)$$
    $$= \left(\dfrac{1}{2}, -\dfrac{9}{2} \right)$$

12. Replace x with 5 and y with -3, and see if the equation is true.

$-x + 2y = -11$	This checks which means
$-5 + 2(-3)\,?-11$	$(5,-3)$ is a solution for
$-5 + (-6)\,?-11$	$-x + 2y = -11$.
$-11 = -11$	

13. Replace x with -3 and y with -4, and see if the equation is true.

$y = \dfrac{2}{3}x + 1$	This does not check
	which means $(-3,-4)$
$-4\,?\dfrac{2}{3}(-3) + 1$	is not a solution for
$-4\,?-2+1$	$y = \dfrac{2}{3}x + 1$.
$-4 \neq -1$	

14. To find the *x*-intercept, replace *y* with 0 and solve the equation for *x*. To find the *y*-intercept, replace the *x* with 0 and solve the equation for *y*.

x-intercept: $3x - 5y = 6$ $\left(2, 0\right)$
$$3x - 5 \cdot 0 = 6$$
$$3x - 0 = 6$$
$$\frac{3x}{3} = \frac{6}{3}$$
$$x = 2$$

y-intercept: $3x - 5y = 6$ $\left(0, -\dfrac{6}{5}\right)$
$$3 \cdot 0 - 5y = 6$$
$$0 - 5y = 6$$
$$-5y = 6$$
$$\frac{-5y}{-5} = \frac{6}{-5}$$
$$y = -\frac{6}{5}$$

15. To find the *x*-intercept, replace *y* with 0 and solve the equation for *x*. To find the *y*-intercept, replace the *x* with 0 and solve the equation for *y*.

x-intercept: $y = 3x - 4$ $\left(\dfrac{4}{3}, 0\right)$
$$0 = 3x - 4$$
$$4 = 3x$$
$$\frac{4}{3} = x$$

y-intercept: $y = 3x - 4$ $\left(0, -4\right)$
$$y = 3 \cdot 0 - 4$$
$$y = 0 - 4$$
$$y = -4$$

16. To find the *x*-intercept, replace *y* with 0 and solve the equation for *x*. To find the *y*-intercept, replace the *x* with 0 and solve the equation for *y*.

x-intercept: $y = 3$
$$0 \neq 3$$
Since the equation cannot be solved for *x*, there is no *x*-intercept.

y-intercept: $y = 3$ $\left(0, 3\right)$

$y = 3$ is the equation of line that is parallel to the *x*-axis and passes through the point (0, 3).

17. Individual solution points may vary but should all lie on the same line. Some possible solutions are $(-1, 3)$, $(0, 0)$, and $(1, -3)$.

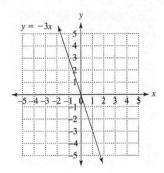

18. Individual solution points may vary but should all lie on the same line. Some possible solutions are $(-5, -4)$, $(0, -3)$, and $(5, -2)$.

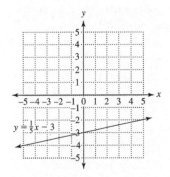

19. Individual solution points may vary but should all lie on the same line. Some possible solutions are $(2, 5)$, $(3, 4)$, and $(4, 3)$.

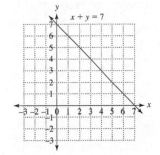

20. Individual solution points may vary but should all lie on the same line. Some possible solutions are $(0, -2)$, $(1, 1)$, and $(2, 4)$.

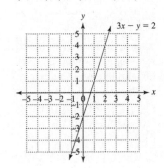

21. The graph of $y = 1$ is a line drawn parallel to the x-axis having a y-intercept of 1.

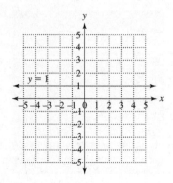

22. The graph of $x = -2$ is a line drawn parallel to the y-axis having a x-intercept of -2

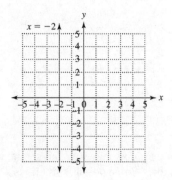

23. a)   Initial velocity means the velocity at the time the rocket is released, which is 0 seconds. Let $t = 0$.

$v = -9.8t + 40$

$v = -9.8(0) + 40$

$v = 0 + 40$

$v = 40$ m/sec.

b)   We are given a time of 3 seconds. Let $t = 3$.

$v = -9.8t + 40$

$v = -9.8(3) + 40$

$v = -29.4 + 40$

$v = 10.6$ m/sec.

c)   We must calculate the time at which the rocket stops in midair. After stopping, the rocket will then reverse direction and begin falling back toward the ground. When an object is at a stop, it is obviously not moving and therefore has zero velocity. We must solve for t when the velocity, v is 0. Let $v = 0$ and solve for t.

$v = -9.8t + 40$

$0 = -9.8t + 40$

$\underline{-40 =} \qquad \underline{-40}$

$-40 = -9.8t + 0$

$\dfrac{-40}{-9.8} = \dfrac{-9.8t}{-9.8}$

$4.1 \text{ sec.} \approx t$

d)

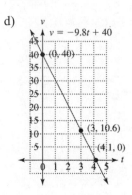

24.

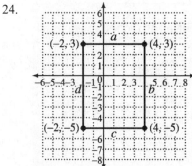

centroid

$= \left(\dfrac{x_1 + x_2 + x_3 + x_4}{4}, \dfrac{y_1 + y_2 + y_3 + y_4}{4} \right)$

$= \left(\dfrac{-2 + 4 + 4 + (-2)}{4}, \dfrac{3 + 3 + (-5) + (-5)}{4} \right)$

$= \left(\dfrac{4}{4}, \dfrac{-4}{4} \right)$

$= (1, -1)$

25. The figure is a parallelogram. The formula for area of a rectangle is $A = bh$, so we need the base and height. For the base, we can subtract the x-coordinates.

base $= |2 - (-5)| = |2 + 5| = |7| = 7$

For the width, we can subtract the y-coordinates.

height $= |4 - (-4)| = |4 + 4| = |8| = 8$

Now we can calculate the area.

area $= bh$

$= 7 \cdot 8$

$= 56$ square units

Chapter 9 Cumulative Review Exercises

1. false   2. false   3. false   4. true

5. true   6. false   7. true   8. true

9. false   10. true   11. true   12. true

13. 1) grouping symbols
   2) exponents or roots from left to right
   3) multiplication or division from left to right.
   4) addition or subtraction from left to right.

14. Add the absolute values and keep the same sign.

15. Subtract the absolute values and keep the sign of the number with the larger absolute value.

16. Change the operation from $-$ to $+$. Change the subtrahend to its additive inverse.

17. positive   18. negative

19. 1) Simplify by clearing parentheses, fractions, and decimals and combining like terms.
   2) Use the addition/subtraction principle of equality so that all variable terms are on one side of the equation and all constant terms are on the other side.
   3) Use the multiplication/division principle of equality to clear any remaining coefficients.

20. 1) Find at least two solutions.
   2) Plot the solutions as points on the rectangular coordinate system.
   3) Draw a straight line through the points.

21. $24{,}607 = 2 \times 10{,}000 + 4 \times 1000 + 6 \times 100 + 7 \times 1$

22. four thousand six hundred seven and nine hundredths

23. a) 80,000: 7 is in the ten thousands place. Look to the 9 on the right. Since 9 is 5 or more, we round up.

   b) 80,000: 9 is in the thousands place. Look to the 8 on the right. Since 8 is 5 or more, we round up.

   c) 79,805: 4 is in the ones place. Look to the 6 on the right. Since 6 is 5 or more, we round up.

   d) 79,804.7: 6 is in the tenths place. Look to the 5 on the right. Since 5 is 5 or more, we round up.

   e) 79,804.65: 5 is in the hundredths place. Look to the 2 on the right. Since 2 is 4 or less, we round down.

24. a) $\begin{array}{r} 21{,}000 \\ +7{,}000 \\ \hline 28{,}000 \end{array}$ b) $\begin{array}{r} 729{,}000 \\ -5{,}000 \\ \hline 724{,}000 \end{array}$

c) $90 \times 20 = 1800$; Think 9×2 and add 2 zeros.

d) $2200 \div 100 = 22$

25. 2.75×10^{10} 26. 3,530,000

27. a)

   b)

   c)

28. a) 1   b) 2   c) 4

29. The angle is acute because its measure is between $0°$ and $90°$.

30. $(-3)^4 = (-3)(-3)(-3)(-3) = 81$

31. $\left(\dfrac{1}{6}\right)^2 = \dfrac{1}{36}$ 32. $\left(-\dfrac{1}{2}\right)^3 = -\dfrac{1}{8}$

33. $\begin{aligned} 9 - 4\left|3-8\right| - \sqrt{49} &= 9 - 4\left|-5\right| - \sqrt{49} \\ &= 9 - 4 \cdot 5 - \sqrt{49} \\ &= 9 - 4 \cdot 5 - 7 \\ &= 9 - 20 - 7 \\ &= -11 - 7 \\ &= -18 \end{aligned}$

34. $\begin{aligned} 19 - (6-8)^3 + \sqrt{100-36} &= 19 - (-2)^3 + \sqrt{64} \\ &= 19 - (-8) + 8 \\ &= 19 + 8 + 8 \\ &= 27 + 8 \\ &= 35 \end{aligned}$

35. $\begin{aligned} \dfrac{14 + (18 - 2 \cdot 4)}{4^2 - 10} &= \dfrac{14 + (18 - 8)}{16 - 10} \\ &= \dfrac{14 + 10}{6} \\ &= \dfrac{24}{6} \\ &= 4 \end{aligned}$

36. $9\dfrac{3}{4}-\left(-2\dfrac{2}{3}\right)=9\dfrac{3}{4}+2\dfrac{2}{3}$

$$=9\dfrac{3\cdot3}{4\cdot3}+2\dfrac{2\cdot4}{3\cdot4}$$

$$=9\dfrac{9}{12}+2\dfrac{8}{12}$$

$$=11\dfrac{17}{12}$$

$$=11+1\dfrac{5}{12}$$

$$=12\dfrac{5}{12}$$

37. $-6\dfrac{1}{8}\div\dfrac{3}{16}-\dfrac{3}{4}=-\dfrac{49}{\cancel{8}}\cdot\dfrac{\cancel{16}^{2}}{3}-\dfrac{3}{4}$

$$=-\dfrac{98}{3}-\dfrac{3}{4}$$

$$=-\dfrac{392}{12}-\dfrac{9}{12}$$

$$=-\dfrac{401}{12}$$

$$=-33\dfrac{5}{12}$$

38. $\sqrt{(27)(0.03)}=\sqrt{0.81}=0.9$

39. $-11.7\div1.8+2.6(7.5)=-6.5+2.6(7.5)$

$$=-6.5+19.5$$

$$=13$$

40. $\left(\dfrac{3}{4}\right)^{2}(-8.4)=(0.75)^{2}(-8.4)$

$$=(0.5625)(-8.4)$$

$$=-4.725$$

41. $\dfrac{3}{5}-(0.2)^{3}=0.6-0.008$

$$=0.592$$

42. $\sqrt{126}\approx11.22$

43. a) $\dfrac{3}{8}=3\div8=0.375$

   b) $\dfrac{1}{3}=1\div3=0.\overline{3}$ c) $8\%=8\div100=0.08$

   d) $24\dfrac{2}{5}\%=24.4\div100=0.244$

   e) $4.5\%=4.5\div100=0.045$

44. a) $5\dfrac{1}{6}=\dfrac{6\cdot5+1}{6}=\dfrac{30+1}{6}=\dfrac{31}{6}$

   b) $0.145=\dfrac{145}{1000}=\dfrac{29}{200}$

   c) $6\%=\dfrac{6}{100}=\dfrac{3}{50}$

   d) $5\dfrac{1}{2}\%=\dfrac{5\dfrac{1}{2}}{100}$ e) $16.4\%=\dfrac{16.4}{100}$

$$=5\dfrac{1}{2}\div100\qquad\qquad=\dfrac{16.4(10)}{100(10)}$$

$$=\dfrac{11}{2}\cdot\dfrac{1}{100}\qquad\qquad=\dfrac{164}{1000}$$

$$=\dfrac{11}{200}\qquad\qquad\qquad=\dfrac{41}{250}$$

45. a) $\dfrac{1}{4}=0.25(100\%)$ b) $\dfrac{4}{9}=0.44\overline{4}$

$$=25\%\qquad\qquad\qquad=0.44\overline{4}(100\%)$$

$$\qquad\qquad\qquad\qquad=44.\overline{4}\%$$

   c) $0.65=0.65(100\%)$ d) $0.035=0.035(100\%)$

$$=65\%\qquad\qquad\qquad=3.5\%$$

   e) $2.3=2.3(100\%)$

$$=230\%$$

46. a) $(-4)^{3}-5(-4)=-64-5(-4)$

$$=-64-(-20)$$

$$=-64+20$$

$$=-44$$

   b) $-(-4)(20)+2\sqrt{-4+20}=-(-80)+2\sqrt{16}$

$$=80+2\cdot4$$

$$=80+8$$

$$=88$$

47. $9x^{3}-15.4+8x-12x^{3}+4x^{2}+x^{4}-4.9-x$

$$=x^{4}+9x^{3}-12x^{3}+4x^{2}+8x-x-15.4-4.9$$

$$=x^{4}-3x^{3}+4x^{2}+7x-20.3$$

48. $\left(5n^{2}+8n-9.3\right)+\left(4n^{2}-12.2n-3.51\right)$

$$=5n^{2}+4n^{2}+8n-12.2n-9.3-3.51$$

$$=9n^{2}-4.2n-12.81$$

49. $\left(x^4 - \dfrac{1}{3}x^2 + 9.6\right) - \left(4x^3 + \dfrac{3}{5}x^2 - 14.6\right)$

$= \left(x^4 - \dfrac{1}{3}x^2 + 9.6\right) + \left(-4x^3 - \dfrac{3}{5}x^2 + 14.6\right)$

$= x^4 - 4x^3 - \dfrac{1}{3}x^2 - \dfrac{3}{5}x^2 + 9.6 + 14.6$

$= x^4 - 4x^3 - \dfrac{5}{15}x^2 - \dfrac{9}{15}x^2 + 24.2$

$= x^4 - 4x^3 - \dfrac{14}{15}x^2 + 24.2$

50. a) $\left(\dfrac{5}{8}a^2\right)\left(-\dfrac{4}{9}a\right) = \left(\dfrac{5}{8}\right)\left(-\dfrac{4}{9}\right)a^{2+1}$

$= -\dfrac{5}{18}a^3$

  b) $(m-9)(m+9) = m^2 - 81$

  c) $(4.5t - 3)(7.1t + 6)$
  $= 4.5t \cdot 7.1t + 4.5t \cdot 6 - 3 \cdot 7.1t - 3 \cdot 6$
  $= 31.95t^2 + 27t - 21.3t - 18$
  $= 31.95t^2 + 5.7t - 18$

51. $630 = 2 \cdot 3^2 \cdot 5 \cdot 7$

52. $30n^4 = 2 \cdot 3 \cdot 5 \cdot n \cdot n \cdot n \cdot n$
  $45n^2 = 3 \cdot 3 \cdot 5 \cdot n \cdot n$
  $\text{GCF}\left(30n^4, 45n^2\right) = 3 \cdot 5 \cdot n \cdot n = 15n^2$

53. $36x^3 = 2^2 \cdot 3^2 \cdot x^3$
  $24x = 2^3 \cdot 3 \cdot x$
  $\text{LCM}(36x^3, 24x) = 2^3 \cdot 3^2 \cdot x^3$
  $= 8 \cdot 9 \cdot x^3$
  $= 72x^3$

54. $\dfrac{30y^5}{-6y^2} = \dfrac{30}{-6}y^{5-2}$
  $= -5y^3$

55. $28b^5 + 24b^3 - 32b^2$
  $= 4b^2\left(\dfrac{28b^5 + 24b^3 - 32b^2}{4b^2}\right)$
  $= 4b^2\left(\dfrac{28b^5}{4b^2} + \dfrac{24b^3}{4b^2} - \dfrac{32b^2}{4b^2}\right)$
  $= 4b^2\left(7b^3 + 6b - 8\right)$

56. a) $-\dfrac{9x^3}{10} \cdot \dfrac{25}{12x} = -\dfrac{\cancel{3} \cdot 3 \cdot \cancel{x} \cdot x \cdot x}{2 \cdot \cancel{5}} \cdot \dfrac{\cancel{5} \cdot 5}{2 \cdot 2 \cdot \cancel{3} \cdot \cancel{x}}$
  $= -\dfrac{15x^2}{8}$

  b) $-\dfrac{10}{9n^2} \div \left(-\dfrac{5}{12n}\right) = -\dfrac{10}{9n^2} \cdot -\dfrac{12n}{5}$

$= \dfrac{-2 \cdot \cancel{5}}{3 \cdot \cancel{3} \cdot \cancel{n} \cdot n} \cdot \dfrac{2 \cdot 2 \cdot \cancel{3} \cdot \cancel{n}}{-\cancel{5}}$

$= \dfrac{-8}{-3n}$

$= \dfrac{8}{3n}$

  c) $\dfrac{3}{5} - \dfrac{y}{4} = \dfrac{3 \cdot 4}{5 \cdot 4} - \dfrac{y \cdot 5}{4 \cdot 5}$

$= \dfrac{12}{20} - \dfrac{5y}{20}$

$= \dfrac{12 - 5y}{20}$

57. $m - 25 = -31$ Check: $-6 - 25 = -31$
  $m = -31 + 25$ $-31 = -31$
  $m = -6$

58. $-6x = 54$ Check: $-6(-9) = 54$
  $\dfrac{-6x}{-6} = \dfrac{54}{-6}$ $54 = 54$
  $x = -9$

59. $-4x + 19 = -1$ Check: $-4 \cdot 5 + 19 = -1$
  $\underline{\quad -19 \quad \underline{-19}}$ $-20 + 19 = -1$
  $-4x + 0 = -20$ $-1 = -1$
  $\dfrac{-4x}{-4} = \dfrac{-20}{-4}$
  $x = 5$

60. $\dfrac{3}{5}k - 7 = \dfrac{1}{4}$

$20 \cdot \left(\dfrac{3}{5}k - 7\right) = 20 \cdot \dfrac{1}{4}$

$\dfrac{\overset{4}{\cancel{20}}}{1} \cdot \dfrac{3}{\cancel{5}}k - 20 \cdot 7 = \dfrac{\overset{5}{\cancel{20}}}{1} \cdot \dfrac{1}{\cancel{4}}$

$12k - 140 = \quad 5$

$\underline{\quad +140 = +140}$

$12k + 0 = 145$

$\dfrac{12k}{12} = \dfrac{145}{12}$

$k = 12\dfrac{1}{12}$

Check: $\dfrac{\cancel{8}}{\cancel{8}}\left(\dfrac{\overset{29}{\cancel{145}}}{\underset{4}{\cancel{12}}}\right)-7=\dfrac{1}{4}$

$$\dfrac{29}{4}-7=\dfrac{1}{4}$$

$$\dfrac{29}{4}-\dfrac{28}{4}=\dfrac{1}{4}$$

$$\dfrac{1}{4}=\dfrac{1}{4}$$

61. $6.5h-16.4=9h+3.2$

$10\left(6.5h-16.4\right)=10\left(9h+3.2\right)$

$10\cdot6.5h-10\cdot16.4=10\cdot9h+10\cdot3.2$

$65h-164=90h+32$

$\underline{-65h\qquad\quad=-65h}$

$0-164=25h+32$

$\underline{\quad-32=\qquad-32}$

$-196=25h+0$

$\dfrac{-196}{25}=\dfrac{25h}{25}$

$-7.84=h$

Check: $6.5(-7.84)-16.4=9(-7.84)+3.2$

$-50.96-16.4=-70.56+3.2$

$-67.36=-67.36$

62. $2.4\left(n-5\right)+8=1.6n-14$

$2.4n-12+8=1.6n-14$

$2.4n-4=1.6n-14$

$10\left(2.4n-4\right)=10\left(1.6n-14\right)$

$10\cdot2.4n-10\cdot4=10\cdot1.6n-10\cdot14$

$24n-40=16n-140$

$\underline{-16n\qquad\quad=-16n}$

$8n-40=0-140$

$\underline{+40=\qquad+40}$

$8n+0=-100$

$\dfrac{8n}{8}=\dfrac{-100}{8}$

$n=-12.5$

Check: $2.4\left(-12.5-5\right)+8=1.6\left(-12.5\right)-14$

$2.4\left(-17.5\right)+8=1.6\left(-12.5\right)-14$

$-42+8=-20-14$

$-34=-34$

63. $\dfrac{u}{14}=\dfrac{-3}{8}$ Check: $\dfrac{-5.25}{14}=\dfrac{-3}{8}$

$8\cdot u=14\cdot-3$ $8\cdot-5.25=14\cdot-3$

$8u=-42$ $-42=-42$

$\dfrac{8u}{8}=\dfrac{-42}{8}$

$u=-5.25$

64. a) $12\text{ ft.}=\dfrac{12\ \cancel{\text{ft.}}}{1}\cdot\dfrac{12\text{ in.}}{1\ \cancel{\text{ft.}}}=144\text{ in.}$

b) $78\text{ ft.}=\dfrac{78\ \cancel{\text{ft.}}}{1}\cdot\dfrac{1\text{ yd.}}{3\ \cancel{\text{ft.}}}=\dfrac{78}{3}\text{ yd.}=26\text{ yd.}$

c) $2.4\text{ mi.}=\dfrac{2.4\ \cancel{\text{mi.}}}{1}\cdot\dfrac{5280\text{ ft.}}{1\ \cancel{\text{mi.}}}=12{,}672\text{ ft.}$

d) Because cm is 2 units to the right of m, we move the decimal point 2 places the right in 4.8.
4.8 m = 480 cm

e) Because m is 3 units to the right of km, we move the decimal point 3 places to the right in 12.5.   12.5 km = 12,500 m

f) Because m is 3 units to the left of mm, we move the decimal point 3 places to the left in 4500.   4500 mm = 4.5 m

65. a) $30\text{ ft.}^2=\dfrac{30\ \cancel{\text{ft.}^2}}{1}\cdot\dfrac{1\text{ yd.}^2}{9\ \cancel{\text{ft.}^2}}=\dfrac{30}{9}\text{ yd.}^2=3\dfrac{1}{3}\text{ yd.}^2$

b) $4\text{ yd.}^2=\dfrac{4\ \cancel{\text{yd.}^2}}{1}\cdot\dfrac{9\text{ ft.}^2}{1\ \cancel{\text{yd.}^2}}=36\text{ ft.}^2$

c) $504\text{ in.}^2=\dfrac{504\ \cancel{\text{in.}^2}}{1}\cdot\dfrac{1\text{ ft.}^2}{144\ \cancel{\text{in.}^2}}$

$=\dfrac{504}{144}\text{ ft.}^2=3.5\text{ ft.}^2$

d) $0.6\text{ m}^2=\dfrac{0.6\ \cancel{\text{m}^2}}{1}\cdot\dfrac{10{,}000\text{ cm}^2}{1\ \cancel{\text{m}^2}}=6000\text{ cm}^2$

e) $0.085\text{ km}^2=\dfrac{0.085\ \cancel{\text{km}^2}}{1}\cdot\dfrac{1{,}000{,}000\text{ m}^2}{1\ \cancel{\text{km}^2}}$

$=85{,}000\text{ m}^2$

f) $42{,}000\text{ cm}^2=\dfrac{42{,}000\ \cancel{\text{cm}^2}}{1}\cdot\dfrac{1\text{ m}^2}{10{,}000\ \cancel{\text{cm}^2}}$

$=\dfrac{42{,}000}{10{,}000}\text{ m}^2$

$=4.2\text{ m}^2$

66. a) $9.5\text{ lb.}=\dfrac{9.5\ \cancel{\text{lb.}}}{1}\cdot\dfrac{16\text{ oz.}}{1\ \cancel{\text{lb.}}}=152\text{ oz.}$

b) $4500 \text{ lb.} = \dfrac{4500 \ \cancel{\text{lb.}}}{1} \cdot \dfrac{1 \text{ T}}{2000 \ \cancel{\text{lb.}}}$

$= \dfrac{4500}{2000} \text{ T}$

$= 2.25 \text{ T}$

c) Because g is 3 units to the right of kg, we move the decimal point 3 places to the right in 0.23. $0.23 \text{ kg} = 230 \text{ g}$

d) Because kg is 1 units to the left of hg and t is 3 units to the left of kg, we must move the decimal point 4 places to the left in 9600. $9600 \text{ hg} = 0.96 \text{ t}$

67. a) $4 \text{ gal.} = \dfrac{4 \ \cancel{\text{gal.}}}{1} \cdot \dfrac{4 \ \cancel{\text{qt.}}}{1 \ \cancel{\text{gal.}}} \cdot \dfrac{2 \text{ pt.}}{1 \ \cancel{\text{qt.}}} = 32 \text{ pt.}$

b) $68 \text{ c.} = \dfrac{68 \ \cancel{\text{c.}}}{1} \cdot \dfrac{1 \ \cancel{\text{pt.}}}{2 \ \cancel{\text{c.}}} \cdot \dfrac{1 \text{ qt.}}{2 \ \cancel{\text{pt.}}} = \dfrac{68}{4} \text{ qt.} = 17 \text{ qt.}$

c) Because l is 3 units to the left of ml, we move the decimal point 3 places to the left in 400. $400 \text{ ml} = 0.4 \text{ l}$

d) $1 \text{ cm}^3 = 1 \text{ ml}$: Because ml is 1 units to the right of cl, we move the decimal point 1 place to the right in 0.9. $0.9 \text{ cl} = 9 \text{ cc}$

68. a) $C = \dfrac{5}{9}(F - 32)$ b) $F = \dfrac{9}{5}C + 32$

$C = \dfrac{5}{9}(77 - 32)$ $F = \dfrac{9}{\cancel{5}_1} \cdot \dfrac{-\cancel{20}^4}{1} + 32$

$C = \dfrac{5}{\cancel{9}} \cdot \dfrac{\cancel{45}^5}{1}$ $F = -36 + 32$

$C = 25°$ $F = -4°$

69. $15 \cdot 24 = 360$ windows

70. old balance + payment = new balance
$-1245 + 375 = -870$

Her new balance is –$870.

71. Net worth is the sum of all assets and debts. Write an addition statement adding all assets and debts.

$1282 + 3548 + 21,781 + 5745$
$+ (-2942) + (-83,605) + (-4269)$
$= 32,356 + (-90,816)$
$= -58,460$

Since the total debt –58,460 is a number with larger absolute value than the total assets 32,356, the family's net worth was a debt/negative amount of $58,460.

72. a) The figure is a parallelogram. The measures of the opposite sides of a parallelogram are equal. Perimeter is the sum of the measures of the sides of the figure.
$P = (h + 1) + (h + 3) + (h + 1) + (h + 3)$
$= 4h + 8$

b) Substitute 5 for h.
$4h + 8 = 4 \cdot 5 + 8 = 20 + 8 = 28 \text{ cm}$

c) The formula for area of a parallelogram is $A = bh$. The base is $(h + 3)$ and the height is h.
$A = bh = (h + 3)h = h^2 + 3h$

d) Substitute 7 for h.
$A = h^2 + 3h = 7^2 + 3(7) = 49 + 21 = 70 \text{ cm}^2$

73. a) $P = R - C$
$P = (3.5x + 600) - (0.2x + 4000)$
$P = (3.5x + 600) + (-0.2x - 4000)$
$P = 3.3x - 3400$

b) $P = 3.3(1500) - 3400$
$P = 4950 - 3400$
$P = \$1550$

74. $SA = 2(lw + lh + wh)$
$SA = 2(4.5 \cdot 6 + 4.5 \cdot 3 + 6 \cdot 3)$
$SA = 2(27 + 13.5 + 18)$
$SA = 2(58.5)$
$SA = 117 \text{ ft.}^2$

75. one integer $= 2x + 3$
other integer $= x$
Sum means to add. $(2x + 3) + x = 123$
$3x + 3 = 123$
$\underline{-3 = -3}$
$3x + 0 = 120$
$\dfrac{3x}{3} = \dfrac{120}{3}$
$x = 40$

One integer is $2(40) + 3 = 83$. The other integer is 40.

76. The key word *sum* means to add. We must find the relationships between the three angles in a triangle.
If we choose the first angle to be x, we can say:
first angle: x
second angle: $10 + x$
third angle: $x - 7$

The sum of the angles in any triangle is 180°.
first angle + second angle + third angle = 180°

$$x + (10 + x) + (x - 7) = 180$$
$$x + x + x + 10 - 7 = 180$$
$$3x + 3 = 180$$
$$\underline{-3 = -3}$$
$$\frac{3x}{3} = \frac{177}{3}$$
$$x = 59$$

Because x represents the first angle, we can say:

first angle: $x = 59°$

second angle: $10 + x = 10 + 59° = 69°$

third angle: $x - 7 = 59° - 7 = 52°$

77.

Categories	Value	Number	Amount
$5 bills	5	$17 - x$	$5(17 - x)$
$10 bills	10	x	$10x$

$$5(17 - x) + 10x = 140$$
$$85 - 5x + 10x = 140$$
$$85 + 5x = 140$$
$$\underline{-85 \qquad = -85}$$
$$0 + 5x = 55$$
$$\frac{5x}{5} = \frac{55}{5}$$
$$x = 11$$

Jeremy has 11 ten dollar bills and $17 - 11 = 6$ five dollar bills.

78. The fraction representing the long-distance calls to her parents is $\dfrac{3}{4} \cdot \dfrac{24}{32} = \dfrac{72}{128} = \dfrac{9}{16}$. We will translate *of* to mean multiplication. We must find $\dfrac{9}{16}$ of the 32 calls. $\dfrac{9}{16} \cdot 32 = \dfrac{9}{\cancel{16}} \cdot \dfrac{\cancel{32}^{2}}{1} = 18$

Sandra made 18 calls to her parents.

79. $A = \dfrac{1}{2}bh$

$$= \frac{1}{2} \cdot 17 \cdot 9$$
$$= \frac{1}{2} \cdot 153$$
$$= \frac{153}{2}$$
$$= 76\frac{1}{2} \text{ in.}^2$$

80. $\qquad C = 2\pi r$

$$12\frac{1}{2} = 2 \cdot \frac{22}{7} \cdot r$$
$$\frac{25}{2} = \frac{44}{7} r$$
$$\frac{7}{44} \cdot \frac{25}{2} = \frac{\cancel{7}}{\cancel{44}} \cdot \frac{\cancel{44}}{\cancel{7}} r$$
$$\frac{175}{88} = r$$
$$1\frac{87}{88} \text{ ft.} = r$$

81. a) $\dfrac{1}{6} + \dfrac{3}{5} = \dfrac{1 \cdot 5}{6 \cdot 5} + \dfrac{3 \cdot 6}{5 \cdot 6}$

$$= \frac{5}{30} + \frac{18}{30}$$
$$= \frac{23}{30}$$

$\dfrac{23}{30}$ said "excellent" or "good."

b) $\dfrac{1}{6} + \dfrac{3}{5} + \dfrac{1}{8} = \dfrac{1 \cdot 20}{6 \cdot 20} + \dfrac{3 \cdot 24}{5 \cdot 24} + \dfrac{1 \cdot 15}{8 \cdot 15}$

$$= \frac{20}{120} + \frac{72}{120} + \frac{15}{120}$$
$$= \frac{107}{120}$$

$\dfrac{107}{120}$ said "excellent", "good" or "fair"

$\dfrac{13}{120}$ said "poor"

82.

Categories	Rate	Time	Distance
Alan	3	t	$3t$
Jessica	2	t	$2t$

$$3t + 2t = \frac{1}{4}$$
$$5t = \frac{1}{4}$$
$$\frac{1}{5} \cdot 5t = \frac{1}{4} \cdot \frac{1}{5}$$
$$1t = \frac{1}{20}$$
$$t = \frac{1}{20} \text{ hr.} = 3 \text{ min.}$$

83. a) $14.50 - 11.90 = 2.60$
The closing price was $2.60 more on April 5 than on April 3.

b) $\dfrac{12.80 + 11.90 + 13.20 + 14.50}{4} = \dfrac{52.40}{4} = 13.10$
The mean closing price was $13.10.

c) Arrange the data from least to greatest and take the mean of the two middle numbers, which are 12.80 and 13.20.

11.90, 12.80, 13.20, 14.50

$\dfrac{12.80 + 13.20}{2} = \dfrac{26}{2} = 13$

The median closing price was $13.00.

84. We must subtract the area of the circle from the area of the trapezoid.

$A = \dfrac{1}{2} \cdot h \cdot (a + b) - \pi r^2$

$A = \dfrac{1}{2} \cdot 10 \cdot (7.5 + 12) - \pi \cdot 4^2$

$A = 0.5 \cdot 10 \cdot 19.5 - \pi \cdot 16$

$A = 97.5 - 16\pi$

$A \approx 97.5 - 16 \cdot 3.14$

$A \approx 97.5 - 50.24$

$A \approx 47.26 \text{ cm}^2$

85. $V = \pi r^2 h$
$V = \pi \cdot 3^2 \cdot 3.5$
$V = \pi \cdot 9 \cdot 3.5$
$V = 31.5\pi$
$V \approx 31.5 \cdot 3.14$
$V \approx 98.91 \text{ in.}^3$

86. Since the diameter is 12 in., the radius is 6 in.

$V = \dfrac{4}{3}\pi r^3$

$V = \dfrac{4}{3} \cdot 6^3 \cdot \pi$

$V = \dfrac{4}{\cancel{3}} \cdot \dfrac{\cancel{216}^{72}}{1} \cdot \pi$

$V = 288\pi$

$V \approx 288 \cdot 3.14$

$V \approx 904.32 \text{ in.}^3$

87. $V = \dfrac{1}{3}\pi r^2 h$

$V = \dfrac{1}{3} \cdot 0.8^2 \cdot 12 \cdot \pi$

$V = \dfrac{1}{\cancel{3}} \cdot \dfrac{0.64}{1} \cdot \dfrac{\cancel{12}^4}{1} \cdot \pi$

$V = 2.56\pi$

$V \approx 2.56 \cdot 3.14$

$V \approx 8.04 \text{ cm}^3$

88. We must use the Pythagorean Theorem.
$a^2 + b^2 = c^2$
$8^2 + 20^2 = c^2$
$64 + 400 = c^2$
$464 = c^2$
$\sqrt{464} = c$
$21.54 \approx c$

The beam is about 21.54 ft.

89. There are 41 favorable outcomes out of 650 possible outcomes. $P = \dfrac{41}{650}$

90. Let F represent the unit price of the 15.5-oz. box of cereal and T represent the unit price of the 16-oz. box of cereal.

$F = \dfrac{\$2.89}{15.5 \text{ oz.}}$ $T = \dfrac{\$3.45}{20 \text{ oz.}}$
$F \approx \$0.19 / \text{oz.}$ $T \approx \$0.17 / \text{oz.}$

Because $0.17 is the smaller unit price, the 20 oz. box of cereal is the better buy.

91.

Smaller triangle		Larger triangle
9 cm	↔	12 cm
16.2 cm	↔	x
20 cm	↔	y

$\dfrac{9}{12} = \dfrac{16.2}{x}$ $\dfrac{9}{12} = \dfrac{20}{y}$

$x \cdot 9 = 12 \cdot 16.2$ $y \cdot 9 = 12 \cdot 20$

$9x = 194.4$ $9y = 240$

$\dfrac{9x}{9} = \dfrac{194.4}{9}$ $\dfrac{9y}{9} = \dfrac{240}{9}$

$x = 21.6 \text{ cm}$ $y = 26.\overline{6} \text{ cm}$

92. a) $24 + 40 + 64 + 26 + 22 = 176$ students

b) $\dfrac{40}{176} = \dfrac{40 \div 8}{176 \div 8} = \dfrac{5}{22}$

c) Basic percent sentence:
What percent is 24 out of 176?

$$\frac{P}{100} = \frac{24}{176}$$
$$176 \cdot P = 24 \cdot 100$$
$$176P = 2400$$
$$P = 13.\overline{63}$$

$13.\overline{63}$ % of the students earned an A.

d)

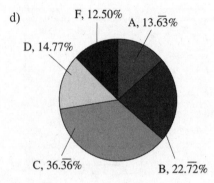

F, 12.50%
A, 13.$\overline{63}$%
D, 14.77%
C, 36.$\overline{36}$%
B, 22.$\overline{72}$%

93. The discount amount is 20% of the initial price.
What amount is 20% of 75.95?

$$d = 20\% \cdot 75.95$$
$$d = 0.2(75.95)$$
$$d = 15.19$$

The discount amount is $15.19. The amount that Cody will pay is 75.95 − 15.19 = $60.76.

94. amount of increase = new price − initial price
$$= 9.75 - 8.95$$
$$= \$0.80$$

What percent of 8.95 is 0.8?
$$\frac{P}{100} = \frac{0.8}{8.95}$$
$$8.95 \cdot P = 100 \cdot 0.8$$
$$8.95P = 80$$
$$\frac{8.95P}{8.95} = \frac{80}{8.95}$$
$$P \approx 8.9\%$$

95. $I = Prt$
$$I = (680)(4.2\%)\left(\frac{1}{2}\right)$$
$$I = \frac{680}{1} \cdot \frac{0.042}{1} \cdot \frac{1}{2}$$
$$I = \frac{28.56}{2}$$
$$I = \$14.28$$

96. $B = P\left(1 + \dfrac{r}{n}\right)^{nt}$
$$B = 3200\left(1 + \frac{0.06}{2}\right)^{2 \cdot 2}$$
$$B = 3200(1 + 0.03)^4$$
$$B = 3200(1.03)^4$$
$$B = 3200(1.12550881)$$
$$B = 3601.63$$

The balance will be $3601.63.

97. **Understand:** The variable expressions represent the measures of angles that are consecutive interior angles, so they are supplementary.

Plan: Translate to an equation, then solve. The sum of x and $2x - 21$ is 180.

Execute: $x + 2x - 21 = 180$
$$3x - 21 = 180$$
$$3x = 201$$
$$x = 67$$

Answer: The measure of the smaller angle is 67°. To find the measure of the smaller angle, we replace x in $2x - 21$ with 67.
$$2(67) - 21 = 134 - 21 = 113$$

The two angles are 67° and 113°. We can find all the other angles using the rules concerning angles formed by a transversal intersecting two parallel lines.

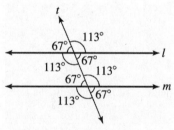

Check: The sum of the measures of two consecutive interior angles is 67° + 113° = 180°.

98. A: (5, 1), B: (−3, 4), C: (−1, −3), D: (3, −2), E: (−4, 0), F: (0, 3)

99. a) Individual solution points may vary but should all lie on the same line. Some possible solutions are (1, −5), (2, −4), and (4, −2).

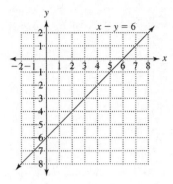

b) Individual solution points may vary but should all lie on the same line. Some possible solutions are (0, 2), (1, –1), and (2, –4).

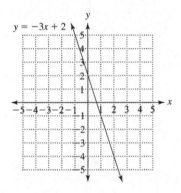

c) Individual solution points may vary but should all lie on the same line. Some possible solutions are (– 2, 1), (0, 0), and (2, –1).

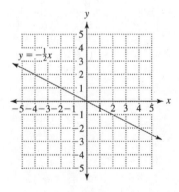

d) Individual solution points may vary but should all lie on the same line. Some possible solutions are (–5, 2), (0, 2), and (4, 2).

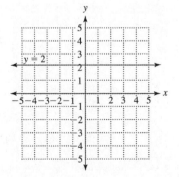

100.a) Initial velocity means the velocity at the time the rocket is released, which is 0 seconds. Let $t = 0$.

$$v = -32.2t + 580$$
$$v = -32.2(0) + 580$$
$$v = 0 + 580$$
$$v = 580 \text{ ft./sec.}$$

b) We are given a time of 5 seconds. Let $t = 5$.

$$v = -32.2t + 580$$
$$v = -32.2(5) + 580$$
$$v = -161 + 580$$
$$v = 419 \text{ ft./sec.}$$

c) We must calculate the time at which the rocket stops in midair. After stopping, the rocket will then reverse direction and begin falling back toward the ground. When an object is at a stop, it is obviously not moving and therefore has a velocity of 0. We must solve for t when the velocity, v is 0. Let $v = 0$ and solve for t.

$$v = -32.2t + 580$$
$$0 = -32.2t + 580$$
$$\underline{-580 = -580}$$
$$-580 = -32.2t + 0$$
$$\frac{-580}{-32.2} = \frac{-32.2t}{-32.2}$$
$$18 \text{ sec.} \approx t$$